D0456245

Where the truth lies

Where the truth lies

Franz Moewus and the origins of molecular biology

Jan Sapp

Department of History and Philosophy of Science
University of Melbourne

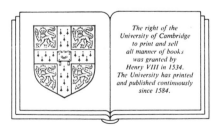

The right of the
University of Cambridge
to print and sell
all manner of books
was granted by
Henry VIII in 1534.
The University has printed
and published continuously
since 1584.

Cambridge University Press

Cambridge
New York   Port Chester   Melbourne   Sydney

Published by the Press Syndicate of the University of Cambridge
The Pitt Building, Trumpington Street, Cambridge CB2 1RP
40 West 20th Street, New York, NY 10011, USA
10 Stamford Road, Oakleigh, Melbourne 3166, Australia

First published 1990

Printed in the United States of America

Library of Congress Cataloging-in-Publication Data
Sapp, Jan.
Where the truth lies : Franz Moewus and the origins of molecular
biology / Jan Sapp.
p.   cm.
"Publications of Franz Moewus" : p.
Bibliography: p.
Includes index.
ISBN 0-521-36550-3. – ISBN 0-521-36751-4 (pbk.)
1. Molecular biology – History.   2. Molecular genetics – History.
3. Moewus, Franz, 1908 – Career in molecular biology.   I. Title.
QH506.S27   1989
574.8'8'09–dc20   89-32582
                                                                         CIP

British Library Cataloguing in Publication Data
Sapp, Jan
Where the truth lies : Franz Moewus and the
origins of molecular biology.
1. Molecular biology, history
I. Title
574.8'8'09

ISBN 0-521-36550-3 hard covers
ISBN 0-521-36751-4 paperback

For Will

Contents

Acknowledgments

Many people generously provided useful leads and/or materials for discussion in this study. Their specific contributions are noted in the text itself. It is enough to say here that this book could not have been written without their understanding and support. It is a pleasure to thank again Arthur Birch, Elof Carlson, Bernard Davis, Karl Grell, Norman Horowitz, Reinhard Kaplan, Liselotte Kobb, Joshua Lederberg, Ralph Lewin, John A. Moore, David Nanney, Robert Olby, Elizabeth Ryan, Ruth Sager, Ruth Sonneborn, Herbert Stern, and James D. Watson.

Richard Ziemacki of Cambridge University Press and Homer Le Grand provided relentless encouragement for writing this book when it was in its early stages. As a referee for Cambridge University Press, Edward Yoxen offered constructive criticisms, advice, and reassurance when they were needed. Liselotte Kobb and Ralph Lewin offered friendship, support, and valuable criticisms of several chapters. I was also fortunate to capture the interest and suggestions of Bruno Latour, who was *la tour de force* in this department for three months when I was completing the final draft. Carole McKinnon, David Turnbull, Camille Limoges, Steve Cross, and my colleagues and students in the Department of History and Philosophy of Science, Melbourne University, have also been rich sources of inspiration. Bernd Bartl, Doug McCann, and Jane Alvarez helped in various ways, at different times during this project.

I would also like to thank Helen Wheeler and Edith Feinstein of Cambridge University Press, New York, and especially Rosalind Corman, copy editor, for skillful and meticulous attention to the text. The reader will also appreciate the beautiful drawings of microorganisms by David Hill of the Botany Department, University of Melbourne. Research for this book was generously supported with funds from the Australian Research Grant Scheme.

1. Is science fiction?

In October 1981, I was searching through the papers of the late T. M. Sonneborn at Indiana University. At that time, I was a doctoral student in the history of science at the University of Montreal. I was working on the history of genetics, more precisely, the study of cytoplasmic inheritance, an area that had been virtually ignored by historians of modern biology. Sonneborn was one of the central figures in the development of modern genetics; he was widely respected as a brilliant experimentalist, with the broadest grasp of fundamental biological problems. I had some correspondence with him, and after his death in 1981, his wife, Mrs. Ruth Sonneborn, generously invited me into her home and permitted me to search through his unpublished papers and professional correspondence. It was there that I found a file marked "Moewus" that held together a great body of correspondence which told a remarkable story. The letters told of a major controversy in the origins of what is now molecular biology surrounding the work of the German biologist Franz Moewus.

I had never heard of Moewus. But as I read on, I learned that during the late 1940s and early 1950s, Moewus was hailed by many biologists as one of the outstanding leaders in biological research of this century and one of the principal architects of the revolution in modern biology. Moewus was a pioneer in the development of microbial genetics. When, in the early 1930s, geneticists did not know if microorganisms had genes like higher organisms, Moewus provided some of the first demonstrations that microorganisms did possess genes that were inherited in the classical Mendelian way. During the late 1930s, when geneticists did not know what a gene was or how it worked, Moewus led the way again by providing basic concepts and important methodologies. He also provided the experimental "facts" concerning the biochemical means by which genes affect sexuality in one microorganism.

Yet, during the late 1930s, 1940s, and 1950s, Moewus did not always receive credit for his insightful concepts and pathbreaking methodolo-

1

gies. Instead, he was relentlessly criticized and defamed by many others. There were charges that Moewus's data were "too good to be true," that he had "polished off" his data, and that some of his interpretations were faulty. There were stories about Moewus's refusal to send his cultures to others or his sending dead cultures to those who wanted to repeat and extend his work. During the early 1950s, there were several failed attempts to repeat some of his experiments. Some geneticists interpreted these failures to be clear disconfirmations of Moewus's results and further indication that his reports were unreliable. Others dismissed the disconfirmations of Moewus's work as being faulty and argued that Moewus deserved the highest recognition regardless of the criticism that had been made against his work. Even when the controversy came to a close, leading geneticists believed they still had no direct evidence for invalidating any specific observation, experiment, or idea set forth in Moewus's publications. They had only circumstantial evidence. Although Moewus himself continually denied the charges made against him, he was ultimately judged guilty of perpetrating one of the most ambitious cases of fraud in the history of science. Shortly thereafter, Moewus died of a heart attack on May 30, 1959.

The story, as I first learned of it in Sonneborn's letters, caught my imagination. It conflicted with the little I knew of the history of genetics and of human nature. I, along with a whole generation of scientists and historians, had been brought up believing that George Beadle and Edward Tatum (1941) were the first to show how microorganisms could be used for investigating genic action. I believed they had laid the foundations for the biochemical genetics of microorganisms. In 1958, they were awarded a Nobel Prize, which they shared with Joshua Lederberg. I wanted to know more. I consulted texts on the history of genetics. I soon found that Moewus's work was excluded from all of the many texts celebrating the historical development of modern genetics and the triumph of molecular biology (see, for example, Dunn, 1965; Sturtevant, 1965; Olby, 1974; Allen, 1978a; Judson, 1979). Those that did mention his name (Sturtevant, 1965; Olby, 1974) alluded to him only in passing.

The Moewus story, like the research on cytoplasmic inheritance, represented another case of historians' "neglect." But the telling of the story had to wait until I completed my thesis on the history of cytoplasmic genetics. By that time, I hoped I would know more about the history of genetics. When I finished my doctoral dissertation, in 1984, coming to grips with the historical neglect of Moewus was not that difficult. I came to recognize more and more the biases of scientists and historians which shape how the "past" is constructed. These biases stem not only from "reading into" the literature by superimposing modern ideas on past scientific work; they result also from omitting historical facts that conflict with the historians' preconceived views of the nature of science. In history, the unconscious biases of the writer in selecting and interpreting data is not called fraud. We have another word for it. These kinds of accounts are quaintly labeled "Whiggish."

The claim that Moewus had polished off his data did not bother me. As I will explain momentarily, I suspect that all scientists do this. Moreover, I knew that Mendel's results were also "too good to be true," but no one denied his contributions and excluded him from history. Indeed, he is universally hailed as the "founding father" of the entire science of genetics. What I found difficult to digest was that Moewus had deliberately fabricated his results in a wholesale way. Moewus's biochemical genetic results seemed to be too elaborate and consistent to be fabricated in a wholesale way. For me to believe *that* would be similar to the reader's believing that I am fabricating the entire story that will unfold before you. As you will soon recognize, to do this would be very clever indeed. Moreover, the breadth of Moewus's accomplishments – his ideas and methodology – seemed to be too sophisticated to have originated with someone who would completely fabricate his data. And why would anyone who was obviously as intelligent as Moewus go to the trouble of faking his experiments when he could actually do them?

I turned to accounts of other controversies involving fraud charges to help unravel these questions. I read Koestler's (1971) account of Paul Kammerer and the midwife toad, and Weiner's (1955) account of the Piltdown forgery. But it seemed that each of these cases represented a hoax – a not-so-funny joke. Moreover, it was perfectly clear that someone had put the India ink on the toads to prove or disprove the inheritance of acquired characteristics. And someone deliberately had placed the skull of a human with the jaw of a monkey to be later discovered as the "missing link." The crime in these two cases is clear. The remaining mystery is a "Who dunnit?" The Moewus case was different. The suspected fraud was much more elaborate, and there was no clear-cut evidence suggesting that he had committed a crime. This problem warranted investigation. It was clear to me that I needed to know more about how scientists assess and evaluate knowledge claims.

In the meantime, I needed a job. And I soon found one. After working for nine months as visiting assistant professor in the History Department of the University of Arizona at Tucson, I flew back to Montreal to defend my doctoral dissertation. The next morning I was on a plane bound for Australia, where I was to take up a position as lecturer in the History and Philosophy of Science Department of Melbourne University. Here, I found excellent teaching and research conditions.

I immediately began to work on the Moewus story again, this time in earnest. My work was interrupted only by having to polish up for publication my thesis on cytoplasmic inheritance (Sapp, 1986, 1987a). As I began to collect the published literature concerning Moewus, I came to realize that the story was much more complex than I had imagined. There seemed to be an inside story that could not be fully revealed by studying the published scientific literature alone. It was clear that if this story was to be told in sufficient detail, I needed help from those scientists who had participated in the controversy. I immediately began to write them, asking for literature and advice, and inquiring whether I could arrange an interview.

My inquiries were met with overwhelming help and encouragement. Virtually all of the participants I contacted cooperated in any way they could. All of them wanted this story to be told as thoroughly as possible. Many sent letters of correspondence they had with or about Moewus, as well as names and addresses of others I should be sure to contact. All of those who were asked agreed to be interviewed. Others, including Bernard Davis, D. L. Nanney, and R. W. Kaplan, wrote me lengthy letters. Moewus's scientific career spanned three continents – Europe, Australia, and North America. In order to retrace his steps, I needed travel money. The Australian Research Grant Scheme generously provided funds for this project. My German is poor, so I hired Bernd Bartl, a German doctoral student, to help with translation.

This book could not have been written without the help of Mrs. Ruth Sonneborn, and many others. Upon arriving in Australia, I interviewed the leading Australian biochemist, Arthur Birch. Birch and Moewus had written important papers together in the early 1950s. Soon after, through the help of David Nanney, I learned that the German protozoologist Karl Grell often spent the winter working at the Australian Institute for Marine Sciences in Townsville, off the Great Barrier Reef. I happily flew to Townsville to talk with him about the conditions in Germany, where Moewus had done most of his work. My next travel leave took me to the East and West Coasts of the United States, where I met John A. Moore and his wife, Betty Moore, at the University of California, Riverside. John Moore was instrumental in bringing Moewus to the United States in the 1950s. No one was more helpful and encouraging than Ralph Lewin, a leading algologist at the Scripps Institution of Oceanography, La Jolla. In the early 1950s, Lewin had been working along lines similar to Moewus's; he became, in effect, one of his chief adversaries. Lewin had followed the Moewus stories closely, and he had long hoped that someone would write a detailed account.

In New York City, I was able to talk with Joshua Lederberg, president of Rockefeller University. Lederberg told me about the political gossip about Moewus that he had heard as a young pioneer in the development of microbial genetics. On the outskirts of Columbia University, I spent an evening with Mrs. Elizabeth Ryan, wife of the late Francis Ryan. Francis Ryan was a leading bacterial geneticist at Columbia University who had invited Moewus into his laboratory in an attempt to repeat some of his work in order to help bring the controversy to a close. James D. Watson, director of the Cold Spring Harbor Laboratory on Long Island, generously provided lodging for me and told me how stimulating he had found Moewus's work on the eve of his own discovery, with Francis Crick, of the structure of DNA. Ruth Sager, director of the Division of Cancer Genetics at Harvard University, told me of her meetings with Moewus at the Marine Biological Laboratory, Woods Hole, Cape Cod.

While on the East Coast of the United States, I took the opportunity of searching through the archives of the Rockefeller Foundation in Tarrytown, New York, for information on Moewus and the institutions in which he worked. The Natural Science Division of the Rockefeller Foundation, headed by Warren Weaver, played a leading role in fostering the development of molecular biology. Rockefeller officials kept detailed diaries of interviews with scientists and their visits to scientific institutions. I learned of the attitude of Rockefeller officials toward providing support for Moewus and others who attempted to repeat his work and bring the controversy to a close. I also learned about the political basis of their funding policies during the decade following World War II and gained some insight into the thinking of the directors and the internal workings of the two main institutions involved in Moewus's work.

The following year, I was able to travel to Heidelberg to spend several hours talking with Franz Moewus's widow, Mrs. Liselotte Kobb. Mrs. Kobb was a respected scientist in her own right and had frequently assisted her husband in the laboratory and in writing his scientific papers. She is completely convinced that there was no intended deception in her husband's reported observations. Mrs. Kobb generously shared with me some of her happiest and saddest memories of her life with her husband. She allowed this story to be written with the hope that the lessons we all might learn from it will not be lost.

When my journey was ended, I returned with a story that is, above all, one about the human side of science – a story about scientific truth, authority, war, racism, sexism, national pride, and individual dignity. This was hardly a simple story of a psychopathic scientist who managed to fool a naive community of researchers. Moreover, one point was certainly clear: When the controversy ended in the mid-1950s, much was unsettled in the minds of those scientists who dismissed Moewus as the perpetrator of a fraud. Moewus's judges based their opinions on a diverse and scattered body of circumstantial evidence. No one was certain about how much data Moewus "falsified," what could be retained as "true," and what had to be discarded as "false." Nor was anyone certain about Moewus's motives for perpetrating a "fraud." But, if it was fraud, as many participants believe, then it was indeed one of the most ambitious of its kind in the history of science. On the other hand, it seemed to be entirely possible that Moewus's "crime" may have been constructed by the scientific community itself. Again, I turned to the secondary literature on "fraud" for some insights to help unravel this mystery.

The normal and the pathological

In recent years, there has been a spate of disclosures of fraud in science. Indeed, with so many public disclosures of fraud occurring constantly, it is proving difficult to keep up with them. Fraud has been uncovered in all areas of scientific activity and all levels of the scientific community, from the "hired hand," to industrial scientists, to those in the "mainstream" of the academic scientific community (Bridgstock, 1982). In each of these areas, investigators have explored the social conditions that might encourage fraud and the motives of the researchers. Generally, they have traced the occurrence of fraud to a consideration of three factors: the rewards, the perceived risks of getting caught, and the honesty and integrity of the individual. Hired hands are often willing to take shortcuts to obtain the specific answers sought by supervisors. They are also generally interested in making money, and usually there is no career to be threatened – at most, a temporary job (Roth, 1966). In industrial science, fraud frequently occurs when scientific tests are performed solely to satisfy a government bureaucracy of a safety requirement, or when research is done for publicity purposes. The case of thalidomide

clearly illustrates what can happen when scientists become totally committed to a company that has pinned all its hopes on a product currently
undergoing safety tests (Knightly et al., 1979).

Many cases of fraud have been recently disclosed also in the mainstream of scientific research. They have been widely publicized by the
science journalists Broad and Wade (1982). Two of the best-known recent examples are the so-called painted-mice affair (Hixson, 1976) and
the case surrounding Cyril Burt (Kamin, 1974; Hearnshaw, 1979). William Summerlin, a scientist at the prestigious Sloan Kettering Institute,
painted the skin of two mice to demonstrate the success of his newly
developed techniques of skin transplantation. Summerlin confessed to
his fraudulent act. Whereas the research administrators claimed that
Summerlin's behavior was the result of a deranged mind (temporary
insanity), Summerlin claimed that he was put under a great deal of pressure from his superiors to produce positive results (Broad and Wade,
1982: 153–157).

Sir Cyril Burt died before his work came under serious scrutiny. Burt,
one of the pioneers of applied psychology in England, invented in a
wholesale way his I.Q. test data and even the very existence of his co-
workers in order to support his theory that intelligence is determined by
heredity. His official biographer, Hearnshaw (1979), believes that Burt's
research reports from 1943 onward have to be regarded with suspicion.
But his fraud went undetected for 31 years (Kamin, 1974). Two circumstances have been proposed to explain how Burt was able to pass off his
elitist hereditarian opinions as fact without severe scrutiny. First, Burt
held such a powerful position in the psychological establishment that
he became immune to scrutiny. He was the editor of the *British Journal
of Statistical Psychology* and used his position to publish numerous articles under pseudonyms. Because of Burt's prestige in the field, those
who were critical of his work were afraid to make their views public.
Second, Burt's data fit the dominant views of his times. People believed
what they wanted to believe. Kamin, the first to attack publicly the legitimacy of Burt's data, was a socialist who adopted an environmental
view of human intelligence. Burt's fraud is particularly disturbing since
his data had a serious effect on education policy in Britain and the
United States.

These and some other less publicized cases have led several writers
to believe that the known cases of fraud may represent only the tip of
the iceberg (see Rensberger, 1977; Gould, 1978; Weinstein, 1979; Broad
and Wade, 1982; Martin Bridgstock, 1982; Chubin, 1985). In recent

years, members of almost every major scientific institution have been forced to acknowledge that fraud occurs and have come to realize they must deal with it in some sensible way. In the relations between science and society, professional scientists are concerned about maintaining the reputation of scientists as purveyors of truth. In their relations with funding agencies, members of scientific institutions need to protect their reputations against charges of misusing funds. In their relations with each other, scientists want to rely on the trustworthiness of their colleagues' data. Lastly, scientists want to protect their intellectual property rights and guard against plagiarism. Formal guidelines have been proposed by scientific organizations to prevent fraud and to protect the innocent from irresponsible charges, while simultaneously encouraging individuals with certain knowledge of wrongdoing to make it known in appropriate ways.

Any attempt to understand fraud in science necessarily reflects one's view of how science properly functions, just as the pathological reflects the normal. Virtually all discussion of misconduct in science begins with the basic "rules" of science – the norms of science first put forward by Robert Merton, a pioneer in the sociology of science. In 1942, Merton briefly summarized a series of institutional imperatives for science. Along with technical or cognitive norms, such as requirements of logical consistency and empirical confirmability, Merton's rules of science consisted essentially of four moral normative requirements which he believed comprised the ethos of modern science. The moral norms that Merton held were necessary for the extension of certified scientific knowledge may be listed as follows (see Merton, 1973):

1. *Universalism:* This norm requires that knowledge claims are evaluated in terms of cognitive criteria; not in terms of personal attributes of their authors. In other words, the social standing of the scientist making a claim (i.e., whether he or she is an assistant researcher or Nobel Prize winner) should not significantly affect the judgments of others toward the knowledge being assessed.
2. *Communism:* The findings of scientists are a product of social collaboration and thus belong to the scientific community as a whole. Scientists do not *own* their work; intellectual property is limited to peer recognition. All information is made public, and secrecy is avoided.
3. *Organized skepticism:* Knowledge claims must be subjected to "detached scrutiny of beliefs in terms of empirical and logical criteria."
4. *Disinterestedness:* A "distinctive pattern of institutional control of a wide range of individual motives characterizes the behavior of scientists" such that it is "to the interest of scientists to conform" by engaging in disinterested activity directed toward the extension of scientific knowledge. Of all Merton's norms,

sociologists have found disinterestedness to be the most difficult to understand. For our purposes it is sufficient to know that most interpret it as a motivational requirement: that scientists are engaged in the pursuit of truth, not prestige or financial gain (see Weinstein, 1979). All have understood fraud in science in terms of a violation of the norm of disinterestedness.

Zuckerman (1977), Weinstein (1979), Bridgstock (1982), Broad and Wade (1982), and others have all shown that studies of fraud and "deviant behavior" are very useful for studying the system of social control in science. If the norms defined by Merton were operative, then certified knowledge would be increased and fraudulent assertions would not be made. Fraud can therefore be used as a probe to investigate how well the so-called "self-regulating structure" of science is functioning. It should be stressed that Merton himself set up the framework for these studies when he argued that the absence of fraud in science was due, not to the personal virtues of scientists, but instead to institutionalized mechanisms of social control. In fact, Merton (1942) made "the virtual absence of fraud in the annals of science" a hallmark of the uniqueness of scientific activity and a necessary corollary of his belief in the "verifiability of results," in "the exact scrutiny of fellow experts" (organized skepticism), and in "rigorous policing to a degree perhaps unparalleled in any other field of activity" (Merton, 1973: 276).

This statement of Merton's about the absence of fraud in science was wrong, even at the time he made it. In fact, in the early nineteenth century, Charles Babbage, the famous English mathematician and inventor of the first modern calculating machine, believed that fraud was prevalent enough in science to classify various kinds. In his well-known diatribe against the elitist nature of the Royal Society, *Reflections on the Decline of Science in England*, Babbage (1830: 174–183) listed various kinds of "frauds of observers":

Hoaxing: The scientist's "deceit is intended to last for a time, and then be discovered, to the ridicule of those who have credited it." The affairs of the Piltdown man and Kammerer's toad may be placed in this category.

Forgery: "The forger is one who, wishing to acquire a reputation in science, records observations which he never made." William Summerlin was found guilty of this practice; Sir Cyril Burt also forged his data on the inheritance of intelligence. However, it is not clear at all, in these two cases, whether the forgers wished only "to acquire a reputation" and were not, in fact, committed to the "truth."

Trimming: The data are manipulated so as to make them look better: "Trimming consists in clipping off little bits here and there from those observations which differ most in excess from the mean, and in sticking them on to those which are too small; a species of 'equitable adjustment,' as a radical would term it, which cannot be admitted in science." In modern terminology this is referred to as "massaging data" or "fudging."

Cooking: According to Babbage, cooking means choosing only those data that fit the researcher's hypothesis, and discarding those that do not; telling half-truths.

In direct conflict with Mertonian views, recent investigations of fraud in science have been devoted largely to showing that adequate policing in science is impossible and that fraud is likely to be endemic in modern science. Pressure to cheat has been traced to the reward system of contemporary science, with its emphasis on the quantity, as opposed to the quality, of publications. The quest for individual recognition and prestige has led scientists to violate the professed norms of science. But this is only half the story. The nature of big science – the quest for big money to finance huge laboratory factories – also seems to be playing a role in eroding the foundations upon which sound and useful knowledge rests. It appears that the success of biotechnologies, such as those associated with recombinant DNA, have set a precedent for science funding policies. They have set unreasonable expectations on other domains of scientific inquiry that are not yet ripe for exploitation. The scientific establishment is beginning to realize how the lure of big money leads them to abandon the ideals of the profession. Investigations of several cases of fraud that have recently appeared in big research laboratories suggest that they may often result from pressure to publish successful results in order to meet the demands set by large funding agencies that want immediately applicable results (see Broad and Wade, 1982).

In principle, replication of experiments, the "Supreme Court" of the scientific system (Collins, 1985), should be a powerful deterrent to fraud. However, in contrast to the common image of science which portrays replication as standard practice, many science analysts claim that replication of another's findings is seldom done in practice since there is little incentive for doing it. Recognition in science is accorded for originality; reward for repeating another's results is granted only in extraordinary circumstances.

One can easily understand how science writers could find a major place for fraud, once barriers to the enforcement of Mertonian norms were detected. These criticisms of Merton's theories generally have been a major source of anxiety and have contributed to the belief that fraud is likely to be endemic in modern institutionalized science. Yet, estimates of the prevalence of fraud in science vary from author to author and even within accounts of the same author. The following passages will highlight some of the contradictions in the literature pertaining to fraud:

Broad and Wade (1982: 87) have written:

> We would expect that for every case of major fraud that comes to light, a hundred or so go undetected. For each major fraud, perhaps a thousand minor fakeries are perpetrated. The reader can supply his own multiplication factors; ours would indicate that every major case of fraud that becomes public is representative of some 100,000 others, major and minor combined, that lie concealed in the marshy wastes of the scientific literature.

Yet, later, in the same book they claim that

> fraud . . . is a small, but not insignificant, endemic feature of the scientific enterprise. (Broad and Wade 1982: 219)

Everything relies on a definition of fraud, and again, any attempt to define it necessarily reflects one's view of science, just as the pathological reflects the normal. For those who believe that the norms described by Merton are, or *should be* the operative rules of science, fraud is understood in terms of violation of these norms (e.g., see Zuckerman, 1977). Plagiarism violates the norm of communism. The contriving of fraudulent evidence is understood to be in direct violation of the norm of disinterestedness. Nonetheless, there is a great inconsistency in the use of the term "fraud." Some use it to mean deliberate deception; others use it to embrace only deliberate deception for personal gain; still others use it to embrace any form of bias, whether deliberate or not. Even within the writings of the same author, there are inconsistencies. The following passages are illustrative:

> Cases of fraud in science – the outright fabrication of data – do indeed seem to be rare in science. (Zuckerman, 1977: 100)

> Fraud, the generic term for deliberate deception in science, occurs in three principal forms: the fabrication, fudging, and suppression of data. (Zuckerman, 1977: 113)

> The law differentiates "actual fraud", the intentional perversion of truth . . . or false representation of a matter of fact from "constructive fraud" which "although not originating in any evil design or contrivances to perpetrate a positive fraud . . . [has] led to the tendency to deceive or mislead . . . [and] is deemed equally reprehensible with actual fraud". . . . In other words, fraud is fraud whether intended or not. The same is true in science. (Zuckerman, 1977: 115)

> The scientist who is cheating knowingly, who falsifies or invents research data, or who lies about them, is not strictly fraudulent as long as he is not using the false data to obtain financial support from public or government agencies, or from private funds. Fraud is also committed when, on the basis of false data, the scientist is trying to secure a research job, to prove that public funds have been properly used, or to convince the public or the grantors that a certain procedure, material or drug is acceptable and safe. (Kohn, 1986: 2)

> Fraud, of course, is deliberate and self-deception unwitting. . . . (Broad and Wade, 1982: 20)

"Fraud" then is a catch-all term; it seems to be used to embrace almost every kind of behavior that might lead scientists away from the *truth* and the moral norms of science as first outlined by Merton. There are serious problems with this view, and, as the above conflicting statements suggest, with any approach that attempts to define fraud on a priori grounds. If fraud in science is understood in terms of Merton's norms, or in reference to some objective methodological "truth," then just about every leading scientist would have to be considered guilty. Galileo is held to have exaggerated the outcome of his experimental results (Koyré, 1968). Isaac Newton introduced a "fudge factor" into his work on the velocity of sound and gravitation to make it agree precisely with his theory (Westfall, 1973). The so-called father of modern atomic theory, John Dalton, published experimental results that probably could not have been obtained as described (Nash, 1956). The Nobel Prize–winning physicist Robert Millikan kept out of publication results that were unfavorable to his theories on the electric charge of the electron (Holton, 1978). Sigmund Freud is said to have "lied" about, or perhaps massively repressed, aspects of the discovery of the Oedipus complex (Cioffi, 1976). Mendel published results that statistically were "too good to be true" (Fisher, 1936).

If these cases represent crimes against science, as Broad and Wade and others claim, then we are left with the paradox of the great scientist–great fraud. To call such cases crimes against scientific truth is problematic indeed, and results from a failure to recognize science as a social process through which specific contributions gain the status of discoveries. To call Mendel's work, for example, a crime against science and the truth, is to contradict and disregard his universally recognized contributions that have become and remain central to modern biology. There is no question that these cases conflict with the conventional view of science as an arena where objectivity and impartiality rule. However, because this view of science is mythical, any view of fraud made in reference to it is also mythical. The above cases represent crimes only against the professed norms or ideology of science; they are not crimes when considered against the background of how science actually operates as a social activity.

Those who analyzed scientific fraud were not the first to criticize the normative structure of science as first proposed by Merton. Since about the mid-1970s, many investigators who have written about the social nature of science, especially those who have examined scientific controversies, have come to realize more and more that science does not operate in the manner that Merton had first envisaged. At the frontiers of science, correct and incorrect behavior are not as clearly defined as they are with respect to crude plagiarism and wholesale fabrication. Merton's views seemed to be based on what scientists *say* – at least, what they say to the public – not on what scientists *do*. They comprised the professed norms of science, not the statistical norms (see Barnes and Dolby, 1970; Mulkay, 1976). The actual operation of modern science bears little resemblance to the normative structure Merton described. Merton had constructed an idealistic view which today even few scientists would accept. Scientists are human; their views are often colored by personal, institutional, and larger political biases.

In fact, Merton himself later recognized that the reality represented a significant departure from his proposed norms, and he attempted to account for it by introducing the notion of "counternorm." He argued that social institutions such as science tended to be built on conflicting pairs of norms. His call for counternorms was answered by Mitroff (1974) in a detailed study of scientists involved in research concerning the moon. Mitroff's study showed that scientists did use variants of the norms described above in accounting for the actions of their colleagues and in their descriptions of how scientists ought to behave. But he also detected an opposing set of norms which scientists insisted were essential to the progress of science. For example, Mitroff suggested that the norm of "emotional neutrality" (organized skepticism) is countered by a norm of "emotional commitment." The latter was necessary because without it researchers would be unable to bring to fruition lengthy and laborious projects and tolerate many setbacks along the way. Similarly, the norm of "universalism" is balanced by a norm of "particularism." Scientists frequently regard it as perfectly acceptable to judge knowledge claims on the basis of personal criteria. The central point is that, according to Mitroff, there are at least two sets of norms functioning together in a dynamic way.

But is there a compelling reason to believe that there exist "operating rules of science" that are institutionalized in the form of norms? The most critical examination of the concept of norms in science, by Barnes and Dolby (1970) and Mulkay (1976), suggests that there is not. As these writers have emphasized, if social norms are to be considered institutionalized, they have to be positively linked to the reward system of science. Conformity to them would be maintained because it would be regularly rewarded. However, as Mulkay (1976) has argued, the primary conclusion of the body of recent social studies of science is that scientists are rewarded for communicating information that their colleagues perceive to be useful in pursuit of their own research. In other words, rewards are allocated overwhelmingly in response to the perceived quality of scientific findings, irrespective of the professional ethics of their producer. The moral norms of science have to be understood as rhetoric that scientists use in flexible ways when accounting for their own activity and that of their competitors (see, e.g., Mulkay and Gilbert, 1982).

If this view were to be accepted, then one would expect a greater degree of tolerance toward the presentation of biased data. What counts as fraud in science would be a much more complex issue than is often assumed. Not all "lies" would be considered fraud; it would depend on the claim in question, the social context in which it was made, and who was making it. For example, scientists who "lied" for the sake of truth and were considered "right" would normally be forgiven. Their flights of fancy would be attributed to the great personal commitment great scientists need to convince lesser scientists of their great ideas.

In fact, when the historian Alexandre Koyré (1968) pointed out that Galileo never made certain observations he claimed to have made, Koyré did so, not to discredit Galileo, but to celebrate his great mind – to show how Galileo had the truth worked out in his mind and used rhetoric to persuade his readers of his truth. To Koyré, Galileo's methodology was exemplary for understanding how the Scientific Revolution occurred, and epitomized the priority of theory over observation. Similarly, when Richard Westfall revealed Newton's "fudge factor," he did so, not to condemn Newton because he violated "moral norms"; rather, he praised the great scientist's achievement in helping to establish "the quantitative pattern of modern science," and celebrated Newton's genius, claiming that "no one could manipulate the fudge factor quite so effectively as the master mathematician himself" (Westfall, 1973: 752). As will be discussed in Chapter 5, when Fisher (1936) claimed that Mendel's data were statistically too good, again, he did so to praise Mendel, not to bury him. Fisher was interested in showing that Mendel had worked out his laws before carrying out his experiments, which, Fisher claimed, were done solely for the benefit of illustrating them to others. All of these cases, which marked significant revolutions in scientific thought, were originally raised to point out the power of theory over observation in science.

The theory-ladenness of observations mandates serious objection to any claim that cases in which scientists consciously or unconsciously tidy up the data a little, or report only favorable data, necessarily represent crimes against science and the truth. Such a simplistic view of fraud entails the assumption that when a scientist makes an observation or carries out an experiment, his or her mind *should be* an empty vessel ready to receive whatever information the reality of nature reveals to it. This view is often referred to as "naive empiricism." However, science simply does not and cannot work this way. Yet, this view is implicit in any account which claims that unconscious bias in reporting data is fraudulent in science. Scientists, historians, sociologists, and many philosophers of science have long realized that all scientific work of an exploratory or experimental nature begins with some expectation of the outcome and that all scientific observations are embedded in theory. It is theory that makes sense out of observations. It is one's expectations and theoretical understanding of the laws of nature that tell the observer what is a good experiment and what is a failed experiment, what are good data and what are bad or insignificant data that can be ignored and kept unpublished.

There is yet a second major objection to any attempt to define fraud on a priori grounds. Any attempt to define fraud in this way relies on the assumption that there exists a unique, timeless, and efficacious scientific method. However, today, many historians and sociologists, and some philosophers of science, recognize that there is no universal, ahistorical scientific method that purports to offer criteria for the genesis or evaluation of knowledge. Scientists' appeals to formal scientific method doctrines (inductivism, hypotheticodeductivism, falsificationism, etc.) have to be understood as rhetoric, as argumentative resources in scientific practice and debate (see Schuster and Yeo, 1986). Science is not the unique, socially autonomous activity that at one time had been imagined.

The scientific field is normally a place of conflict and competition. What is at stake in this struggle is a definition of science itself – what questions are important, what phenomena are interesting, what techniques are suitable, and what answers are acceptable. There is no neutral or value-free way of engaging in this activity. When a scientist attempts to evaluate knowledge, each one tends to uphold those values that are most closely related to him or her personally or institutionally. And it is against this background that scientists negotiate what counts as good reason.

Ernst Haeckel and his critics

A brief glance at the controversy surrounding the work of the German biologist Ernst Haeckel might help to clarify these points. Haeckel was a towering figure in nineteenth-century biology, a champion of Darwinism and materialistic interpretations of nature. Haeckel's formulation of the theory of recapitulation (that ontogeny is the recapitulation of philogeny) ranks as one of the most influential theories of nineteenth-century science. "Philogeny," according to Haeckel, was "the mechanical cause of ontogeny"; one no longer needed to invoke vitalistic principles, as was common in his day. The concept of evolution was just beginning to transform human thought, and ways were being sought to trace the course of evolution. Haeckel's theory of recapitulation suggested that the paths taken by evolution might readily be observed by following stages in the embryonic development of higher organisms. Haeckel argued that similarities could be observed among the embryonic forms of very different animals. For example, very early stages in the development of humans corresponded to the structure of and conditions for life in lower fishes. The next phase of development presented as a transformation of the fishlike organism into an amphibious animal and so on.

To support his recapitulation theory, Haeckel published various drawings to illustrate the striking similarities among different types of embryos. He was soon attacked by critics who claimed that many of the diagrams printed in a number of his books were completely fabricated, and that others had been tampered with to make them support the theory he was advancing (Figure 1). In arguing their case, Haeckel's critics described what they considered to be normal procedure in science. They claimed that in scientific texts, if a picture

> is not a representation of an object really seen, but merely an illustration of the author's view or theory, it must be so described in express terms, (e.g., *"Schematic Figure"*), so as to prevent the reader from imagining that it represents a real object.
>
> Similarly with regard to the written text of scientific books. Only ascertained facts must be put forward as such. Theories must be clearly labelled as theories, and hyptheses as hypotheses. Moreover, nothing must be

stated as universally true, if reliable authorities have put on record observations proving that the thing asserted is *not* universal. In any case, the existence of investigations giving results contradictory to those put forward by the author ought to be mentioned.

The infringement of these principles, when arising from carelessness, haste, looseness of mind, or prejudice and special pleading, tends to deprive the scientific writer of all reliability. And if the infringement has been deliberate, without the foregoing excuse, it opens him to the charge of fraud or forgery in greater or less degree, according to the degree or kind of misrepresentation, whether in plates or in texts, of which he has been guilty. (Assmuth and Hull, 1915: 3)

The controversy surrounding Haeckel's drawings began to come to a head around 1908, when it reached the German newspapers. On December 29, 1908, Haeckel himself tried to end the unpleasant affair by writing a confession to his forgery in the *Berliner Volkszeitung*. He admitted that, in his critics' terms, he was indeed guilty of "forgery." But at the same time, he was quick to argue that his scientific work was by no means a deviation from normal scientific practice in his field. Haeckel wrote:

To cut short this unsavory dispute, I begin at once with the confession that a small fraction of my numerous drawings of embryos (perhaps 6 or 8 percent) are really, in Dr. Brass's sense, falsified – all those, namely, for which the present material of observation is so incomplete or insufficient as to compel us, when we come to prepare a continuous chain of the evolutive stages, to fill up the gaps by hypothesis, and to reconstruct the missing links by comparative synthesis. . . . After this compromising confession of "forgery," I should be obliged to consider myself "condemned and annihilated," if I had not the consolation of seeing side by-side with me in the prisoner's dock hundreds of fellow-culprits, among them many of the most trusted observers and most esteemed biologists. For the great majority of all the figures – morphological, anatomical, histological, and embryological – that are widely circulated and valued in the best text- and handbooks, in biological treatises and journals, would incur in the same degree the charge of "forgery." All of them are inexact, and more or less "doctored," schematised, or "constructed." Many unessential accessories are left out, in order to render conspicuous what is essential in form and organization. (quoted in Assmuth and Hull, 1915: 14–15)

Haeckel was quite right. As mentioned above, most scientists, historians, sociologists, and philosophers of science realize that data are fitted into a theoretical structure the moment a problem is chosen, a specific technique is used, and an observation is made, for these choices reflect the scientist's intention to interpret "nature" in a certain way. One

PLATE I.
SPECIMENS OF KEIBEL'S EXPOSURES.

(1) HAECKEL'S "COPY."   (2) SELENKA'S ORIGINAL.

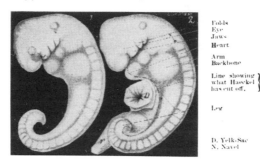

Folds
Eye
Jaws
Heart

Arm
Backbone

Line showing }
what Haeckel }
has cut off. }

Leg

D. Yelk-Sac
N. Navel

Haeckel takes Selenka's genuine figure of a *Macaque* embryo, cuts off such essential parts as the arms, legs, heart, navel, yelk-sac, so as to make it as much like a fish-embryo as possible, and then labels it "Embryo of a *Gibbon* in the fish-stage." Haeckel excuses himself by pretending that the omitted parts are *not* essential.

(3) HAECKEL'S "COPY."   (4) HIS'S ORIGINAL.

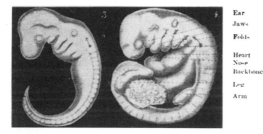

Ear
Jaws
Folds

Heart
Nose
Backbone

Leg
Arm

Compare the full embryo of a man, figured by His from life, with Haeckel's diminished figure. Haeckel omits such essential parts as the arms, legs, heart, so as to make it as much like a fish-embryo as possible, and then labels it "Embryo of a man in the fish-stage." This is not merely a mutilated copy, but a free invention, and is absolutely unlike the reality as observed by others

Figure 1. Comparison of Haeckel's drawings with those of Selenka and His (Plate I) and Huxley (Plate II). (From J. Assmuth and E. R. Hull, *Haeckel's Frauds and Forgeries*, Examiner Press, Bombay, 1915.)

cannot make a hard and fast distinction between fact and theory. The "theory-ladenness of observations" in the broadest sense can be readily recognized by any lay person who looks down a microscope at a cell and is asked to describe what he or she sees and compare that with textbook descriptions of a cell. What one should include in the reported observations is not at all clear. After drawing what one sees, one will be quickly told that much is artifact of the technical procedures: The dyes

PLATE II.
SKELETONS OF APES AND. OF MAN.

I. HUXLEY'S ORIGINAL PLATE (REVERSED.)

Man.   Gorilla.   Chimpanzee   Orang.   Gibbon.

Huxley's plate displays many differences between man and the apes:
e. g.. the bent posture of the apes, the turned-up position of their feet, etc.

II. HAECKEL'S MODIFICATION OF THE ABOVE.

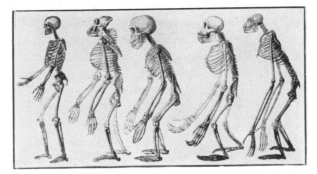

Haeckel substitutes two new figures, makes the feet of the apes flat like
those of man, straightens up their backs, cuts off the neckbones of the gorilla,
and creates an impression of sequence altogether untrue to nature, so as to
support his theory of descent.

used to stain the cell, the magnification by the microscope and so forth,
all have to be taken into consideration. One will be told that one part of
the cell is much more important than the others, that there is a hierarchy
of cellular constituents laid down by accepted theory. One might also
be told that not all cells are alike; one has to know which species one is
observing.

What counts as correct data in science is negotiated through theory,
techniques, and social interaction among scientists. To know how to use
the microscope properly, to know which cells are alike and which are

different, to know what can be left out and what has to be detailed in one's reports, are socially acquired skills. A scientist's judgment of these issues is influenced by his or her understanding of the field itself, by his or her years of training, by attending conferences, by writing letters, by working in the laboratories of others, and so on. It also requires judgment as to whose work is reliable, who is important, and whose views are authoritative. Scientific knowledge-making is a social activity.

Once we understand that all knowledge about nature is "fabricated" (i.e., constructed out of theory, techniques, and social interaction), and that all scientific observations and reports are based on conscious and unconscious biases, then one cannot judge fraud in terms of timeless truth. It is not a matter of science and truth on the one hand, versus politics and bias on the other. It is also not a matter of "betrayers of the truth." Once we understand that what counts as evidence is negotiated, we can see that what counts as a crime in science is also negotiated. The problem of fraud in science cannot be reduced to the motives of the perpetrator, and to the social barriers against the enforcement of norms. One cannot assume that the fraud or crime simply lies hidden in scientific texts, waiting to be detected. We need to know, for example, under what conditions is deliberately fudging or doctoring the data considered fraudulent in science. If fraud, like all scientific knowledge, is constructed within specific scientific contexts, then we need to know why charges are made – that is, What are the motives of those making the charge?

In Haeckel's case, the motives of his critics were clear. Many were creationists and political conservatives, who were attacking his diagrams in order to discredit the theory of evolution. Haeckel was an ardent defender of Darwinism. In Darwinism, Haeckel and many other scientific leaders in Germany found not only a means to understand "nature." They saw in Darwinism also principles for a new social order. Through his scientific writings, Haeckel took up the struggle against dogmatic conservativism in both the social and religious spheres. From the 1880s onward, he rapidly gained a following of people with similar ideas, and they employed Darwinism as their principal weapon. Haeckel brought his political, social, and biological views together into his "monistic" philosophy, the unity of mind and matter, in contrast with dualism, the separation of mind and matter.

Haeckel devoted a large proportion of his writings to methodological considerations. In his view, the correct method of research was "philosophical empiricism," the interaction of induction and deduction. His mechanical – causal approach was to take the place of any vitalistic – teleological way of viewing nature. Haeckel's major popular work, *Welt-ratsel* (the Riddle of the Universe), was among the most spectacular successes in the history of publishing. It sold 100,000 copies in its first year, went through 10 editions by 1919, and was translated into 25 languages.

It is little wonder that those who opposed Darwinism in Germany attacked Haeckel's work. They claimed that Haeckel's diagrams gave the theory of evolution a veneer of legitimacy (see Assmuth and Hull, 1915: 16). Haeckel's public statement, suggesting that he had done only what other biologists have done, caused considerable unrest among the biological community of Germany. "If true," his adversaries argued, "we have here the greatest blow ever struck against the reliability of scientific men, and against the current method of pushing the evolution theory" (Assmuth and Hull, 1915: 17). To Haeckel's adversaries it seemed that Haeckel had "cast a slur" upon the scientific community of Germany. In order to show that Haeckel's defense was gratuitous, the Kepler-Bund – perhaps the largest natural history association in Germany – sent out a questionnaire to zoologists in German universities, asking if they approved of Haeckel's methods. The effect of the circular backfired.

Many evolutionists were annoyed by the steps taken by the Kepler-Bund. They saw it as an unwarranted interference in scientific inquiry and professional freedom. In their view, Haeckel had merely "let his imagination run away from him"; he had "overstepped the bounds which separated his facts from his deductions." They refused to attack him since the concept of evolution could not be disparaged by his "inaccurately drawn diagrams." The real crime, they implied, was committed by those who sought to campaign against Haeckel in order to discredit the theory of evolution. In 1909, 46 leading zoologists and anatomists (including Theodor Boveri, Alexandre Goethe, Richard Hertwig, and August Weismann) drew up and signed collectively a declaration which they published in *Die Münchener Allgemeine Zeitung*, one of Munich's leading newspapers. They were careful to tell the public that "the undersigned professors herewith declare that they do not approve of the method of schematizing which Haeckel has in some instances made use of." But they added, "At the same time, in the interests of science and professional freedom, they condemn the warfare waged against Haeckel by Brass and the Kepler-Bund. They declare moreover that the evolutionist idea can suffer no detriment from a few inaccurately produced embryo-diagrams" (quoted in Assmuth and Hull, 1955: 19).

Those who attacked Haeckel in order to discredit the theory of evolution lost the battle. Their mistake resulted from not recognizing that Haeckel, a major theoretician and popularizer of evolutionary theory, was, in fact, on the side of the angels! Indeed, today, Haeckel's "frauds" and "forgeries" are not mentioned in any of the modern books and papers discussing the triumph of evolutionary theory. Instead, he is typically hailed as one of the outstanding theoreticians of nineteenth-century biology, as the man who coined the terms "ecology," "ontogeny," and "philogeny"; and as the "founding father" of "protistology." Nordenskiöld (1929) suggested that Haeckel was more influential than Darwin in convincing the world of the truth of evolution. And, as Nordenskiöld (1929: 505, 506) put it, "There are not many personalities who have so powerfully influenced the development of human culture – and that, too, in many spheres – as Haeckel." Many biologists and historians have highlighted Haeckel's theory of recapitulation. They have argued that it proved to be a useful heuristic which led to a splendid flowering of comparative embryology between 1870 and 1910 (Mayr, 1982: 474–475).

Given Haeckel's stature within the history of biology, it is not surprising that his so-called frauds and forgeries have been overlooked by all modern commentators on fraud in science. Had this case been considered, it may well have restrained analysts from viewing fraud in static terms of operative norms of science, or as perpetrated by "betrayers of the truth." Once we come to terms with recent social studies of science and recognize that the conventional view of science as a cooperative, disinterested activity where knowledge is certified by an objective method is pure myth, the problem of fraud appears to be more complex. It is not simply a matter of investigating the mechanisms of social control in science that prevent fraud, and commenting on the obstacles to its detection. We have to take a more dynamic approach to what counts as fraud in science, and investigate the contexts in which fraud charges are made, the negotiations that often ensue over what counts as evidence and what counts as good reason.

Fraud and the knowledge-making process

Fraud charges in scientific controversies are often made to discredit a competing individual or theory. And there is no doubt that they may damage the integrity of the individual and the claims that he or she is putting forward. For example, there is no question that the charges made against Kammerer, and his subsequent suicide, generally helped to discredit the concept of the inheritance of acquired characteristics. As will be discussed later in this book, the charge of fraud certainly made those scientists who were sympathetic to the idea more reluctant to make similar claims. This is not to suggest that no line can be drawn between real misdemeanor and "mere rhetoric." I am only insisting that the line cannot be drawn in terms of institutionalized moral norms of science or in terms of an ahistorical scientific method. The line between "rhetoric" and misdemeanor, like all scientific knowledge, has to be placed in its technical and social contexts.

This investigation will lead us deep into what is usually called the scientific method – to examine the difficulties scientists have in evaluating and assessing the validity of knowledge claims. If scientific observations are theory-laden and often require special personal skills, then disconfirmation of another's work through replication becomes a difficult and precarious task (see Rosenthal, 1966; Collins, 1985). Scientists often find it difficult to distinguish between a "failed replication" due to the replicator's error, and a "disconfirmation." They often find it equally as difficult to differentiate simple error from intended deception. Scientists' assessment of the validity of scientific knowledge claims extends well beyond experimental evidence alone. As we shall see when examining the controversy surrounding the work of Franz Moewus, it involves an understanding of the novelty of the claims in question, their fit with existing knowledge, judgments about the competency and integrity of the individual in question, and the credibility of others who bear witness and give testimony to their validity.

When investigating the controversy surrounding Moewus, one cannot simply examine the biases of those who believe in the reality of his reports, and celebrate the "integrity and courage" of those who attacked him. We have to examine where the truth lies on both sides of scientific controversy. The scientists' conclusion that Moewus was wrong all along does not help us in the slightest as we search for truth in the midst of controversy. It does not help us to understand how the controversy originated, how it was sustained for two decades, how it was closed, and how the stamp of fraud was constructed and placed on him. We have to let Moewus's fraud be the result of controversy, not its cause (cf. Latour, 1987). Haeckel, Mendel, Millikan, and Newton were winners in the controversies surrounding them, and no historical information we can offer to show that they consciously and/or unconsciously faked data will change that. Perhaps, what makes the case of Moewus so different was that he lost.

In saying this, I do not mean to imply that "reality" plays no role in the construction of scientific knowledge. "Material reality" imposes an important constraint on the production of scientific knowledge. But it never acts alone in the construction of knowledge-making and knowlege-breaking claims. One has to take into consideration the "domestic politics" of science – the strength relations among competitors in science resulting from their socially recognized technical achievements and institutional positions in science. One also has to consider the constraints resulting from the "external politics" of science – the relations of scientists and the knowledge they produce, to society at large. Scientific knowledge results from the interaction of all three constraints.

Once we come to grips with recent social studies of science and recognize that science is not the unique activity that had previously been imagined, the question "Is science fiction?" should not appear to be such a radical suggestion. Certainly the conventional view of science as a disinterested activity where knowledge is certified by an objective method is pure myth. But by posing the above question, I mean more than this. I mean that science may be fiction in the sense that arguments put forward for the existence and nature of things, events, and processes are creatively constructed, invented, and composed by scientists. Scientists are creative writers, not unlike fiction writers, who write their accounts in such a way as to persuade the reader of the truth that they are attempting to illustrate.

In the controversy surrounding Franz Moewus, we shall follow the scientists as they themselves explore and discuss every aspect of the knowledge-making process. We shall see that the scientists in this controversy have raised many issues that have only recently begun to be recognized by those historians, sociologists, and philosophers whose work is dedicated to understanding scientific method. We shall see why they find difficulties in judging whether data are correct or incorrect on the basis of experimentation alone; in discussing their own theoretical and political biases; in questioning each other's motivations and interpreting each other's experimental data. Yet, we shall see how when the controversy came to a close, scientific writers omitted all the nonrational elements of the scientific process – "all the human passions and prejudices that shape the original findings" – and presented "a desiccated residue of knowledge" that only *appears to be* so "distant from its human originators that it at last acquires the substance of objectivity" (Broad and Wade, 1982: 217).

The production of scientific knowledge is based on a series of "stories" intricately spun together in unique ways. Our aim is to try to unravel them. The knowledge-making process begins with the theories and laws one attempts to illustrate, or that one is using for explanation and prediction. It moves to the technical procedures employed, to the informal recording of data which is then reorganized, refined, polished, and formally presented in original published reports. We will explore the rhetorical nature of the "story" told in scientific papers – that is, how it often conceals the motivations and thought processes of the author.

The production of scientific knowledge progresses from the writing of original scientific reports to the composition of review papers. Although review papers also marshal data in order to persuade the reader of a particular view, they often differ in their structure. Review papers are often explicitly polemical. They often contain discussions of opposing theoretical views, expose conflicts between competing groups, and explicitly appeal to authority figures. Other review papers simply present a chronology of research reports on a particular theoretical problem. We shall see that the literature cited in such review papers does not simply reflect a disinterested attempt to recognize the contributions of others. Instead, the practice of citing others is an attempt to enlist other scientists and their work in support of a particular view.

We shall see, too, that the process of enlisting others is done selectively according to the theoretical and social biases of the author of the review paper. Quite frequently, the authors of such papers enlist the names and evidence of other scientists to support their views without regard for, and sometimes in virtual conflict with, the original intentions of those scientists. What frequently results is not only a "misrepresentation" of the original authors' views; a pseudohistorical continuity of research and rational argument is also constructed. This rational reconstruction gives the illusion of intellectual coherence, social order, and inherent logic and progression in the field. These intellectual syntheses are again reprocessed and further refined and reorganized in scientific texts.

The final stage of this knowledge-making process, as is well known, ends with mythical and semimystical accounts of the actual discoveries that are frequently found in the "historical" introductions of scientific texts. Indeed, the historical myths perpetuated by scientists are often embedded in the very narrative of scientific texts. These discovery accounts are frequently recited and recorded in the proceedings of scientific meetings. They usually take the form of celebrations of the victors. The historical record resulting from such accounts is often tidied up by omitting the losers in scientific controversy. They are normally doctored again by omitting any "false" views of the individuals one wishes to celebrate and focusing only on their "correct" views. They are heroic tales in which the views of the celebrated predecessor(s) are often taken out of their own context and placed into a contemporary one, where new meaning is bestowed upon them. These discovery accounts are not frauds, nor are they simply the writings of ahistorically minded and naive scientists. On the contrary, we shall see that these "folk histories" play important roles in science. They provide historical and cultural identities for different scientific groups. They also are important argumentative resources in scientific knowledge-making itself. And it is here that we shall begin our story.

Before proceeding directly to the controversy surrounding Moewus, then, we shall start our journey where scientists end theirs. We shall begin with their account of the origins of biochemical genetics. Our investigation of scientists' accounts will take us back to the origins of genetics when crucial questions about the gene and how it controls inherited characteristics were first being posed. Their discussion will take us up to the 1940s and 1950s and provide the necessary historical background for understanding the ensuing controversy surrounding the work of Franz Moewus.

2. Founding-father fables

Like Mendelism, biochemical genetics began with a remarkable investigator whose work remained unappreciated for a third of a century. I refer to A. E. Garrod, later Sir Archibald, English physician and biochemist who, shortly after the turn of the century, proposed that the disease alkaptonuria in man is a Mendelian recessive trait. (Beadle, 1958: 156)

The work of Garrod was certainly known among men of medicine. Nevertheless, its pregnant insight into the nature of gene action was disregarded among geneticists. . . . Why was this development, so surely foreshadowed in Garrod's work, postponed for nearly forty years?
(Glass, 1965: 231)

Garrod's work, like Mendel's, was largely ignored for 30 years. Why no one picked up Garrod's lead is one of those difficult historical questions that deserve more study. (Allen, 1978a: 199)

The pattern of discovery, neglect, and rediscovery is often repeated in accounts of the history of genetics. The classic story of this kind is that surrounding Gregor Mendel. It is almost universally accepted today that Mendel's discovery of the principles of heredity was resisted from the time of its announcement in 1865, until the end of the century. Mendel's conception of the separate inheritance of characters ran counter to the predominant conception of joint and total inheritance of biological characteristics. It was not until biologists changed their views and concentrated their research on the inheritance of individual characters that Mendel's discovery was recognized. Mendel's laws are said to have been "rediscovered" independently in 1900 by three botanists – Hugo de Vries working in Amsterdam, Carl Correns working in Tubingen, and Erich Tschermak working in Vienna.

This same historical sequence of discovery and rediscovery has been applied to a second major breakthrough in our knowledge of heredity: the discovery of how genes control hereditary characteristics. During the first four decades of the present century, geneticists demonstrated that Mendelian theory could account for almost every case of the inheritance of characters between interbreeding groups of plants and animals. The work of the *Drosophila* school led by T. H. Morgan at Columbia University gave genes a physical location in chromosomes in the cell nucleus (see Allen, 1978b). However, what the gene was and how it controlled hereditary characters remained unresolved. Nonetheless, it is generally claimed today that during the first decade of the century, Archibald Garrod had made the astonishing discovery of the nature of gene action: What a gene does is to regulate biochemical reactions by controlling the specificity of a single enzyme. However, his discovery was unappreciated and overlooked for several decades until it was re-discovered independently in the 1940s by the American team of George Beadle and Edward Tatum.

Today, the study of gene action is part of the celebrated field of molecular biology. But during the 1940s and 1950s, it was at the center of the domain known as "biochemical genetics." The turning point for investigations of how genes control inherited characteristics is well acclaimed today to have occurred during the 1940s, when microorganisms were first used for biochemical genetic analysis. The biochemical genetic analysis of microorganisms was led by George Beadle and Edward Tatum, who worked on the bread mould *Neurospora*. They proposed that a gene acts by determining the specificity of a particular enzyme and thereby controls, in a primary way, enzymatic synthesis and other chemical reactions in the organism. Throughout the 1940s and 1950s their proposal became well known as the one-gene–one-enzyme hypothesis. For their contributions to biochemical genetics, Beadle and Tatum were awarded a Nobel Prize together with Joshua Lederberg in 1958.

The story of the neglect of Garrod's discovery originated in the 1940s and 1950s. It was popularized by George Beadle himself, who upheld the work of Garrod as having laid the foundation of biochemical genetics. During the 1950s and 1960s, Beadle persistently referred to Archibald Garrod as the actual "father of chemical genetics" (Beadle, 1951: 222). He acknowledged his predecessor succinctly in his Nobel Prize lecture when he claimed that he and his collaborators had only "rediscovered what Garrod had seen so clearly so many years before" (Beadle, 1958: 156). Since that time, it has come to be universally accepted that (1) Garrod was the founder of biochemical genetics; that (2) his discovery was unappreciated, almost wholly ignored, by geneticists for some 30 or 40 years; but that (3) his contributions were "rediscovered" independently by Beadle and Tatum, who were working on the same problem.

The stories written about Mendel and Garrod are extraordinary. They present us with the patriarchal image of how "the founding fathers" penetrate Mother Nature to leave a child destined for greatness. However, the child is born "premature" in an alien world. Underdeveloped and weak, it is left to die because no one is prepared or willing to take care of it. This view was well captured by Beadle himself when he wrote:

> It seems to me that, like Mendel, Garrod was so far ahead of his time that biochemists and geneticists were not ready to entertain seriously his gene–enzyme–reaction concept. Like Mendel's, Garrod's work remained to be rediscovered independently at a more favorable time in the development of biological sciences. (Beadle, 1966: 34)

The image of scientists as "ahead of their time," of discoveries as "premature," and of the "neglect stories" spun around them are strange indeed. In effect, Garrod, like Mendel is considered to be a classic case of what historians of science have called a "precursor." A precursor is someone who runs ahead of his contemporaries, someone who sets out on a journey that he or she never completes, a journey that is only later completed by another (Clark, 1959: 103; Koyré, 1973: 77; Canguilhem, 1979: 20–22; Barthélemy-Madaule, 1982: xi). Historians have had harsh words about the precursor: The precursor is difficult to conceptualize as anything but a phantom that appears only to strip science of its historical dimension. Precursors are timeless, ahistorical figures. Thanks to their almost prophetic vision, they are uprooted from their own contexts; they are said to be "ahead of their time." Yet, because they came too soon, "before their time," they do not belong to the period to which it is held they belong. "Caught between 'not yet' and 'no longer', where is the precursor's time? Uchronia, utopia!" (Barthélemy-Madaule, 1982: xi).

How have scientists constructed such strange notions as discoveries and discoverers being "ahead of their time"? Part of the answer stems from how scientists examine the past. Later generations of scientists are often concerned with those who precede them only insofar as they see in them traces of what they regard as truth. When searching for this truth, they see only what they want to see; they often impose their own meaning on what they observe, irrespective of the intentions of the original worker.

This was indeed the case for accounts concerning "the long neglect" of Mendel. Olby (1979), Brannigan (1979, 1981), and Callender (1988) have shown clearly in detailed studies that Mendel was no Mendelian. Brannigan has shown further how Mendel's discovery was attributed to him by biologists who gave new meaning to his work in a new context, at the turn of the century. Mendel's "rediscovery," Brannigan has argued, has to be understood in terms of a priority dispute among De Vries, Correns, and Tschermak. The labeling of the discovery as "Mendel's laws" was a strategy to neutralize the dispute. "This is perhaps the single most important fact in the reification of Mendel as the founder of genetics" (Brannigan, 1981: 94).

To Brannigan, the case of Mendel's "rediscovery" is a good example of his social attributional model of discovery which he juxtaposes with mentalistic models. That is, instead of viewing discovery in terms of the creative genius of scientists, Brannigan argues that discovery should be treated as a process of social recognition which only later appears to be mentalistic or independent. Within this framework, the great problems presented by scientific discovery are not simply who said, did, or "found" something first, or how several scientists sometimes almost simultaneously converge on a single theoretical model or technical procedure; the question is not how ideas come to mind, but how specific contributions come to be regarded as discoveries. As Brannigan (1979: 448) put it, "A theory of discovery should concern itself *not* with determining what makes certain discoveries happen, but with what makes certain happenings discoveries."

When reconsidering the work of Garrod and placing him in his own historical setting, I will offer several alternative suggestions for understanding the history of biochemical genetics, namely that (1) Garrod's work cannot be understood in terms of a discovery in biochemical genetics that lay waiting to be detected or ignored by others; (2) he did not make a conceptual advance that was neglected for 30, 40, or 50 years; and that (3) his work was not simply "rediscovered" by geneticists of the 1940s and 1950s. On the contrary, Garrod, who was not a geneticist, did not offer a theory of genic action. When casting him as the founder of biochemical genetics, geneticists have bestowed their own meaning on his work.

The suggestion that the long-neglect story of Archibald Garrod is indeed a myth does not explain why scientists repeatedly tell such stories. To answer this question fully one must first take into consideration that scientists' accounts of history play various roles in their knowledge-making process. They surround experimental evidence itself as part of the art of persuasion in science. "Neglect" stories have, as already mentioned, an underlying moral element; they clash with the stereotype of the scientist as the "open-minded man" (Barber, 1962). It is self-evident that persons should be excited by discoveries, intensely interested in the detailed working of nature, and committed to the elaboration of theories. This issue alone has attracted considerable attention among geneticists.

Geneticists have devoted a great deal of historical commentary in trying to explain the "long neglect" of Garrod's work. The case of Garrod, like that of Mendel (see Brannigan, 1979, 1981), has been used by scientists as a lesson to illustrate almost every condition that might contribute to the resistance of scientists to new scientific discoveries. Attention has been drawn to such apparent obstacles as (1) the overwhelming technical difficulties of doing biochemical genetic work on higher organisms; (2) the preoccupations of individuals with their own work, and the resulting lack of interest in that of others; (3) barriers between disciplines imposed by specialization; and, more specific to Garrod's case, (1) the "prematurity" of Garrod's publications, (2) his inconclusive data, (3) the "inadequately prepared biological world", (4) the demand for social applications of human genetics and the alleged resulting interference with the development of research on fundamental problems (see Haldane, 1954; Beadle, 1958; L. C. Dunn, 1965; Glass, 1965).

To understand fully Garrod's significance in the origins of biochemical genetics, one would also have to recognize the importance of "founding father mythologies" in the social and intellectual construction of science. The aggrandizement of past scientists through stories of their heroic insights and subsequent "neglect" may play an important social role in strengthening emergent scientific research traditions. Paul Forman (1969) has exposed various myths in scientists' accounts of the discovery of x-ray crystallography. He interpreted these myths as attempts to strengthen the tradition of x-ray crystallography by "tracing it to a higher, better, more supernatural reality of initial events." He argued (1969: 68) that "the traditional account may be regarded as a "myth of origins," comparable to those which in "primitive" societies recount the story of the original ancestor of a clan or tribe." Olby (1979) has put forth the same hypothesis to explain the aggrandizement of Mendel in scientists' accounts of the origin of genetics (see also Sapp, 1989).

However, the myth surrounding Archibald Garrod has been left untouched; it stands as originally fabricated by scientists in the 1950s and 1960s. The particular polemical roles played by "the long neglect of Archibald Garrod" in the foundations of biochemical genetics will be treated later when we discuss why Beadle originally constructed the myth. First, however, it is necessary to divest Garrod, the precursor, of his fictitious cloak and assign to him his own historical place as a creator of his own products with their own destinies. We shall see that the meaning of Garrod's work changed over time with a shift in contexts. It began with a recognition of the importance of enzymes for biochemistry and pathology. In genetics, it shifted to a concern with the nature of hereditary variations in order to illuminate evolutionary mechanisms. Garrod's work was given new meaning again in the field of biochemical genetics after World War II.

The physiological context

Archibald Garrod was a lecturer in chemical pathology at St. Bartholomew's Hospital, and senior physician at the Hospital for Sick Children. Around the turn of the century, he was interested in a group of congenital metabolic diseases in humans, which he later named, in the title of his now classic book, *Inborn Errors of Metabolism* (1909, 1923). Alkaptonuria, one of the first inborn errors studied by Garrod, has as its most obvious symptom blackening of urine on exposure to air. It had been recorded medically long before Garrod became interested in it, and important aspects of its biochemistry had long been understood. The substance responsible for blackening of the urine is alkapton or homogentisic acid. Homogentisic acid is oxidized to a black pigment on exposure to air, thus leading to darkening of the urine. Although alkaptonuria's biochemistry was known, very little was understood regarding its inheritance. The condition was known to be very rare, and, though found in several members of the same families, it had only once been known to be directly transmitted from parent to offspring. Garrod consulted with William Bateson and R. C. Punnet, leaders of the new Mendelian movement in England. In 1902, Bateson diagnosed the case in terms of a Mendelian recessive character (see Garrod, 1908: 5).

In 1908, Garrod discussed the case in more detail and suggested that the condition was due to the blocking of an enzymatically controlled reaction; alkapton is not further oxidized and as a consequence accumulates and is excreted in the urine. He also drew similar conclusions concerning other congenital abnormalities occurring in humans, such as albinism, cystinuria, and porphyrinuria. In his second edition of *Inborn Errors of Metabolism* (1923), Garrod extended his discussion to two other metabolic defects: steatorrhea and pentosuria. He was also able to offer further evidence for his interpretation of alkaptonuria. In 1914, a German chemist, Gross, had reported the presence, in normal blood plasma, of an enzyme capable of oxidizing alkapton and its absence in the blood of alkaptonuric persons. Garrod believed that Gross had provided conclusive evidence to substantiate his view that lack of a particular enzyme in alkaptonurics causes a chemical reaction to be blocked, thereby resulting in the accumulation of alkapton. Garrod's own words are worth quoting:

> In alkaptonuria the failure to break up the benzene ring extends to acids with hydroxyl groups in the 2.5 position other than homogentisic acid, and . . . the essential error resolves itself into an inability to destroy the ring of acids so constituted. Homogentisic acid is apparently the only compound formed in normal metabolism which offers itself for such disruption, and accordingly the alkaptonuric excretes it.
>
> This conception of the anomaly locates the error in the penultimate stage of the catabolism of the aromatic protein fraction. . . .
>
> We may further conceive that the splitting of the benzene ring in normal metabolism is the work of a special enzyme, that in congenital alkaptonuria this enzyme is wanting, whilst in disease its working may be partially or even completely inhibited. (Garrod, 1923: 85)

This quotation is taken by Beadle (1951, 1958, 1966), Harris (1963), Glass (1965), and others to suggest that Garrod took two of the major intellectual steps fundamental to biochemical genetics of the 1940s and 1950s, namely (1) recognition that lack of an enzyme is blocking a metabolic step, and (2) recognition that the specific enzyme that is normally present is absent owing to alteration of the controlling gene. Thus, it is claimed that Garrod explicitly offered an interpretation of gene action, which was later rediscovered by Beadle and Tatum and their collaborators and formulated as the one-gene–one-enzyme hypothesis. As Allen (1978a: 204) put it, "Beadle and Tatum pursued their work in complete ignorance of Garrod's similar findings 30 years previously. But in reality, as they acknowledged, they only confirmed what their predecessor had stated so eloquently."

It is clear that Garrod interpreted alkaptonuria and some other "inborn errors of metabolism" to be due to metabolic blocks resulting from a lack of an enzyme. He was also interested in applying Mendelian laws in order to understand the pattern of transmission of the congenital defects 1 – 1 that is, their tendency to occur in several members of one generation of a family, and the rarity of their direct transmission from parent to child. He tried to show that the reappearance of the metabolic anomalies may be predicted in accordance with the behavior of Mendelian recessive characters. However, he showed no interest in using his studies of metabolic anomalies as a basis for developing a theory of gene action. As F. G. Hopkins (1938: 228) wrote, "Garrod remained first and foremost a clinician with a deep sense of dignity and importance of that calling." However, I suggest that it was not solely this commitment that prevented Garrod from developing gene theory.

Garrod made no eloquent statements about gene action. He never offered a theory of gene action or mentioned genes at all. The biochemical genetic concepts which Beadle and others attributed to him were simply not addressed in any of his writings. Indeed, even though there were numerous suggestions of gene action in terms of enzymes in the genetics literature by 1923, when Garrod published the second edition of his *Inborn Errors of Metabolism*, he referred to none of them. This was not a case of a great scientist being ignored by geneticists. On the contrary, one could interpret it as a case of a physiologist ignoring geneticists. Moreover, as will be discussed momentarily, Garrod himself might have opposed the kind of biochemical genetic theory proposed by Beadle and his associates – that is, had he actually expressed any views on the matter.

The chief significance Garrod attributed to his own work was in relation to concepts of metabolism and pathology, not to gene theory. Developing a theory of gene action was not on his working agenda. Physiologists and biochemists of Garrod's day were generally hesitant to relate their work on metabolism to gene theory. A brief glance at the conceptual framework of physiological studies as they related to the metabolism and organization of cells will help us to understand why this was so.

Garrod's writings on metabolic "anomalies" were part of a conceptual transformation occurring in what in fact soon became the new discipline of biochemistry. Around the turn of the century, great strides in chemical physiology, pathology, the newly acquired knowledge of the constitution of proteins and especially of the part played by enzymes in chemical changes within the organism had profoundly modified conceptions of the nature of metabolic processes. As Garrod himself explained in 1908:

> It was formerly widely held that many derangements of metabolism which result from disease were due to a general slackening of the process of oxidation in the tissues. The whole series of catabolic changes was looked upon as a simple combustion, and according as the metabolic fires burnt brightly or burnt low, the destruction of the products of the breaking down of food and tissues was supposed to be complete or imperfect. . . .
>
> Nowadays, very different ideas are in ascendance. The conception of metabolism in block is giving place to that of metabolism in compartments. The view is daily gaining ground that each successive step in the building

up and breaking down, not merely of proteins, carbohydrates, and fats in general, but even of individual fractions of proteins and of individual sugars, is the work of special enzymes set apart for each particular purpose. (Garrod, 1908: 1–2)

Garrod's announcement was among many programmatic statements of the early 1900s which heralded the importance of intracellular enzymes in physiological processes (see Kohler, 1982: 287).

Although physiologists recognized the important role of enzymes in biochemical reactions, they were quite reluctant to relate their work on metabolic action to genetics. This reluctance was not simply a matter of technological obstacles that would have to be overcome to investigate genic action in humans and other higher organisms. One also has to recognize that there were major theoretical differences between the physiologists' conception of "life" and that of the geneticists. During the second decade of the century, the *Drosophila* school at Columbia University, led by T. H. Morgan A. H. Sturtevant, C. B. Bridges, and H. J. Muller, had lined up genes on the chromosomes, like beads on a string. Muller argued that the ability of genes to vary (mutate) and reproduce the change in successive generations bestowed upon them the building blocks of evolution. However, as H. J. Muller (1929: 914) wrote in his highly celebrated paper, "The Gene as the Basis of Life": "Just how these genes thus determine the reaction-potentialities of the organism and so its resultant form and functioning is another series of problems; at present a closed book in physiology and one which physiologists as yet seem to have had neither the means nor the desire to open!"

During the first decades of the century, many physiologists actively protested against the genetic view of the organism in terms of discrete organic bodies such as genes. As Joshua Lederberg, who first suggested to me that Garrod may not have anticipated the one-gene–one-enzyme hypothesis, pointed out, "The theoretical milieu of his time was that of 'protoplasm,' an almost mystical living colloid. Genes, when altered, might influence the workings of that protoplasm; they were not yet thought to be the exclusive or nearly exclusive seat of hereditary information (to use an anachronistically modern expression" (Lederberg, 1986: 11). Moreover, some leading physiologists doubted the actual existence of genes. They were inclined to view them merely as hypothetical elements useful for explaining the results of breeding, but of little explanatory value for physiology. The American physiologist Charles Manning Child, for example, compared gene theory to the discredited corpuscular theories of inheritance proposed by morphological theorists of the second half of the nineteenth century. Such theories, he claimed, merely assumed the properties that he wanted to explain – the properties of growth, nutrition, inheritance, etc. In Child's view, they served only as impediments to scientific knowledge:

> It is scarcely necessary to call attention to the fact that these [corpuscular] theories do not help us in any way to solve any of the fundamental problems of biology; they merely serve to place these problems beyond the reach of scientific investigation. The hypothetical units are themselves organisms with all the essential characteristics of the organisms that we know; they possess a definite constitution, they grow at the expense of nutritive material, they reproduce their kind. In other words, the problems of development, growth, reproduction, and inheritance exist for each of them, and the assumption of their existence brings us not a step nearer the solution of any of these problems. These theories are nothing more nor less than translations of the phenomena of life as we know them in terms of the activity of multitudes of invisible hypothetical organisms, and therefore contribute nothing in the way of real advance. No valid evidence for the existence of these units exists, but if their existence were to be demonstrated we might well despair of gaining any actual knowledge of life. (Child, 1915: 11–12)

The physiologists' view of the cell stood in virtual opposition to the claims of leading geneticists about the determining role of genes. From a physiological point of view, the life of a cell was not governed by discrete physical determinants such as genes, nor was it governed by enzymes. Physiologists recognized that enzymes – which were probably organic bodies (of unknown chemical composition) – controlled the speed of chemical reactions. It was clear to them that, were it not for catalytic enzymes, reactions would go on so slowly that the phenomena of life would be quite different from what they are. However, they also recognized that living reactions had another peculiarity besides speed, and that is their "orderliness." Indeed, physiologists argued that the cell was not simply a bag of enzymes. The orderliness of reactions in the cell was not due to enzymes but resulted in part from the specialization of the cell in different compartments – nucleus, cytoplasm, and so on.

That countless chemical reactions could be organized in the cell, and that catalysis was possible, was due primarily to the structure of "the cell as a whole," rather than to genes. Physiologists viewed the cell as a chemical factory where enzymes carried out chemical reactions, but the orderliness of their activity was in turn attributed to the structure of the factory. A. P. Mathews (1916: 11–12), professor of physiological chemistry at the University of Chicago, summarized this view concisely in his well-known textbook, *Physiological Chemistry:*

> The cell is not a homogeneous mixture in which reactions take place hap-
> hazard[ly], but [rather] it is a well-ordered chemical factory with special-
> ized reactions occurring in various parts. . . . The orderliness of chemical
> reactions is due to the cell structure; and for the phenomena of life to
> persist in their entirety that structure must be preserved.

Garrod dedicated his book on *Inborn Errors of Metabolism* to J. G. Hopkins, founder of the Biochemistry Department at Cambridge. There is
no reason to believe that his views would have been different had he
expressed them. Hopkins took a view similar to that of Mathews. The
concern with the structural organization of chemical reactions in cells
was one of the hallmarks of Hopkins's school at Cambridge (see Kohler,
1982: 83). In his celebrated address on "The Dynamic Side of Biochemistry" (1913), Hopkins stressed the need to appreciate the structural geog-
raphy of the cell. Hopkins (1913: 150) referred to Garrod to point out
that in modern research "the conception of metabolism in block" is "giving place to that of metabolism in compartments." At the same time,
Hopkins stressed that "life" could not be reduced to any particular molecule, but was a property of the cell as a whole which maintained its
physiological equilibrium. In Hopkins's words:

> On ultimate analysis we can hardly speak at all of living matter in the cell;
> at any rate, we cannot, without gross misuse of terms, speak of the cell
> life as being associated with any one particular type of molecule. *Its life is
> the expression of a particular dynamic equilibrium which obtains in a polyphasic
> system.* Certain of the phases may be separated, mechanically or other-
> wise, as when we squeeze out of the cell juices, and find that chemical
> processes still go on in them; but "life" as we instinctively define it, is a
> property of the cell as a whole, because it depends upon the organization
> of processes, upon the equilibrium displayed by the totality of the co-exist-
> ing phases. (Hopkins, 1913: 152)

This view is in many ways antithetical to the simple "one gene–one enzyme" or "one gene–one primary function" hypothesis of George Beadle (see Beadle, 1951: 228). The physiological view of the cell as a chemical factory, tooled with enzymes and organized by the structure of the
cell, resulting in a self-regulating dynamic equilibrium, was echoed
throughout the physiological community between 1900 and 1930 (see
Williams, 1931: 4). It was also rigorously defended by those physiologists who, unlike Garrod, concerned themselves directly with problems
of development (see Sapp, 1987a).

The problem of development was formidable. If the organism was
made up of discrete particles as Mendelism proposed, how do they
come together at the right time and place so as to constitute the adult

organism? Typical answers of the day were "It is the organism as a whole," it is a "property of the system as such," it is "organization." Some leading embryologists who were concerned with the physiology of development felt compelled to invoke unknowable vitalist principles. Others simply came to assume that "organization" of the cell or organism was a vital property unto itself, irreducible to and inexplicable in terms of the parts that comprised it. Indeed some leading physiologists considered "organization" itself to have finally become a category that stood beside those of matter and energy (see Wilson, 1923: 286).

Certainly, to most biologists the central question was not whether a mysterious unknown force was required to account for the origin, properties, and behavior of organisms. It was whether existing laws pertaining to the analysis of matter were sufficient in themselves, or whether new laws had to be found. As one observer saw the situation:

> The biochemist renders inestimable services in elucidating the chemical mechanism of living organisms but the problem of individuality and specific behavior, as manifested by living things, is beyond the scope of his science, at least at present. Such problems are essentially of distinctive vital nature and their treatment cannot be brought satisfactorily into relation at the present time with the physico-chemical interactions of the substances composing the living body. (Minchin, 1916: 35–36)

The gene theory seemed only to make matters more difficult to those physiologists who were concerned with the problem of making adult organisms out of eggs. The leading American physiologist, Jacques Loeb (1916: v), stated the problem clearly when he wrote:

> Since the number of Mendelian characters in each organism is large, the possibility must be faced that the organism is merely a mosaic of independent hereditary characters. If this be the case the question arises: What moulds these independent characters into a harmonious whole?

The whole materialist interpretation of life to physiologists rested upon the assumption that the specific character of cells must somehow depend on what was called their "organization." As Loeb (1916: 39) himself put it, "Without a structure in the egg to begin with, no formation of a complicated organism is imaginable." Even with synthetic enzymes as a starting point, which might be capable of forming molecules of their own kind, Loeb could still not imagine how a cell or organism could be built up without a structural organization to begin with.

Whereas Child simply doubted the existence of genes, other developmental physiologists attempted to formulate a compromise with the

Mendelian-chromosome theory. Physiologists such as Loeb brought the above theoretical considerations into relation with various experimental observations concerning the behavior of developing eggs and argued for the existence of hereditary determinants in the cytoplasm as well as the nucleus of the cell. They suggested that chromosomal genes controlled only such traits as eye color, skin color, and sexual differences which topped off, as it were, the fundamental organismic features. Conversely, they proposed that the structural organization and "organ-forming materials" of the cytoplasm (transmitted through the egg) were responsible for large characteristics which distinguished phyla, classes, orders, and perhaps even genera and species (see Sapp, 1987a). This belief was championed by many biologists throughout the first half of the twentieth century.

It should be clear, then, that Garrod and the physiologists of his day had no intention of formulating a general theory of heredity based on genes and enzymes. To some, the reality of the gene was dubious, and to many, genes and enzymes alone could not control the orderliness of metabolic reactions in the cell or organism. But it was not only physiologists who held these views; as we shall see, similar views were held by many leading geneticists of Garrod's day.

Presence and absence: the first genetics context

William Bateson gave Garrod's work its first genetics context when he introduced it to biologists in his two widely read texts, *Mendel's Principles of Heredity* (1909) and *Problems of Genetics* (1913). At first glance it might seem that the fact that Bateson had publicized Garrod's work should have at least heralded the association of genic action and enzymes to geneticists. Glass (1965: 231) expressed this expectation succinctly when he wrote that Garrod's "pregnant insight into the nature of gene action was disregarded among geneticists even though Garrod had relied upon the help of William Bateson in his genetic analysis." Dunn (1965: 68) wrote that "Bateson . . . helped to put Garrod on the right track in human biochemical genetics"; and Sturtevant (1965: 100), suggested, in his discussion of the case, that Bateson was the first to propose that gene action is expressed in terms of enzymes.

All this might make a case of "neglect" of Garrod's work by geneticists even more significant. However, here, too, the case of neglect loses its meaning once we take into account three issues: (1) the actual significance Bateson accords to Garrod's work; (2) the ubiquity of the gene –

enzyme relation in genetics generally; and (3) Bateson's failure to identify Mendelian factors with material parts of chromosomes.

Bateson publicized Garrod's work, not in terms of its significance for
a biochemical theory of genic action, but in the general context of evolutionary theory, and as a means toward elucidating the nature of genetic
mutations. More specifically, Bateson employed Garrod's results to support his "presence and absence" theory (which, in fact, may have done
more harm than good for publicizing the possible significance of Garrod's work for a theory of gene action). According to Bateson, recessive
Mendelian characters were merely the result of the absence of a factor
responsible for the dominant Mendelian character. This idea conflicted
with the alternative hypothesis – that separate factors (alleles) were responsible for dominant and recessive traits. Bateson's hypothesis was
simpler, and it was supported by two important features of Mendelian
mutations: Mutations were almost invariably recessives, and they often
involved the loss of a phenotypic character, or they were harmful in
some way. (Bateson had difficulty explaining the origin of dominant
characters, but suggested that they might be somehow acquired by the
organism through pathogenic organisms and integrated into the germ
by some unknown process; see Bateson, 1913: 90.)

Bateson's arguments about gene mutations led to his "unpacking theory." If most mutations were recessives and if the recessive is a gene
lacking some element that the dominant has, then evolution must proceed by the loss of elements of the genotype. Bateson's theory was severely criticized by many geneticists, especially those led by T. H. Morgan and his *Drosophila School* (Swinburne, 1962). Many rejected his
theory on the grounds that it led to absurd evolutionary conclusions.
They argued, for example, that if evolution proceeded by loss of genes,
then this would imply that simpler organisms such as worms or microorganisms would have a more complex genetic makeup than higher organisms such as humans. Bateson's hypothesis also was seen to smack
of preformationism, evolution through an unpacking.

Swinburne (1962: 141) argued, "The detailed criticisms, which when
finally developed and substantiated killed the theory, were in the main
provided by two discoveries – first, the existence of multiple allelo-
morphs, and secondly, the existence of reverse mutations." However,
Swinburne overestimated the power of observation and experimentation and underestimated the power of theory. As the geneticist Char-
lotte Auerbach (1967: 68–70) pointed out, the presence–absence theory
persisted in a modified and enlarged form throughout the 1920s and

1930s. Auerbach (1967: 68–69) put the issues clearly when she wrote about the presence–absence theory:

> Although this theory was based largely on the confusion between gene and character that is found in many of Bateson's writings, and although it is obvious that evolution could not have proceeded as a mere succession of loss mutations, experimental refutation of the presence–absence theory was extraordinarily difficult. It is true that observations soon turned up which contradicted the theory, but all these would be explained away by additional assumptions, which might have seemed plausible but were not testable. This has remained a popular way for protecting cherished theories against adverse experimental evidence.

Bateson clung to his theory and supported it with evidence drawn from Garrod's work on alkaptonuria. Indeed, this first example of Mendelism in man supported Bateson's theory perfectly. Alkaptonuria, it will be recalled, is a recessive trait and is due to the loss of an enzyme necessary to break down alkapton. Bateson referred to the work of Garrod as providing an example of his presence–absence model. In *Mendel's Principles of Heredity*, in a chapter entitled "Heredity and Pathology," he wrote:

> If, for example, a disease descends through the affected persons, as a dominant, we may feel every confidence that the condition is caused by the operation of a factor or element added to the usual ingredients of the body. In such cases there is something present, probably a definite chemical substance, which has the power of producing the affection. . . . On the contrary, when the disease is recessive we recognize that its appearance is due to the absence of some ingredient which is present in the normal body. So, for example, albinism is almost certainly due to the absence of at least one of the factors, probably a ferment which is needed to cause the excretion of pigment; and, as Garrod has shown, alkaptonuria must be regarded as due to the absence of certain ferments which have the power of decomposing the substance alkapton. In the normal body, that substance is not present in the urine, because it has been broken up by the responsible ferment; but when the organism is deficient in the power to produce that ferment, then the alkapton is excreted undecomposed and the urine is coloured by it. (Bateson, 1909: 232–233)

Bateson discussed alkaptonuria again in his book *Problems of Genetics* (1913) in a chapter on "The Classification of Variations." The context again was that hereditary "changes must occur by either the addition or loss of factors" (Bateson, 1913: 86).

The discovery Bateson saw in Garrod's work was not that genes control metabolic reactions by way of enzymes. This was not the issue. In fact, the general relationship between genes and enzymes was ubiquitous in genetics from its very beginnings. However, it was not at all considered to be a discovery of great significance. There is no question that Bateson recognized that factors or genes operated through enzymes (Bateson, 1913: 86). Bateson himself suggested that the enzymes should be thought of as gene products, not as the genes themselves. The idea that genes were themselves enzymes or in some way controlled enzymes did not represent any great intellectual leap on the part of geneticists such as Bateson and T. H. Morgan, who were originally steeped in embryology of the late nineteenth century. Although it is not usually mentioned by those who have commented on the history of biochemical genetics, embryologists in the last decades of the nineteenth century recognized that the units of heredity – if they existed – must control elementary chemical or physiological events of development, not the final characteristics as ordinarily observed. Indeed, Hans Driesch (1894), for example, is well known for his discussions of the interaction between the nucleus and cytoplasm in development in terms of ferments or enzymes.

From the very beginning of Mendelism, it seemed obvious that heredity would, at least in part, have to involve control of metabolic reactions by way of enzymes. The idea itself was fundamental but unquestioned in genetics. It was neither defended nor opposed. It was nonproblematic. As Leonard Thompson Troland wrote in 1917:

> The conception of enzyme action is, of course, one with which all biologists, including students of genetics, are extremely familiar. Probably there is no student of morphogenesis who would not consider it absurd to deny that enzymes play a very important role in individual development. In a number of cases such participation has been clearly demonstrated by experiment, and the suggestion that the germ-cell contains "determiners" for the production of enzymes, which in turn, regulate certain aspects of the development, is a common one. Several Mendelians have even hinted that the "unit characters" themselves are enzymes, but so far as I am aware, no worker in genetics, with the exception of Goldschmidt, has regarded this conception as an important one. (Troland, 1917: 327–328)

Troland was not exaggerating. In fact, the relationship of Mendelian factors to enzymes was discussed in the first studies showing that these principles applied to animals. Thus, in his first studies pertaining to Mendelism, the French geneticist Lucien Cuénot (1902), for example, had proposed that in mice hair color is determined by a chromogene and two enzymes. A gray mouse has a chromogene and both enzymes; a black mouse has a chromogene but only one enzyme; an albino has both enzymes but no chromogen. The idea that genes controlled reactions through their control of enzymes was suggested or implied by many geneticists throughout the century. O. Riddle (1908) and A. R. Moore (1912) suggested that Mendelian determiners might be enzymes. Richard Goldschmidt (1916) attempted to relate pigmentation of moths to genes through their control of enzymes. The gene–enzyme relationship was mentioned by J. B. S. Haldane (1920), H. J. Muller (1922), Bridges (1923), and Sewall Wright (1917). Morgan (1926: 257) himself also recognized that enzymes were products of genes rather than being the genes themselves.

There were various interpretations of the relationship between Mendelian factors and enzymes in the analysis of pigmentation in plants and animals. Color differences are almost always due to biochemical rather than structural variations, and biochemical information on the nature of the pigments was often known. As a result, there were sometimes rather detailed descriptions of the effects of particular genes, couched in biochemical terms, with enzymes regularly assigned major roles. In fact, Bateson's belief that genes made enzymes was strengthened by the work of Muriel Wheldale Onslow, who had worked with Bateson on the inheritance of flower pigments. Wheldale, appointed lecturer at University College, Cambridge, in 1927, taught advanced courses in plant biochemistry and carried out pioneering research on the biochemical genetics of plant pigments until her untimely death in 1932 (Kohler, 1982: 83).

The idea that hereditary factors controlled metabolic activity by way of enzymes was thus pervasive; it was not seen as a discovery at all. Why was it considered to be unimportant? In order to understand the reaction, or rather lack of reaction, to the concept that genes acted through enzymes, it is necessary to take a closer look at the attitude of geneticists toward the problems of growth and development. A brief glance at Troland's ideas and their reception by biologists is helpful. Troland (1917) attempted a sweeping identification of gene action with enzymatic control. He believed that the concept of enzyme action, or of specific catalysis (he distinguished between "heterocatalysis" and "autocatalysis" and suggested that hereditary determinants possessed both abilities), provided a general solution to many, if not all, of the fundamental problems of theoretical biology – the mysteries of the origin of life, the source of variations, and the mechanisms of heredity and ontogeny and of general organic regulation.

> It is an answer, moreover, which links these great biological phenomena directly with molecular physics, and perfects the unity not only of biology, but of the whole system of physical science, by suggesting that what we call life is fundamentally a product of catalytic laws acting in colloidal sys-

> tems of matter throughout long periods of geological time. . . . Catalysis
> is essentially a determinative relationship, and the enzyme theory of life,
> as a general biological hypothesis, would claim that all intra-vital or "he-
> reditary" determination is, in the last analysis, catalytic. (Troland, 1917:
> 327)

Troland went further and linked his theory of the chemical nature of
heredity with the chromosome theory, which was just emerging into
prominence at the time.

It is striking that, if Beadle wanted to construct a precursor, Troland
was not chosen. Troland's ideas were very similar to those that emerged
with the rise of biochemical genetics under the aegis of Beadle and his
school. But whatever merit we may want to grant to Troland's grand
ideas, they ran counter to the prevailing opinion of his time. Some geneticists, like the physiologists already mentioned, refused to accept the
chromosome theory. The hypothesis that genes were discrete entities
located on chromosomes was anathema to Bateson, the leader of the
British Mendelians (Coleman, 1970). It was opposed by Carl Correns,
one of the so-called rediscoverers of Mendel's laws in Germany. Wilhelm Johannsen, who in 1909 coined the primary genetics terms – "genotype," "phenotype," and "gene" – also opposed defining the gene
as a material particle (see Sapp, 1987a). Bateson's views toward the chromosome theory were shared by many European geneticists when he
wrote, in 1916:

> It is inconceivable that particles of chromatin or of any other substance,
> however complex, can possess those powers which must be assigned to
> our factors. . . . The supposition that particles of chromatin, indistin-
> guishable from each other and indeed almost homogeneous under any
> known test, can by their material nature confer all the properties of life
> surpasses the range of even the most convinced materialism. (Quoted in
> Sturtevant, 1965: 49)

Troland's idea that genes were enzymes, or any view that sought a
one-to-one relationship between gene and function (i.e., gene and enzyme), seemed simplistic and naive to many biologists of the 1920s,
even to those who supported or developed the chromosome theory.
During the first decade of the century, geneticists had entertained the
idea that there existed a simple one-to-one relationship between factor
and character. Eventually, in the second decade of the century, apologists of the chromosome theory were driven to abandon this idea because of accumulated evidence that a single individual gene may affect
the production not merely of one character but of many. Conversely, it

became apparent that the production of a single character may require the cooperation of many genes.

Morgan (1919: 241–246) clearly delineated this view when discussing the "organism as a whole" in terms of the "collective action of genes." In short, he argued that one-to-one theories were to be abandoned because of evidence of two phenomena: (1) multiple-factor inheritance – that is, the control of many single traits by more than one gene; and (2) pleiotrophy – that is, the multiple effects of single genes. Other leading biologists of the 1920s took the implication of this evidence a step further than Morgan and suggested that production of any single character may involve the interaction of all genetic units. The celebrated biologist E. B. Wilson expressed this view clearly when endorsing the statement of the biochemist J. G. Hopkins (1913: 152) – that cell life could not be associated with any particular type of molecule:

> Genetic evidence is now opening far-reaching horizons of future discovery by the accumulating discovery that no one of the nuclear units plays an exclusive role in the determination of any single character. . . . I believe it is not a great overstatement when I say that every unit may affect the whole organism and that all units may affect each character. We begin to see more clearly that the whole cell-system may be involved in the production of every character. (Wilson, 1923: 285)

The immediate limitations of Troland's "enzyme theory of life" became even more apparent when one considered the larger developmental problem of how the hereditary characters were woven together in a typical order of space and time to make an organism. However, to Wilson, a recognition of the problem of organization and order in the cell and complex organism did not mean that mechanistic methods and conceptions should be abandoned. It did not mean that one had to appeal to vitalism. It meant only that biologists were still far from providing solutions to such problems in mechanistic terms:

> Who will set a limit to their future progress? But I am not speaking of to-morrow but of to-day; and the mechanist should not deceive himself in regard to the magnitude of the task that still lies before him. Perhaps, indeed, a day may come (and here I use the words of Professor Troland) when we may be able "to show how in accordance with recognized principles of physics a complex of specific, autocatalytic, colloidal particles in the germ-cell can engineer the construction of a vertebrate organism"; but assuredly that day is not yet within sight of our most powerful telescopes. (Wilson, 1923: 285)

Morgan, who was the first person not a member of the medical profession to be awarded the Nobel Prize for medicine (1933), avoided the pitfalls of grand speculative unifying theories. He advanced general ideas about the gene with great caution. Morgan defined the gene only in terms of the experimental operations by which the existence of genes may be demonstrated. A Mendelian gene, in his view, was a block of self-duplicating chromosomal material that remained intact and indivisible – at least insofar as can be detected – in the process of crossing-over or during the chromosome breakage associated with rearrangement. The following quotation from his well-known book, *The Theory of the Gene*, in which his complete theory is formulated, describes only the transmission of the gene. Morgan did not attempt to define the gene conceptually nor in any way relate the gene to the character which it controls:

> The theory states that the characters of the individual are referable to paired elements (genes) in the germinal material that are held together in a definite number of linkage groups; it states that the members of each pair of genes separate when the germ cells mature in accordance with Mendel's second law; it states that an orderly interchange – crossing-over – also takes place, at times, between the elements in corresponding linkage groups, and it states that the frequency of crossing-over furnishes evidence of the linear order of the elements in each linkage group and of the relative position of the elements with respect to each other. (Morgan, 1926: 25)

Geneticists lacked the means for investigating what the genes were and how they actually directed the characters said to be under their control. Genetics, in the United States, led by the Morgan school, bypassed developmental questions of how genes control inherited characters. The predominant objective of the genetic research led by the Morgan school was the study of the transmission of visible differences between organisms capable of being crossbred. These investigations entailed a study of the sexual transmission of differences between individuals, a study of their manifestations, and an analysis of the relations of the genes in the chromosomes.

The rhetoric of "rediscovery" and the "founding father"

The question of the exact nature of genic action – how genes control enzymes – did not become a major research problem for geneticists until the 1940s. By that time, enzymes had been crystallized and biochemists had identified them as proteins. And the time had long passed since one heard major protests against identifying genes with regions of concrete chromosomes. Biochemists and geneticists began to collaborate, and new techniques emerged for investigating how genes control biochemical reactions. The American geneticist Sewall Wright signaled the academic stakes when he collected the scattered literature relating to gene action for the first time in 1941:

> There is a certain similarity in all cells in gross chemical constitution and physical organization. Yet there is the greatest diversity in the substances produced. The usual interpretation is that there is an almost infinite field of possible syntheses of which protoplasm is capable, but that in each case the course of metabolism is guided along particular paths by the particular assemblage of specific catalysts. The demonstration that several enzymes are proteins capable of repeated crystallization without loss of properties has suggested that their specificity rests on the same basis as that of other proteins or protein compounds. The model of inheritance of enzyme differences is obviously a question of great importance in physiological genetics. (Wright, 1941: 503)

Garrod's work found its third meaning in this new context with the emergence of the biochemical genetics of microorganisms.

After World War II, the face of genetics quickly changed. Many young geneticists came to concern themselves with the study of genic control over the physiological properties and chemistry of the organism. This shift in interest was conditioned by the domestication of microorganisms for genetic use. The higher organisms of classical genetics such as *Drosophila* and maize had proven to be very useful for establishing the chromosome theory of inheritance. However, when geneticists attempted to bridge the gap between gene and character, the utility of higher organisms of classical genetics was overwhelmingly surpassed by that of rapidly reproducing microorganisms such as bacteria, yeast, algae, fungi, and protozoa. The higher organisms retained their usefulness as favorite organisms for population genetics and cytogenetics.

Microorganisms allowed geneticists to avoid the complexity of tissue differentiation and integration of multicellular organisms when investigating the chemical nature of the gene and the nature of genic action. Indeed, the domestication of an organism for biochemical genetics required that not only its sex life but its growth as well be brought under meticulous control. The program of biochemical genetics that developed during the 1940s and 1950s relied on joint collaboration of a geneticist and a biochemist. The geneticist isolated mutations that were found to be unable to grow or grew poorly on a well-defined medium, and the biochemist sought the reason for this inability.

The turning point in biochemical genetics is well acclaimed today to have occurred in 1941, when Beadle and Tatum published their first paper on the genetic control of biochemical reactions in the bread mould *Neurospora* (see Chapter 3). The results of the previous 30 years, suggesting genic control of enzymes, were reported in rabbits, guinea pigs, *Drosophila*, and higher plants. However, it was not until after the work of Beadle and Tatum was published that biochemical genetics became a normative practice with more or less standardized procedures. The *Neurospora* work, led by Beadle, distinguished itself from all previous work in two significant ways: Although other investigations had shown in special instances similar gene relations with enzymes and elementary reactions, they usually had been confined to one or a few reactions in any organism. In this respect, the *Neurospora* work differed in both scope and magnitude. It also proved to be of great importance in biochemistry and the drug industry, providing a powerful tool for the analysis of the pathways by which vitamins, amino acids, and other compounds are synthesized.

However, as will be discussed in more detail in Chapter 8, Beadle himself was more interested in academic or intellectual stakes than purely financial ones. He was concerned with fundamental theoretical problems of the nature of the gene and gene action. By the mid-1940s, the team of Beadle and Tatum grew into the *Neurospora* school at the California Institute of Technology, with Norm Horowitz, H. J. Mitchell, and David Bonner. Beadle and Tatum assumed that "genes control or regulate specific reactions in the cell system either by acting directly as enzymes or by determining the specificities of enzymes." In their celebrated paper of 1941, they had argued that because the cell was a highly integrated system, "there must be orders of directness of gene control ranging from simple one-to-one relations to relations of great complexity" (Beadle and Tatum, 1941: 499). However, by the mid-1940s, they suggested a direct and simple role of genes in the control of enzymes. The numerous instances in which single gene mutations resulted in a block of a single metabolic step led them to hypothesize that many or all genes have single primary functions. One gene–one metabolic block–one specific enzyme deficiency soon became well known as the "one-gene–one-enzyme hypothesis." With respect to the specificities of enzymes, the hypothesis stated essentially that a given gene must be concerned in a primary way with only one enzyme (Beadle, 1951: 228).

The one-gene–one-enzyme hypothesis soon rose to the forefront of controversy in biochemical genetics. It was with respect to this particular hypothesis that Beadle claimed to have "rediscovered" the work of Archibald Garrod. Beadle recalled that he had first learned of Garrod through the reviews of Wright in 1941 and J. B. S. Haldane in 1942. He saw in Garrod's work the idea that each inborn error of metabolism in man, like those in *Neurospora,* could be interpreted as a block at some particular point in the normal course of metabolism, owing to the congenital deficiency of a specific enzyme. In other words, Garrod's work seemed to suggest that genes control *single* metabolic functions. The only difference would be that Garrod had shown this simple relationship to be true only for a few genes and a few chemical reactions in humans, whereas Beadle and his group clearly formulated the hypothesis and provided experimental evidence that it was true for many genes and many reactions in *Neurospora.*

Thus Garrod's work was given special significance by Beadle, and throughout the 1950s and 1960s he promoted Garrod in all of his reviews, claiming that his work had been "neglected" and "rediscovered" just like Mendel's. Beadle's acknowledgment of Garrod was not simply an act of humility; it was not one scientist giving in a disinterested way what he believed to be priority to another. In order to understand the rhetorical nature of the tale of the long neglect of Archibald Garrod's work, one first has to recognize that Beadle's own "one gene–one enzyme" hypothesis was criticized by many geneticists during the 1940s and 1950s. This issue is simply overlooked in all accounts of the long neglect of Garrod's work. All of the accounts begin with the assumption that Beadle's one gene–one enzyme theory is correct. However, as will be discussed momentarily, many geneticists of the 1940s and 1950s believed the "one-gene–one-enzyme" theory to be oversimplified and wrong.

Beadle himself tried to help make a heroic discovery out of his theory of gene action by calling upon the legendary story of Mendel's neglect and rediscovery as a precedent to the case of the correctness of Garrod's rediscovery. As Beadle (1966: 31) remarked about the "neglect" of the "discoveries" of Mendel and Garrod, "I strongly suspect that an important component of the unfavorable climate for receptiveness in these two instances is the persistent feeling that any simple concept in biology must be wrong." Certainly Beadle recognized that the idea that genes might act directly as enzymes had been suggested by many geneticists since about 1910. He also acknowledged that the idea of the one-gene–

one-enzyme hypothesis had been implied in the writings of several geneticists. Beadle also recognized that although Garrod by 1923 had concluded that alkaptonuria resulted from a deficiency in an enzyme, he
had made no statement about the nature of gene action in this case nor
in any other. As Beadle (1951: 222) wrote:

> Thus, while he did not express it exactly so, Garrod was able to conclude
> that in a person homozygous for a particular gene defect, a specific en-
> zyme is absent, alcapton is not further oxidized and as a consequence
> accumulates and is excreted in the urine. The relation gene–enzyme– spe-
> cific chemical reaction was certainly in his mind. It is proper that he should
> be recognized as the father of chemical genetics.
>
> It is a fact of both interest and historical importance that for many years
> Garrod's book had little influence on genetics.

Thus although Garrod himself never expressed the idea that a gene controls the specificity of a particular enzyme, nor referred to the work or
ideas of geneticists as they pertained to gene action, he could be forgiven.
And although he probably would have denied a theory of heredity that
was reduced to genes and enzymes, his own interests and aims could be
overlooked. Indeed, Garrod had said just enough; Beadle could now
supply the "right" interpretation, depicting Garrod as *thinking* the right
conclusions so as to make him an ally in his own intellectual struggles.

Beadle constructed the image of a heroic originator whose theory supported his own views but who was neglected. With the one-gene–one-
enzyme hypothesis widely contested, alkaptonuria became a paradigm
case, an exemplar for Beadle. As Beadle later recollected:

> The one-gene–one-enzyme hypothesis . . . is another example of a con-
> cept that was strongly resisted, again, I believe, because it seemed too
> simple to be true. I recall well the year 1953 when, at the Cold Spring
> Harbor Symposium on Synthesis and Structure of Macromolecules there
> appeared to be no more than three of us who remained firm in our faith.
> (Beadle, 1967: 341)

Indeed, the construction and promotion of the romantic myth of Garrod's creative insight and subsequent neglect by geneticists was a rhetorical device used by Beadle in his attempt to turn the one-gene–one-
enzyme theory into truth, a fact, and a discovery. First, as discussed in
the introduction of this chapter, the notion of the independence and
inevitability of the truth is embedded in the notion of rediscovery. Second, the charge of "neglect" suggests that Garrod's ideas, as well as
Beadle's, were not given a fair hearing; they were not assessed properly
on rational grounds. This moral element of the story is well revealed by

the geneticist Bentley Glass (1965: 233): "The first big steps forward in understanding the nature of gene action were those made by Garrod, so ignored by the fraternity of geneticists who perhaps were too engrossed in their own experiments to read anything not published by another recognized geneticist." One can easily deny the historical merit of this claim, but its message is clear. The truth was overlooked because of the false pretenses and self-interested activity of others.

All writings on Garrod's neglect and subsequent rediscovery were, in effect, designed to suggest that those who protested against the one-gene–one-enzyme theory, for whatever reason, simply ignored the truth of the experimental results. They imply that the experimental evidence – and "proof" – of the one-gene–one-enzyme theory was unquestionable. The theory was simply true, incontestable on the basis of any rational argument. But these claims are flawed on various grounds. First, these accounts suggest that there was no experimental evidence launched against the theory. As discussed earlier, one-to-one theories of gene action had been entertained during the first decade of the century but were dismissed on the basis of experimental evidence: that one gene may be concerned with many characters and that a single character may depend on many genes. The one-to-one theory simply took a new form under Beadle and his followers when it was applied to metabolic processes: It was assumed that a given gene specifies only one enzyme and one chemical reaction. The charge of neglect ignores this earlier work and the experimental objections to one-to-one theories. But those who opposed the one-gene–one-enzyme theory did not ignore the experimental evidence. The conflicting experimental evidence for and against the one-to-one theory suggests that experimentation alone cannot be an objective arbiter in theory choice.

This raises the second question, as to *how definitive* experimental evidence can ever be. As we have seen, Beadle claimed that his theory was true but was rejected because of the "feeling" that it was "too simple." This account ignores prior experimental evidence geneticists had accumulated that contradicted one-to-one theories of gene action, and suggests that the experimental evidence for the one-gene–one-enzyme theory was indisputable and based on pure empiricism. This claim also ignores the biases of the technical procedures used by Beadle and his followers to obtain the experimental results that supported their theory. In fact, the biases imposed by the techniques used by the *Neurospora* school were addressed by prominent scientists of the 1940s and 1950s (see Sapp, 1986, 1987a).

Max Delbrück put forth one of the most significant public criticisms of this kind in the summer of 1946 at the Eleventh Cold Spring Harbor Symposium on Quantitative Biology which was dedicated to "Heredity and Variation in Micro-organisms." The apparent triumph of the one-gene–one-enzyme theory was the topic of greatest interest to the audience. Several speakers presented their analyses of biochemical mutants in fungi and bacteria, in strong support of Beadle's theory. After one of these presentations, Max Delbrück pointed out that the data presented were only *compatible* with the interpretation; they could not be considered to offer *proof* of the validity of the theory.

Second, Delbrück pointed out that the very procedures of isolation of the mutations in the *Neurospora* work predetermined the result by restricting possible alternatives. He argued, in effect, that if one gene normally controlled many enzymes, no mutations in such genes could be detected by the procedures used by one-to-one theorists. Delbrück concluded his criticisms by challenging the champions of the one-to-one theory to devise an experiment by which their interpretation could be disproved:

> To sum up: in order to make a fair appraisal of the present status of the thesis of a one-to-one correlation between genes and species of enzymes, it is necessary to begin with a discussion of methods by which the thesis could be *disproved*. If such methods are not available, then the mass of "compatible" evidence carries no weight whatsoever in supporting the thesis. (Delbrück, 1946: 23)

Delbrück's challenge was easy to make but difficult to meet. Advocates of the one-to-one theory argued that it was difficult, if not impossible, to devise an experiment that would disprove the thesis. The *Neurospora* geneticist David Bonner defended the one-to-one theory by turning the question into one of scientific value and judgment. He saw little value in questioning the one-gene–one-enzyme proposal, at least as a working hypothesis. He argued that it had merit in terms of its simplicity and in terms of its ability to account for the data: "Since the data now available are accounted for on a one-to-one basis, it would be of little value to devise a more complex thesis of the relation of genes to enzymes" (Bonner, 1946: 23).

Beadle (1951) discussed the evidence for the one-gene–one-enzyme hypothesis at length and confessed that he too knew "no way of proving it even a single instance" (Beadle, 1951: 234). He discussed slightly different modifications of the one-gene–one-enzyme hypothesis, but again he could not provide definitive ways to test them. Beadle reasoned that though the one-gene–one-function hypothesis "without doubt served a useful purpose" and there were "no compelling reasons for abandoning it," it should not be accepted without reservation. Perhaps bending under the pressure of his critics, Beadle (1951: 234) concluded, "Even if it should prove correct in principle, like many useful working hypotheses in science it may well be found *to err in the direction of oversimplification*" (italics mine).

The controversy over the one-gene–one-enzyme theory continued into the late 1950s, when Beadle and Tatum were awarded a Nobel Prize, which they shared with Joshua Lederberg. It might appear that the fact that Beadle and Tatum were deserving of a Nobel Prize meant that the one-to-one theory of gene action was socially recognized as a discovery. However, although the Nobel Prize no doubt helped to legitimize the theory, it was by no means decisive. Indeed, it was not clear to many of the leading geneticists, who were not privy to the Nobel Prize councils, what the prize was for – that is, exactly what had been discovered. Some suggested that Beadle and Tatum received only one-half a prize between them because some geneticists, including the influential H. J. Muller, believed the one-gene–one-enzyme relation to be not only unproved, but possibly leading in a direction that would actually obscure progress (see Sonneborn to Ephrussi, November 10, 1958).

If the Nobel Prize was awarded simply for "exposing" the general relations of genes to the enzymatic control of reactions, many geneticists would have had excellent grounds to feel discriminated against. This would be especially true of Boris Ephrussi. As will be discussed in the next chapter, it was Ephrussi who first led Beadle to the study of biochemical genetics. During the mid-1930s, Ephrussi and Beadle had done collaborative work on *Drosophila* eye color mutants – work that was highly regarded by geneticists for showing the general relations of genes to enzymes. Ephrussi himself was hit hard by the Nobel Prize decision. As he wrote to T. M. Sonneborn (October 6, 1958), "I must admit I *was* disturbed these days by the Nobel Prize. I hate to admit it: I suddenly felt my life wasted." Many French-speaking biologists were also shaken by the decision. Sonneborn wrote to Ephrussi (November 10, 1958) a beautiful, consoling letter:

> I hope by now you have gotten over the first rough shock and can not only see the whole thing in its proper perspective, but can feel that your life and work – its value and usefulness – are not to be evaluated as success or failure on the basis of winning or not winning a Nobel Prize. There are a lot of first class researchers in modern biology. . . . I hope they all get top recognition. Of course, who wouldn't like to get a Nobel Prize! But

there are satisfactions in less glamorous rewards, even in the act of discovery or thought itself. At least they have been my chief satisfactions and they are sufficient. I'm sure you too have had them and will also have them as sufficient when the fire dies down a bit and the burn is less acute.

I hope you will forgive me for writing in this way, but your letter revealed you as so miserable that I felt I had to speak up. We all have our secret disappointments and regrets, our moments of doubt about the wisdom or the value or the justification of our labors and our life. For all but a few, the only possible stable source of motivation is one's inner resources, not external recognition or applause, however much they or their desire may add to the motivation of some. Perhaps this is just a compensation on my part for the certainty that top recognition can never be mine. But I don't think so, because I felt the same before and when that certainty was not so certain. I hope you can reconcile yourself to yourself and still preserve the zest and drive for discovery which has in the past resulted in a degree of fame and glory for you which should also be a great satisfaction.

In light of this priority dispute, one might suggest a final reason why Beadle promoted Garrod as a precursor. It will be recalled that Brannigan argued that the so-called rediscovery of Mendel was an attempt to neutralize a priority dispute between de Vries, Correns, and Tschermak. One might suggest another parallel in the story of the "rediscovery" of Garrod. By constructing the long neglect of the work of Archibald Garrod, Beadle himself circumvented all his competitors who had found evidence to support the relation of genes to enzymes. He could now attribute the idea and its illustration to a predecessor who was long dead and out of the contest. However, I do not believe that Beadle's construction of the "profound neglect" of Garrod's work was as simple as that. In making this statement, I fully recognize that when scientists appeal to history, they often do so to bolster their own place in it. However, one cannot always make a clear distinction between scientists' search for personal recognition and their search for recognition of the "truth."

In this chapter, I have tried to show that scientists' historical writings do not simply play a passive role of celebrating the achievements of individuals; nor do they account for error simply by illustrating obstacles to the "truth" and the progress of reason. Scientists' accounts of history have always played a role in the very constitution of scientific knowledge. They surround experimentation to strengthen the validity of knowledge claims. I suggest that the casting of Archibald Garrod as the "father of biochemical genetics" was done by Beadle to supply his experimental evidence for the one-gene–one-enzyme theory with a plot that would parallel the rediscovery of Mendel's work.

But one last riddle remains to be resolved in our understanding of the foundations of biochemical genetics. If the Nobel Prize was not meant to sanction or legitimize the one gene–one enzyme theory as a discovery, then what was the discovery attributed to Beadle and Tatum? Many geneticists suspected that their share of a Nobel Prize was for developing the *Neurospora* methodology which proved to be so fruitful for both biochemistry and biochemical genetics. In fact, some leading geneticists, such as Sonneborn, believed that Beadle and Tatum's technical contributions to the development of the biochemical genetics of microorganisms was in itself worthy of a full prize, regardless of the degree to which the one-gene–one-enzyme theory was correct (see Sonneborn to Ephrussi, November 10, 1958). The discovery was not a theoretical achievement of *solving* the problem of gene action; it was a technical achievement of showing how to use microorganisms for biochemical genetics and for *creating* and investigating various problems.

But were Beadle and Tatum the first to use microorganisms for biochemical genetics? Did they originate the whole idea of using microorganisms for biochemical genetics? It is fair to say that historians and biologists have come to believe they did. But here, too, the historical accounts are contrived. As we shall see in the next chapter, Beadle and Tatum were not the first to initiate the biochemical genetics of microorganisms. The origin of biochemical genetics is both obscure and complex.

3. Sex and the simple organism

> The work of Moewus has placed the genetics of protozoa on a new foot-
> ing. It has brought the phenomenon of inheritance in these organisms into
> the same system that is manifested in the Mendelian inheritance of higher
> organisms. (Jennings, 1941/1964: 750)

> Without doubt the most remarkable series of studies in biochemical genet-
> ics is that of the German investigators Moewus, Kuhn, and co-workers on
> the flagellate *Chlamydomonas*. (Beadle, 1945: 43)

The canonical accounts claim that the turning point in biochemical genetics occurred in 1941 when Beadle and Tatum published their first
paper on the genetic control of biochemical reactions in *Neurospora*.
Beadle and Tatum are acclaimed for the basic experimental design that
characterizes modern biochemical genetics, and for the one-gene–
one-enzyme hypothesis which played a directive role in biochemical genetic theory during the 1940s and 1950s. As discussed in the previous
chapter, by the mid-1930s, many geneticists, from those using guinea
pigs to those investigating the chemical effects of genes on the color of
flower blossoms, were interpreting their results in terms of genes controlling the character of enzymes. At the technical level, these early experiments can be easily distinguished from the work of Beadle and Tatum. They are universally recognized today for initiating the first
extensive study of biochemical genetics in any microorganism.

Yet, Beadle and Tatum were not the first to use microorganisms for biochemical genetics. The German biologist Franz Moewus was the first.
During the decade following World War II, Moewus was celebrated by
many biologists as one of the outstanding scientists of the twentieth century. In fact, some leading geneticists argued explicity for the priority of
Moewus's work over that of Beadle and Tatum in laying the foundations
of the modern biochemical genetics of microorganisms. Yet, since that
time, scientists have written many detailed accounts of the rise of biochemical genetics and the resulting revolution in biology. Their accounts
have been followed by those of many historians and science journalists
who have celebrated the triumph of molecular biology. The work of
Franz Moewus has been excluded from all of them. The myth that Beadle

and Tatum originated the whole idea of using microorganisms for biochemical genetics has been perpetuated among current geneticists.

Moewus began to work on the genetics of microorganisms in the early 1930s. By 1937, he began to carry out intensive studies of the biochemical genetics of sexuality in the unicellular green algae *Chlamydomonas*. This work attracted the interest of Richard Kuhn, Nobel Prize winner in biochemistry, who began to collaborate with Moewus. Moewus's first paper along these lines appeared in 1938, three years before Beadle and Tatum's first paper on *Neurospora*. In 1940, before scientific communication between Germany and the United States was cut off by war, Moewus published a series of four remarkable papers dealing with the general theory of biochemical genetics and with its application to *Chlamydomonas*. He clearly set forth how the genetic mechanism of all aspects of the organism – morphogenetic, physiological, and biochemical – could be worked out. These papers were of even more specific importance because they represented the first extensive systematic investigations of biochemical genetics in any microorganism. They were also the first studies purporting to demonstrate the precise manner in which biochemical activities of the cell were controlled by gene and enzymes. In fact, Moewus had provided the first description of the complete life cycle of an organism in terms of genes and molecules. By 1951, Moewus claimed to have detected the effects of, and mapped more than, 70 genes in *Chlamydomonas*, thus making that organism one of the best-known genetic materials of the time.

In order to appreciate the significance of Moewus's contributions, it is first necessary to examine more closely the birth of modern biochemical genetics. As discussed in the previous chapter, the success of biochemical genetics relied on the domestication of microorganisms for genetic use. However, domesticating microorganisms for genetic use and perceiving their usefulness for investigating the biochemical effects of genes were not straightforward tasks. Throughout the first quarter of the century, microorganisms had been used more and more for the study of metabolism and enzyme activities by physiologists and biochemists. The development of medicine and industry demanded accurate identification of germs and knowledge of their properties. Microbiologists studied the growth and life cycles of microorganisms and succeeded in defining their nutritional requirements, their ability to use certain compounds as a source of carbon, and their sensitivity to antimicrobial substances. In all biochemical respects, microorganisms seemed to be fundamentally similar to complex organisms.

Although microorganisms were often well known from a biochemical point of view, prior to the 1930s and 1940s they had been largely excluded from genetics. In most cases, the mechanism of inheritance in microorganisms had remained a mystery. Part of the reason for their exclusion from genetics was that crossbreeding analysis – a sine qua non of Mendelian genetics – had rarely been achieved in microbes. The role of the chromosomes and the mechanics of heredity had been established by a combination of genetic investigation and cytological observation. But nothing of that kind was easily achieved in microbes. In most instances, microorganisms seemed to reproduce solely vegetatively; they showed little signs of sexuality. Generally, their small size and lack of obvious sexual differentiation made it difficult to distinguish between somatic and germinal elements, between character and factor, between phenotype and genotype. Microbes seemed not to follow the concepts and mechanisms of genetics. In fact, many protozoologists and bacteriologists came to believe that Mendelian principles did not apply at all to microorganisms. Thus, for the most part, microbiology and genetics had developed apart.

The investigations that led Beadle and Tatum to the study of the biochemical effects of genes in *Neurospora* are well recorded. Beadle's work in biochemical genetics began with a series of investigations on eye color mutations in *Drosophila*, which he carried out in collaboration with Boris Ephrussi. The celebrated attempts of Ephrussi and Beadle to construct a chain of reactions connecting the gene with visible character in *Drosophila* have become classics in the history of genetics. Ephrussi was born in Moscow and immigrated to France to take up the study of biology at the Sorbonne after the Bolshevic Revolution. His doctoral dissertation, carried out in the late 1920s under the supervision of the biochemist Louis Rapkine and the well-known protozoologist Fauré Fremiet, was on experimental embryology and tissue culture. At that time, genetics had not been shown to apply to embryology, and the problem of cellular differentiation seemed to be outside the mechanistic confines of genetics. Ephrussi's primary interest in genetics was to formulate "the chain of reactions connecting the gene with the character, this chain [not only] being important . . . as an eventual indicator of the gene, but also having a bearing on the general problem of differentiation" (Ephrussi, 1938: 6).

In the 1930s, *Drosophila* was the only convenient resource, owing to the number of hereditary characters described in that organism. In order to familiarize himself with the genetics of *Drosophila*, he spent the academic year 1934–1935 on a Rockefeller Foundation fellowship in Morgan's laboratory, which in 1928 had moved from Columbia University in New York to the California Institute of Technology in Pasadena. There were essentially two principal ways of approaching the gene and genic action. One could try a direct approach, which involved investigating the gene from the gene end of the chain of reactions connecting the gene with the character. In the 1930s, such studies were carried out in relation to a variety of problems and included such studies of the different effects of a particular gene depending on its location on the chromosome, studies of the process of mutation, and studies of the structure of salivary gland chromosomes. On the other hand, studies of the effects of genes on development, which interested Ephrussi, represented a second approach: starting at the "character end" of the postulated chain of reactions.

Unlike Ephrussi, Beadle was trained, not in biochemistry nor in embryology, but rather in cytogenetic studies of corn under the direction of R. A. Emerson at Cornell. Emerson developed one of the largest schools of genetics in the United States and made corn second only to *Drosophila* as an object for genetic investigation. Many of Emerson's students would later become well known, including Barbara McClintock, Marcus Rhoades, and Milislav Demerec. Beadle was a postdoctoral fellow when Ephrussi arrived in Morgan's laboratory at the California Institute of Technology in the 1930s. When Ephrussi returned to France from Morgan's laboratory in 1935, Beadle followed him. In Ephrussi's laboratory in Paris, they developed a transplantation technique whereby they removed embryonic eye tissue from *Drosophila* larvae of one genetic constitution and transplanted them into the body of larvae of a different genotypic constitution. By using eye colors as an example of a character controlled by genes and by applying the transplantation procedure, it was possible to begin to elaborate a chain of reactions leading to pigment formation. This technique was applied to a series of eye color mutations in *Drosophila* – "vermillion" and "cinnabar."

Beadle and Ephrussi soon reported that the two eye-color mutants carried two different blocks in the reaction leading to the development of the normal deep red color of wild-type flies. On the basis of these studies, Beadle and Ephrussi (1936) concluded that two specific gene-controlled substances (eye hormones) intervened in the formation of eye pigment. The next task was to try to establish the chemical identity of the two substances. Ephrussi continued to work on the problem in Paris until the German army invaded. He then flew to America to take up a post as refugee scholar at Johns Hopkins University, which he held until 1944 before becoming active in Les forces françaises libres (see Sapp, 1987a). Beadle, on the other hand, had returned to the United States in 1937 to take up a position at Stanford University. He hired Edward Tatum as a research associate. Tatum was not trained in genetics, but in bacteriology and biochemistry. His doctoral dissertation, completed in 1935, was part of a pioneering wave of research on the physiology and nutrition of bacteria, especially the role of vitamins in bacteria nutrition.

The efforts of Ephrussi and co-workers and Beadle and Tatum were only partially successful. They were able to identify one of the eye color hormones as the amino acid tryptophan. However, after three years of systematic work, all efforts to isolate and characterize the second hormone were unsuccessful. Beadle (1951: 224) later remarked that the great difficulty of identifying the second hormone was "a blessing in disguise" as far as he and Tatum were concerned, for in frustration they were led to abandon *Drosophila* as an object for biochemical studies, and turned their attention instead to the bread mould *Neurospora crassa*.

According to Beadle (1958), the idea of switching to *Neurospora* occurred to him in the middle of a series of lectures given by Tatum on comparative biochemistry. In the course of his lectures, Tatum recounted the nutritional requirements of yeast and fungi, some of which exhibited well-defined blocks in vitamin biosynthesis (see Lederberg, 1986). According to Beadle, while listening to one of Tatum's lectures, it occurred to him that they ought to be able to use microorganisms to reverse the procedure they had been following for investigating the general problem of genic action. Instead of attempting to work out the chemical basis of known genetic characters, they should be able to select mutants in which known chemical reactions were blocked (Beadle, 1966). The idea simply was to select mutants unable to synthesize known metabolites, such as vitamins and amino acids, which could be supplied to the medium. A mutation unable to make a given vitamin could be grown in the presence of that vitamin and classified on the basis of its differential growth response in media lacking or containing it.

Neurospora (Figure 2a) seemed to be an ideal organism for this kind of work. A study of its life cycle and genetics had been begun by B. O. Dodge (1927) of the New York Botanical Garden. When, in 1931, Beadle had arrived in Morgan's laboratory at Cal Tech, Morgan had already assigned a doctoral student, Carl Lindegren, to work on the genetics of *Neurospora* (see Lindegren, 1932, 1934). It had also been recently shown that *Neurospora* could be grown in a culture medium of known chemical

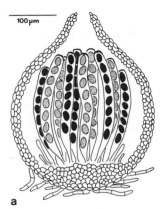

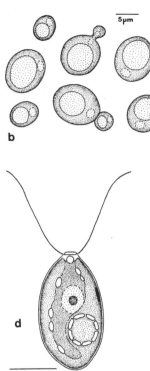

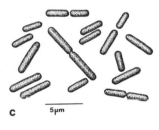

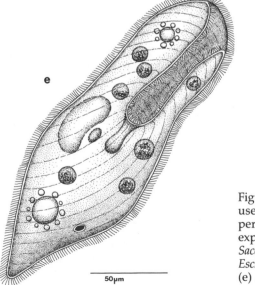

Figure 2. Some microorganisms used in genetic studies: (a) perithicum of *Neurospora crassi* with exposed asci and ascopores; (b) *Saccharomyces cerevisiae;* (c) *Escherichia coli;* (d) *Chlamydomonas;* (e) *Paramecium.*

composition. It soon proved to be an ideal organism in which to seek mutations with biochemical effects illustrated by nutritional requirements.

In the spring of 1941, Beadle and Tatum were exposing *Neurospora* to x-rays or ultraviolet light and seeking the resultant mutants. *Neurospora* soon moved to center stage as the chief organism for biochemical genetic experiments. Despite the exigencies of the war effort, almost immediately an increasing number of graduate students and postdoctoral fellows hurried to Stanford to learn the new techniques of biochemical genetics and to exploit the use of *Neurospora* mutants for the exploration of biochemical pathways. Beadle developed a strong school which included Tatum, M. B. Mitchell, Norman Horowitz, David Bonner, and others. In 1945, Tatum moved to Yale University and Beadle and his group left Stanford en bloc to reshape the program of biology at the California Institute of Technology in Pasadena. By 1946, the *Neurospora* school at Cal Tech constructed a library of about 100 genes controlling vital synthesis. The majority of the mutants that had been obtained were characterized by loss of the ability to synthesize a vitamin, an amino acid, or a nucleic acid component. The investigations of the *Neurospora* school initiated great advances in the knowledge of metabolic pathways and of the ways in which genes act.

Others soon began to apply similar procedures to the study of other microorganisms, especially yeast and bacteria. Ever since Pasteur's day, yeast had been one of the most frequent objects of biochemical studies largely because of its importance to the beer industry. However, genetic investigations of yeast began to emerge only in the late 1930s, mainly because their life cycle was not understood. The early work on the genetics of baker's yeast (*Saccharomyces cerevisiae*; Figure 2b) was done by Otto Winge and his collaborators at the Carlsberg laboratories in Copenhagen (see Winge, 1935; Winge and Lausten, 1938). Following World War II, yeast genetics was developed most prominently by Carl Lindegren and Sol Speigelman at Washington University, St. Louis, Missouri, and by Boris Ephrussi, Piotr Slonimski, and their collaborators at the Rothschild Institute for Physicochemical Biology in Paris.

The study of biochemical genetics in bacteria began later. By the early 1940s, it began to be clear that the genetic systems of the great majority of organisms were very similar: Their hereditary machinery was organized into discrete chromosomes of definite size, shape, and genic makeup. The chromosomes divided normally by mitosis and, at a certain point in the life cycle, underwent meiosis, which was accompanied by crossing-over. This overall scheme pertained to higher plants and animals and to many microorganisms as well. Bacteria, on the other hand, seemed to be forever excluded from this generalization. They seemed to lack the ability to reproduce sexually and consequently had a different type of evolution. There were many attempts to formulate alternative systems for microorganisms such as bacteria which were less amenable to recombination analysis. For example, in his famous book, *Evolution: The Modern Synthesis*, J. S. Huxley (1942) beautifully summarized what seemed clear to everyone as the major non-Mendelian attributes of the hereditary makeup of bacteria:

> Bacteria (and *a fortiori* viruses if they can be considered to be true organisms), in spite of occasional reports of a sexual cycle, appear to be not only wholly asexual but pre-mitotic. Their hereditary constitution is not differentiated into specialized parts with different functions. They have no genes in the sense of accurately quantized portions of hereditary substance; and therefore they have no need for accurate division of the genetic system which is accomplished by mitosis. The entire organism appears to function both as soma and germplasm, and evolution must be a matter of alteration in the reaction-system as a whole. That occasional "mutations" occur we know, but there is no ground for supposing that they are similar in nature to those of higher organisms, nor since they are usually reversible according to conditions, that they play the same part in evolution. We must, in fact, expect that the processes of variation, and evolution in bacteria are quite different from the corresponding processes in multicellular organisms. But their secret has not yet been unravelled. (Huxley, 1942: 131–132)

Similarly in the first edition of his well-known book, *The Evolution of Genetic Systems*, the leading British cytogeneticist C. D. Darlington (1939) referred to "asexual bacteria without gene recombination" (p. 70) and "genes which are still undifferentiated in the viruses and bacteria" (p. 124). As Joshua Lederberg (1948: 153) wrote in one of his first review papers on microbial genetics, "The lack of outward differentiation of bacteria and viruses does give the appearance of holo-cellular propagation and of identity between direct transmission and inheritance. Geneticists and bacteriologists alike have . . . shown justifiable hesitation in accepting unanalyzed genetic variations as gene mutations." Lederberg played a pioneering role in bacterial genetics by solving the problem of bacterial sex and developing a method for bacterial genetic investigations.

Unlike others who worked on bacteria, Lederberg was not schooled in the existing paradigm which maintained the asexual nature of bacteria. Born in New Jersey in 1925, Lederberg developed a strong interest in classical genetics as a teenager. After completing his Bachelor's degree at Columbia College, New York, in 1944, Lederberg started medical school, but his interest in microbial genetics led him to drop out before the second year. Instead, he turned to the laboratory of Francis Ryan, a young assistant professor at Columbia who had just completed a post-doctoral fellowship at Stanford with Beadle and Tatum. Lederberg had already become convinced that bacteria at times exhibited the phenomenon of "sex," or more specifically, the phenomenon of gene transfer between different strains. And, he had mapped out a plan to demonstrate sexuality in bacteria. Ryan heralded the news about the work on biochemical genetics and the prospects of using microorganisms. He invited young Lederberg into his laboratory. Lederberg would have remained in Ryan's laboratory but soon learned that Edward Tatum was interested in similar problems and was then in the process of moving to Yale. After a brief correspondence, Tatum invited Lederberg to work at Yale. He arrived in March 1946, and by early May, genetic recombination in *Escherichia coli* (Figure 2c) was experimentally demonstrated (see Lederberg and Tatum, 1946a,b). Bacteria had sex and genes like other organisms, and they could be domesticated for genetic use.

After completing his Ph.D. in 1947 under Tatum, Lederberg was appointed assistant professor of genetics at the University of Wisconsin. He quickly emerged as one of the most important leaders in biochemical genetics of microorganisms. Lederberg's investigations took three primary routes, which characterized biochemical genetics of the 1940s and 1950s in general (see Brink to Weaver, December 4, 1947). One line of work was designed to test the one-gene–one-enzyme hypothesis as formulated first by Beadle and Tatum. His second principal investigation dealt with the extracellular transformation of genes in *E. coli* mediated through viruses (bacteriophages). This phenomenon was of fundamental importance for elucidating the chemical nature of the gene. Was it a nucleic acid or protein? A final line of work concerned the nature of mutations in microorganisms, especially drug resistance in bacteria. This problem was important both practically for medicine and immunology, and theoretically for some fundamental problems in genetics.

Relative sexuality

These developments – the demonstration of sexuality in microorganisms and their use in genetic investigations – are usually treated as cornerstones in the foundation of modern biochemical genetics and the study of "the chemical basis of life." Yet, such an account is seriously flawed by its omission of the work of Franz Moewus. Moewus's work, highly celebrated by some, but severely criticized and even literally ignored by others during the 1940s and 1950s, emerged independently of, and preceded the work of Beadle and Tatum and other American geneticists. Moewus began to work on the genetics of microorganisms in the early 1930s – many years before the mass migration of biologists, biochemists, and physicists to the study of bacteria, fungi, and algae.

In order to understand how Moewus came to work on problems of biochemical genetics in microorganisms, one has to consider his own unique early career, and the institutional structure of scientific research and education in Germany which fostered it. Moewus's scientific training and general education was quite different from that of American biologists, especially geneticists. He never was exposed to the formal genetic training and discipline of American geneticists. Indeed, throughout his career he never considered himself a geneticist; he called himself a botanist. There is no published biographical account of Moewus; for this I have had to rely heavily on the recollections of Franz Moewus's widow, Liselotte Kobb.

Mrs. Kobb was 75 years of age when I interviewed her at her home overlooking the Neckar River in Heidelberg. Her memory was vivid, and her recollections of her life with Franz provide the only eyewitness account of his career and laboratory work. Liselotte met Franz when she was a student completing her state exams in biology at Dresden. Moewus was an instructor. They married in December 1934, and during the next 20 years she frequently assisted in his experimental work and the writing of his scientific papers. Moewus was born in Berlin (Spandau) on December 7, 1908. He developed a strong interest in biology as a young high school student, and his father supported his interests.

Mrs. Kobb explained that with the rise of the Nazi regime she and Franz did research on their family trees. One had to trace back one's family tree four generations to show that there were no Jewish ancestors. Moewus had no Jewish relatives. He came from a family with a long tradition as tailors, originating in Potsdam, the seat of the Royal family. For three or four generations, his family had been tailors for the Prussian Guard. Uniform tailoring was highly a respected job; his grandparents had four children – two sons and two daughters – all of whom learned the trade; all were uniform tailors. In 1918, the Royal Guard was wiped out, and with them ended this kind of highly specialized tailoring. The family then moved to Spandau (Berlin), also a garrison; here they continued their work until the First World War broke out the next year. They then turned to ordinary tailoring and developed a lucrative family business. Franz Moewus's father was a master tailor. His family became quite wealthy, until inflation took everything again in the Great Depression. Nonetheless, they worked together, and their business picked up quickly.

In the early 1920s, Franz Moewus was at the *Gymnasium;* he was a loner but had developed a strong interest in biology. His father was generous and supported his interests. When he began his studies, his father bought him a microscope and installed a private laboratory for him at home. He bought handbooks on how to do some routine laboratory work, microscopic studies, and so on. And his father furnished his library with expensive science books. One relative was a druggist and gave him all the drugs and chemicals he needed. Therefore, by the time he was 19 years old and ready to enter the University of Berlin, Moewus had already immersed himself in biology. Moewus did not do undergraduate work in science or any other field at the University of Berlin. From the moment he entered, he wanted only to study and do research in botany and biology. He enrolled immediately for the doctorate. To non-European readers, this may appear to be unusual since the typical scenario in Anglo-American universities is first to complete an undergraduate degree before going on to graduate work. The educational structure of German universities needs some explanation.

In Germany, all universities were state-run; one of their principal functions was to train individuals who would later work in the civil service. After completing high school (*Gymnasium*), students who went on to study science in the universities typically would spend four years or so working toward the *Staats Examen.* These were government-supervised exams that allowed one to work in the civil service – as a high school teacher, for example. (There is no Bachelor's degree in Germany.) The state exams were broad-based, embracing three fields – for example, biology, chemistry, and a third subject, say, philosophy. They entailed two oral and written exams in each field plus a minor thesis. However, one was not obliged to take the state exams before doing a doctorate. In exceptional cases, one could bypass them and go straight on to the doctorate. Moewus was an exception. With great support from home, he could afford not to take out a government position and do only his doctoral work which he completed in only four years at the age of 24. To do doctoral work meant, of course, that Moewus had to find a professor who believed in his abilities and would take him on. He found Hans Kniep, a mycologist in the Botany Institute of the University of Berlin, who first introduced him to the problems of sexuality in micro-organisms. Kniep had been working on sexuality in fungi (see Kniep, 1928), and he encouraged Moewus to do the same with alga.

Moewus was an obsessive worker; indeed, throughout his career he developed and maintained an extraordinary knowledge of the scientific literature. He made a habit of writing to anyone working on related problems to gain additional information, stocks, chemicals, etc. However, his life was characterized by a constant struggle to obtain and maintain suitable research and teaching positions. These difficulties began early, when he was still a university student. Kniep died young of cancer, before Moewus's dissertation was completed. Moewus was "orphaned"; he had to look for a new "father professor," as they were often called. This search led him to the laboratory of the famous proto-zoologist Max Hartmann, at the Kaiser Wilhelm Institute for Biology in Berlin-Dahlem. Moewus had previously heard Max Hartmann give a stimulating lecture at the University of Berlin, and after Kniep's death, he asked Hartmann if he could continue his thesis under his direction. Hartmann agreed, and Moewus was awarded a 125-mark scholarship to finish his thesis with Hartmann.

Hartmann's laboratory was particularly productive during the 1920s and 1930s. Many protozoologists in his division worked out the taxonomy and life cycles of various forms of microorganisms, investigated cell structure, and worked on various novel hereditary phenomena and the difficult problem of sex determination. Several of Hartmann's students of the 1920s and 1930s would later become well-known protozoologists and geneticists – including Victor Jollos and Curt Stern. However, Hartmann himself, like Kniep, never did genetic experiments and never trained students specifically as geneticists. He considered himself a biologist. His textbook, *Allgemeine Biologie: Eine Einfuhrung in die Lehre vom Leben* ("General Biology: An Introduction to the study of Life") (1927), was well read in the biological community of Germany. Hartmann spent a great deal of time writing and theorizing on larger problems of evolution, inheritance, development, and – above all – sexuality (see Hartmann, 1929).

Moewus's early research reflected the interests of Hartmann and his workers. In 1934, he investigated the problem *Dauermodifikationen* in microorganisms (see Moewus, 1934). These were cases in which environmentally directed hereditary changes could be induced in microorganisms. They gave the impression of a Lamarckian principle. But in most cases the acquired changes gradually faded away after many generations. Nonetheless, these lasting environmentally induced changes stood in direct contrast to the usual phenomena of random mutations in higher organisms. *Dauermodifikationen* became well known and was widely discussed by evolutionists and geneticists, owing to the work carried out in Hartmann's laboratory (see Sapp, 1987a).

But experiments on the nature of sexuality were by far the outstanding feature of Moewus's work during the 1930s. And it was this work that led him into a great deal of controversy after World War II. In 1932, Moewus published the first of a long series of papers on the nature of sexuality and the genetics of freshwater algae. These studies represented some of the first systematic genetic investigations in any microorganism. Moewus showed how one could carry out genetic analysis of microorganisms, and he demonstrated that, like higher organisms, they possessed genes that were inherited in the classical Mendelian way.

Sexuality was a major and difficult problem for many biologists of the 1920s and 1930s, not just because its control was a necessary condition for doing genetics research on microorganisms. The motivations for understanding and controlling sexuality were many. In the patriarchal society of the Western world, parents often wanted male children, and they wanted to know the recipe for making them. Doctors and educators wanted to enlighten their adult constituencies on the physiological and psychological aspects of their sexual needs and give "expert" advice on correct behavior. Beginning in the 1920s, committees for research in sex were promoted among physiologists and endocrinologists who first explored hormonal regulation in animals before turning to write books about attaining manhood and womanhood (see Beecher, 1970; Long, 1987). However, the questions and problems of those zoologists who worked on single-celled organisms were far removed from such practical matters, although more fundamental to a biological understanding of sexuality.

How was sex determined? If one sex was "male" and the other was "female," what made them so? What made organisms come together and copulate? Was sex a universal biological phemonenon? Did two organisms have to be sexually different in order to copulate? Were there always two and only two sexes – male and female? Biologists of the nineteenth century had reached the conclusion that the egg and sperm were cells and that fertilization was accomplished through their union. Out of this work grew the opinion that ultimately the problems of sex, fertilization, and development were problems of chemistry and cell structure. During the late nineteenth century, one additional generalization emerged: the distinction between germ cells and body cells. The offspring inherits characters from the parent's germ cell, not from the parent's somatic cells, and the germ cell owes its characteristics not to the body which bears it, but to its descent from a preexisting germ cell of the same kind. The body is thus an offshoot of the germ cell.

By considering the sexual cells from which adult individuals originate, instead of the adult individuals themselves, biologists concluded that the differences between the sexes were merely special adaptations for the purpose of facilitating the union of the spermatozoon with the ovum. The leading German zoologist Oscar Hertwig summarized these views clearly in 1902 when he wrote:

> All contrivances connected with sex are variations upon one and the same theme; firstly, they enable the sex cells to come together, and secondly, they ensure that the egg shall be nourished and kept in safety. We call the one set of contrivances "male", the other "female". All these relationships are of secondary nature, and have nothing to do with the real essence of fertilization; this is the union of two cells, and is therefore purely a cell phenomenon. In these views we agree with Weismann, Rich, Hertwig, Strasburger, and Maupas, who have expressed similar opinions. (Quoted in Radl, 1930: 329)

Sex was reduced to the morphology, behavior, and physiology of cells. The sexual cycle of unicellular organisms was in some ways functionally analogous to that of higher organisms. The life cycle of the green alga *Chlamydomonas*, investigated by Moewus, is exemplary: In this unicellular alga (Figure 2d), the single motile individual is haploid (i.e., it contains only a single set of unpaired chromosomes in its nucleus). In copulation, two such haploid cells unite completely, to form a diploid cell (a cell having its chromosomes in pairs, the members of each pair being homologous). The two haploid cells thus correspond functionally to two gametes, whereas the diploid cell is the zygote. The zygote is inactive; it secretes a wall about itself and becomes a cyst. Later, under favorable conditions, the diploid cell divides twice by two "maturation divisions." At one of these divisions, chromosome reduction occurs and four haploid cells are formed. In some species, at times additional cell divisions occur before the cells emerge from the cyst, and eight haploid cells are formed. The cyst wall dissolves, and the haploid cells are freed. Each develops flagella and swims about as a free individual. On emerging from the cyst, the individuals were often called "swarmers" or swarm cells; each is a potential gamete. These free cells commonly multiply vegetatively for many generations, the descendants of each original swarm cell forming a clone.

Although biologists often compared the gametes of microorganisms to the "male" and "female" of higher organisms, there were difficulties with the analogy. In higher organisms, the characteristics by which the female cell was ordinarily recognized were larger size, lesser activity, greater storage of nutritive reserves, and egglike form; the male cell was recognized by the corresponding opposed characters. However, in many species of unicellular organisms, such as those investigated by Moewus, there were no regular morphological or behavioral differences between copulating gametes. How could one identify one gamete as the "male" and the other as the "female"? Operationally, one could define the sexes of microorganisms in terms of the sexes in which they copulated. Two cultures are of the same sex if they do not copulate with each other; they are of different sex if they copulate with each other. Moewus carried out most of his work on *Chlamydomonas eugametos,* which did not show morphological differences between sexes. He designated the sexual types by using the species *Chlamydomonas braunii,* which had two distinct gametes differing by size. The larger one acts as the receiving female gamete. By crossing *Chlamydomonas eugametos* with *Chlamydomonas braunii,* the two sexes of *Chlamydomonas eugametos* were typed. On this basis, Moewus (1933b) designated the two sexes as plus (+) and minus (−).

How then was the sex of the gametes determined? What, if anything, made them "male and "female"? Physiological differences also had to be considered. Moewus's work provided evidence that physiological sex differences may exist in some cases in which morphological sex differences are lacking. This was most clearly evident in those species and races in which each clone consisted exclusively of one sex type. Here sexual union took place *only* between gametes from different clones.

But the interpretation of sexual relations in microorganisms was complicated by still other strange phenomena. In some species or varieties, both sexes could be found among individuals of a single clone. In other words, sex could occur between two genetically identical gametes. This phenomenon led to a great deal of controversy for sex researchers who worked on microorganisms during the 1920s and 1930s. Could copulation ever take place between cells that were physiologically as well as morphologically and functionally identical? Or was there still some invisible difference that separated gametes into sexual types. The question at issue here was considered to be of great theoretical importance.

Moewus's work on sexuality emerged from within a heated dispute which raged between biologists at Berlin led by Hartmann and those at

Prague led by Mainx (1933), Czurda (1933), and Pringsheim. Hostile letters passed back and forth between Berlin and Prague throughout the 1930s. The biologists at Prague freely accepted the possibility that any two – that is, identical – gametes could copulate. In other words, organisms did not need to be sexually different in order to have sex. In their view, sexual union could be brought about simply by copulation-conditioning factors in the environment. Hartmann (1929, 1932), on the other hand, maintained that there could be no sexual union without sexual differentiation, and that properties of "maleness" and "femaleness" were inherent in each and every organism. From this premise, Hartmann constructed his famous theory of "relative sexuality": that any two gametes can copulate only if they differ sufficiently in strength of their sex tendency. In other words, if gametes could be genetically identical, as in several species and strains of unicellular algae, then, he postulated, there must nevertheless exist some invisible difference between them, and the strain is said to possess "bisexual potency."

In order to appreciate fully Hartmann's views of inherent "maleness" and "femaleness," it is helpful to consider the larger theoretical context of German intellectual life. The unique roles of the sexes and the nature of sexuality were at the center of much of nineteenth-century intellectual writings in Germany. In the nineteenth century, biology was characterized by the rise of academic metaphysics. Out of a response to the mechanistic philosophy of eighteenth-century Enlightenment, a movement called *Naturphilosophie* emerged during the early nineteenth century which attempted to construct holistic and unified concepts of the natural world. It was common for *Naturphilosophen* to refer to undefined natural tendencies which operated in organisms. As is well known, the Romantic German *Naturphilosophen*, led by Goethe and the poets, looked with wonder upon the phenomenon of sex. Inherent properties of "maleness" and "femaleness" were frequently invoked (Geddes and Thomson, 1889: 36). Indeed, the polarity between maleness and femaleness represented one of the fundamental dichotomies in the *Naturphilosophen* dialectical image of the world (see Radl, 1930: 328; Churchill, 1979: 141). The influence of *Naturphilosophie* on biological thought reached well into the twentieth century. It is well known that leading German biologists of the 1920s and 1930s were often inclined to combine their thinking about heredity, development, and evolution with metaphysics coming from *Naturphilosophen* such as Goethe, Carus, and Kant (Hamburger, 1980; Saha, 1984).

Hartmann's (1929) theory of sexuality consisted of an elaborate list of propositions:

1. Sex is a universal biological phenomenon.
2. There are always two and only two sexes.
3. These two sexes are always male and female.
4. Male and female are qualitatively diverse.
5. Every cell has the full potential of both male and female.
6. These potencies are not localized in any one cell component, but are general properties of the living material.
7. The sex manifestation by a cell is the result of a weakening or strengthening of the expression of either the male or the female potency.
8. This weakening or stengthening may be determined by external conditions, by developmental conditions, or by genetic factors.
9. The degree of the weakening or strengthening depends on the effectiveness of the determinants listed in proposition 8.
10. This quantitative variation results in the appearance of each sex in a series of strengths called "valences."
11. Sexual union takes place only under one of two conditions: (a) when gametes differ in sex; that is, when one manifests a stronger male than female potency, the other a stronger female than male potency; (b) when the gametes are all alike in sex, but very different in sex valence; that is, when one is strong female, the other weak female; or when one is strong male, the other weak male.
12. Sexual union equalizes or reduces the tension resulting from a difference in sex or sex valence.

(After Sonneborn, 1941: 678–679).

Moewus's experiments, observations, and interpretations were made in accordance with Hartmann's theory of sexuality, and Hartmann himself upheld Moewus's results as the chief experimental evidence for his theory (see Hartmann, 1943). As mentioned previously, Moewus had provided evidence that in various species and races physiological sex differences existed in cases in which morphological sex differences were lacking. This was most clearly evident in those species and races in which each clone consisted exclusively of one sex type. Here sexual union took place only between gametes from different clones. Moreover, Moewus provided evidence that, in a number of algae, the culture fluid, in which ripe gametes are living, contained material ("sex stuffs") capable of affecting the sexual behavior of other gametes. In *Chlamydomonas*, Moewus (1933b, and later) was able to isolate this material from the organisms that produced it, by means of filtration and centrifugation, and claimed it had striking effects. He held that gametes grown in the dark were incapable of copulating, but treatment with the sex stuff from a suspension of ripe gametes of the same sex rendered them capable of copulation. In other words, gametes that were unreactive could be activated by addition of filtrates from cultures of reactive gametes (see Hartmann, 1934; Moewus, 1934). The chemical nature of the sex stuffs later became the object of exhaustive investigations.

The only cases where doubts continued to be raised about the existence of sex differences were those in which copulation occurred within a single clone: Genetically identical cells seemed to be able to copulate. However, here, too, Moewus and Hartmann continued to maintain that similar sex differences were operative. And Moewus provided evidence for sexual differentiation in these clones. It was based on a study of cells left over after copulation had ceased among certain species of *Chlamydomonas*. Moewus noted that the "leftovers" (those that did not find partners) of any one clone were always of the same sexual type. If these leftover cells were mixed with the leftovers of other clones, some of the mixtures would exhibit typical copulation. Leftovers of one clone mated only with leftovers of another clone. From these observations, Moewus (1934) concluded that each clone produced both types of gametes, but always one in much greater frequency than the other. Some clones regularly produced mostly one kind of gamete; other clones regularly produced mostly the other kind of gamete. Copulation then took place within a clone until all the rarer type of gametes found partners, so that all the leftover cells were of the prevailing type.

However, Moewus's interpretation that copulation was taking place between sexually different types still remained in doubt and was severely criticized by biologists at Prague. The phenomenon of copulation within a single clone was extensively studied by Adolf Pascher (1931) in a species of *Chlamydomonas* and by Pringsheim and Ondratschek (1939) mainly in *Polytoma*. Their interpretations were in fundamental disagreement with those of Moewus; they concluded that these forms showed copulation without any physiological sex differentiation. Further, Pringsheim and Ondratschek (1939) could not confirm Moewus's observations that the cells left over after copulation were unable to copulate with each other because they were all of one sex type. They attributed the cessation of copulation, not to inherent sex differences of the cells, but to a change in the chemical conditions of the culture, rendering it unsuitable for copulation. Appropriate modification of the conditions led to resumption of copulation. They therefore denied the validity of the "leftovers" methodology for the analysis of the issue in question.

Was sex due to inherent properties of "maleness" and "femaleness" or to nutritional requirements? Moewus continued to employ his "father professor's" concept of relative sexuality and found that it could be readily applied to his genetic work as well. Throughout the 1930s, Moewus worked out the breeding relations and carried out extensive crosses between different species and strains of *Chlamydomonas* and *Polytoma*. He soon found that no two of the sexes were exactly alike. In crosses between species and strains, in some cases gametes of like sex could copulate and yield descendants. Plus (+) gametes of one species could unite with either plus or minus (−) gametes from the other species, and (−) gametes from one species with (+) from another. For example, Moewus showed that any gamete of *Polytoma uvella* could copulate with any gamete of *Polytoma pascheri*, irrespective of the sex of the gametes. He suggested that the two sexes of one species were not the same as two sexes from other species.

At least two genetic interpretations were possible. One could draw the conclusion that there were diverse sexes in these interbreeding groups, or one could try to interpret the results in terms of two sexes. Moewus faithfully interpreted his results in terms of Hartmann's theory of relative sexuality, which stated that (1) there are only two sexes, male and female; (2) the gametes of the different species and strains differ only in strength or valence of their sex tendency; and if two gametes differ sufficiently in strength of the sex tendency, they will tend to copulate. On this basis, Moewus assigned the various sex tendencies arbitrary strengths. He argued that copulation could take place either between gametes differing in sex [i.e., between any (+) or any (−)], or between two gametes of the same sex differing in valence by as much as 2.

Moewus (1933a, 1935a) carried out numerous crosses between different species of algae and identified various genes affecting sex. In agreement with Hartmann's theory, he held that the sex genes acted on underlying sexual potencies to produce different strengths in sex tendency. After studying sexuality and its genetics in a number of genera (Moewus, 1933a, 1935a), Moewus soon centered his investigations on *Chlamydomonas*, which he found to be particularly favorable for biochemical investigations. His aim was to determine the precise manner in which genes controlled sexual behavior. By 1937, Moewus already had some evidence that pigments in alga might be involved in sexuality. This interest and the necessity of obtaining continuing support for his research would soon lead him into collaboration with the most celebrated biochemist in Germany, Richard Kuhn.

From relative sexuality to molecular biology

As mentioned earlier, Moewus's early career was characterized by a constant struggle for a secure position to teach and carry out research in the new domain he was pioneering. Certainly, many young scientists all over the world faced similar problems. However, in Germany, the difficulties of obtaining a secure position were extremely severe. This was especially true for those developing new research domains because the German university system was very resistant to change. The first obvious obstacle they faced was the bureaucratic power of the professoriate and the structure of German "Instituten." German institutes were the "feudal" domain of one "chair professor" who possessed the bureaucratic power to allocate facilities and funds to junior staff, thus dictating the direction and nature of research in his institute. Young scientists were forced to rely on the favor of these powerful professors, who tended to be self-serving in their academic interests. Professors also taught the large obligatory courses, for which they were rewarded financially by obtaining part of their income through registration fees. Junior staff often received their income solely from students' fees, and they tended to be left in charge of smaller, more specialized classes. Their only recourse to a better income and security to carry out innovations of any kind was to acquire their own chairs and institutes and train their own following of students. This process took a long time – an average of 16 years by 1909 – and only about half of those who completed their doctorates ever advanced so far (see, e.g., Ash, 1980).

The conservative interests of the German professoriate were strengthened by the Ministries of Culture which legislated over the establishment of new chairs in the universities. Educational officials usually accepted the recommendations of university faculties when they made appointments to existing chairs. But when the establishment of a new chair was at issue, financial and cultural considerations took precedence. The reluctance to establish new chairs was also shared by chair professors, who tended to prefer accommodating new domains within existing institutes by offering temporary teaching contracts or nonbudgetary associate professorships (see Ben-David, 1971). The structure of the German higher education system tended to discourage specialization and the rapid establishment of new disciplines.

A comparison of the institutional structure of German science to that of the United States will help to illustrate the extent of these difficulties. The American "department" stood in striking contrast to the German "institute." The former is characterized by a group of professors of relatively equal security and research autonomy to compete for funds and students. The conservative effect of the German system is readily illustrated in the case of genetics research. The first generation of geneticists in the United States were able to dissociate themselves from the larger problems of biology such as development and evolution. They formed their own semiautonomous discipline with its own journals, societies, doctrines, techniques, and theories. They infiltrated departments of biology, zoology, and botany, creating and solving their own problems, analyzing chromosomes and the sexual transmission of traits from one generation to the next.

This strategy also required support from agriculturalists and the public sectors that found social and economic value in genetic work (see, e.g., Rosenberg, 1976; Kimmelman, 1983). American geneticists enjoyed support from a variety of patrons who provided funds for the establishment of new teaching and research positions in the universities. In the United States, genetics grew rapidly between 1910 and 1930 and outcompeted many of the older disciplines for both students and support (see Sapp, 1983). In Germany, there was only one chair of genetics established before 1945. Genetic investigations were carried out primarily in botany and zoology institutes, where they remained constrained by, and subordinate to, the general interests of the older disciplines. Indeed, many German biologists opposed the disciplinary nature and restricted scope of genetics as developed in the United States. Many European biologists of the 1920s and 1930s claimed that American geneticists studied the easy problems; they focused simply on the transmission of characteristics. Instead, many German biologists worked on difficult problems concerning extragenic inheritance and sexuality, and sought to develop a "unified" theory of "heredity" that would at once embrace ontogenetic development and evolution (see Sapp, 1987a).

The Kaiser Wilhelm Institute for Biology, where Moewus worked with Hartmann, represented an attempt to compensate for the conservative effects of the university system. Founded in 1913, it provided conditions for systematic experimental research, free from the heavy teaching responsibilities in the universities. In 1914, Carl Correns, already famous as one of the "rediscoverers" of Mendel's laws, was appointed its first director. However, much of the elitist and patriarchal structure of the universities was preserved at the Kaiser Wilhelm Institutes. The Kaiser Wilhelm Institutes were not structured around disciplinary commitments or specialties per se. They were structured around institute directors, who still held the kind of power chair professors held in the university.

The difficulties faced by Moewus were typical of those faced by many young German scientists. After completing his thesis in 1933, Moewus received a one-year grant to teach as an instructor at Dresden. It was there that he met Liselotte, who was completing her state exams in biology. The Moewuses then moved to Berlin where Franz returned to work in Hartmann's laboratory on small grants for another two years. But he was soon out of a job again. Hartmann certainly had great admiration for Moewus and his work; Moewus had won the heart of his "father professor." But Hartmann was in his late fifties, a long way from retirement, and he was not immediately concerned with obtaining a permanent job for Moewus. And he showed no immediate interest in developing a new specialty based on biochemical genetics studies of microorganisms.

As Mrs. Kobb explains, Moewus decided to take matters into his own hands. In 1937, to obtain financial aid for his work, he wrote to pharmaceutical and chemical companies: Merck; Schering, a Berlin pharmaceutical firm; and I. G. Farben, one of the largest chemical companies in the world. Moewus sent his thesis on sexual substances in algae which, in his view, contained strong evidence that carotenoid pigments might be involved. Agents from Merck responded that it was not commercial enough; Schering offered him a 125-mark scholarship to be carried out in Hartmann's Institute. But the directors at I. G. Farben showed the greatest interest. They said that although this kind of research was not yet ready for commercial exploitation, Moewus should work with the best biochemist, Richard Kuhn. Kuhn was director of the Chemistry Institute of the Kaiser Wilhelm Institute for Medical Research in Heidelberg and professor at the University of Heidelberg. He was well known for his work on the properties of carotenoid pigments and vitamins, for which he would ultimately be awarded a Nobel Prize in 1939.

The director of I. G. Farben told Moewus that he was meeting with Kuhn the next week at a convention and would bring up the issue of some sort of collaboration. A meeting was planned between Kuhn and Moewus in an elegant hotel in Heidelberg. Kuhn paid the expenses for Moewus's return train fare. The Moewuses were very excited indeed. Mrs. Kobb vividly recalled the episode. She prepared Moewus for his meeting, dressing him up to "look nice"; she even ironed the brim of his hat. But all was to no avail. She later learned that Moewus got caught in a major downpour and arrived at the hotel soaking wet. In spite of his appearance, Kuhn took to Moewus immediately. Moewus returned that night to tell Liselotte what a marvelous man Kuhn was. Kuhn had various stipends he could give; he had money from industry and offered Moewus a one-year contract to begin as soon as possible. The Moewuses packed their belongings and moved to Heidelberg. Franz Moewus and Kuhn got along marvelously. Mrs. Kobb explains that both were quiet men, slow talkers and slow but methodical thinkers; they soon became close colleagues.

With the rise of the Nazi regime, many leading German scientists of Jewish ancestry had been forced to leave their native country. Many of them had taken refuge in the United States, as will be discussed later. With the outbreak of war, many non-Jewish scientists were drafted into the army. Many laboratories were closed down, and communication among the remaining scientists was limited within Germany; by 1942, communication with scientists overseas had ceased. As will be discussed in more detail later (see Chapter 5), Moewus was among the exceptional scientists in Germany, who were able to continue their work, at least to some degree, throughout the war. Moewus's contracts with Kuhn were repeatedly renewed for 15 years.

Moewus had little knowledge of biochemistry, but he bought Kuhn's textbook on biochemistry and immersed himself in it. Immediately, he began a more extensive study of sexuality in algae, isolating and examining the effects of sex substances in filtrates of *Chlamydomonas.* He quickly developed what some geneticists later regarded as the basic concepts and methods for the biochemical genetics of microorganisms. Kuhn soon became a leading partner in the investigations and also published joint papers with Moewus from 1938 to 1949. He and his students were responsible for most of the biochemical work, whereas Moewus concentrated on the biological and genetic aspects.

A brief discussion of the results published by Moewus and his collaborators, pertaining to the sexual cycle of *Chlamydomonas,* will give some indication of the extraordinary sophistication of his work. The first stage of Moewus's work, leading to an analysis of sex substances, was familiar to algologists. Moewus grew *Chlamydomonas* in pure culture by means of the well-known agar plate method. When grown on the surface of agar, the algae forms an immobile green jellylike mass or "palmella stage," but if the cells are flooded with water the cells become motile. The species studied by Moewus for discerning sex substances were of the type in which every motile cell is a potential gamete and capable of fusing with any cell of the opposite sex swimming in its vicinity. When

studying these species, Moewus used a procedure developed by Hartmann (1929) for observing gametic union. Two drops of water, one containing male gametes and the other containing female gametes, are
placed side by side on a slide without a cover glass. After focusing under
the microscope, the two drops are mixed together. Within a matter of
seconds, there is a very marked aggregation of the gametes into clumps.

Using this method Moewus began to explore the effect of light and
sugars on sexual activity. He reported, for example, that when cultures
in the "palmella stage" were placed in darkness for 24 hours and then
flooded with water, the cells remained immobile, but if he flooded the
cultures with a 1% glucose solution, the cells became motile and gametic
union occurred (Moewus, 1933b). He also noted that cells grown in
darkness could be made *motile* (but not capable of copulation) by subjecting them to certain kinds of illumination (visible light). But they were
capable of *gametic union* only if illuminated by light from the blue end
of the spectrum (Moewus, 1939a: 486). Moewus flooded palmelloid cultures that had been standing in light; after the gametes had become motile, he removed all gametes by filtering or centrifuging. He reported
that when palmelloid cultures grown in darkness were flooded with
these filtrates, the cells became motile and sexually functional. Female
cells grown in darkness were made sexually functional by filtrates from
female cultures grown in light; and male cultures grown in darkness
were made sexually functional by filtrates from male cultures grown in
light.

Moewus (1938a) continued to analyze the nature of the substances
excreted by gametes by studying the effects of light on *filtrates* of *Chlamydomonas eugametos*. He claimed, for example, that when female gametes
were made motile in the dark by means of a glucose solution, and the
filtrates were exposed to daylight, the filtrates had no effect on sexuality
of motile cells grown in darkness. In other words, the sex substance
causing gametic union was not formed in darkness. On the other hand,
filtrates from gametes swimming in red light contained a precursor (V)
of a substance causing gametic union. If filtrates from female gametes
swimming in red light were exposed to blue light for no more than 30
minutes, they produced a substance (K♀) capable of making female
gametes functional in darkness. If they were exposed to blue light
for a longer period, the female copulating substance disappeared. But
if illumination was continued for 75–90 minutes, the filtrates produced a substance (K♂) capable of making male gametes functional in
darkness. After illumination of more than 90 minutes, sex substance

(K♂) disappeared resulting in an end product (K$_o$). Moewus (1938b) further claimed that when filtrates from male gametes swimming in red light were exposed to blue light, the female copulating substance was not formed, but the male copulating substance and an end product were both produced. The series of changes may be expressed as follows:

female filtrates $V \rightarrow K_♀ \rightarrow K_♂ \rightarrow K_o$
male filtrates $V \rightarrow K_♂ \rightarrow K_♀ \ K_0$

The climax of the work on sex substances of *Chlamydomonas eugametos* was a demonstration that they were all formed by degradation of the carotenoid pigment, protocrocin. The chemical nature of the sex substances was determined through the efforts of Kuhn and collaborators. However, they were found in filtrates in such small quantities that direct chemical analysis was not possible. Instead they had to be determined by spectroscopic analysis. (Spectroscopic analysis is a powerful tool for recognizing and identifying substances based on their characteristic light absorption.) Kuhn, Moewus, and Wendt (1939) held that the first-formed substance is protocrocin, which is oxidized into two molecules of picrocrocin and one of crocin (see Figure 3). The two molecules of picrocrocin each is cleared to form one molecule of glucose and one molecule of the carotenoid safranal. Each molecule of crocin breaks down into two molecules of gentibiose and one molecule of *cis*-crocetin dimethyl ester, which, in time, rearranges to form *trans*-crocetin dimethyl ester.

By adding each of these carotenoids to the appropriate filtrates, Moewus and his collaborators claimed to be able to show that each has a specific hormonal function in the sex life of *Chlamydomonas eugametos* (see diagram in Figure 3). One hormone, crocin, induced motility in gametes. It was held that one or just a few molecules of crocin per cell could make gametes of *Chlamydomonas eugametos* motile (Kuhn et al., 1938; Moewus, 1938a, 1939a–c). But in other species of *Chlamydomonas* a much richer concentration was necessary (see Moewus, 1939b). Two other carotenoids, called "termones," were necessary for the determination of sex (Kuhn et al., 1939). Picrocrocin was the termone (*gyno*termone) of female gametes; safranal was the termone (*andro*termone) of male gametes. And two other carotenoids, the *cis*- and *trans*-crocetin dimethyl esters, were concerned primarily with the mutual attraction of the gametes. Moewus (1940e) called these attracting substances "gamones" (see Figure 4) after Hartmann and Schartau (1939).

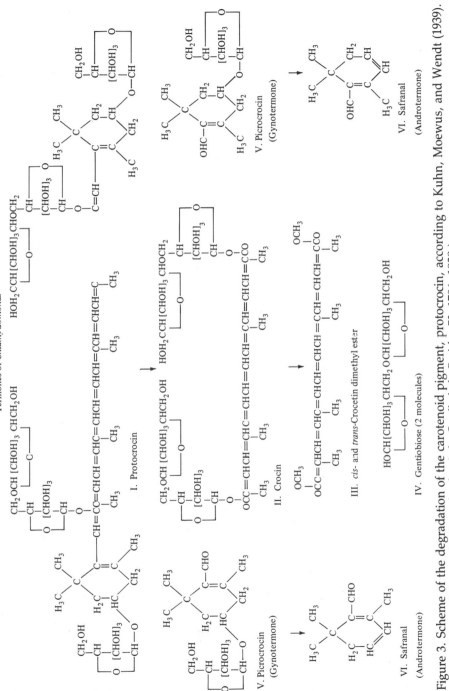

Figure 3. Scheme of the degradation of the carotenoid pigment, protocrocin, according to Kuhn, Moewus, and Wendt (1939). (Adapted from Kuhn et al., *Deutsche chemische Gesellschaft, Berichte* 72: 1706, 1939.)

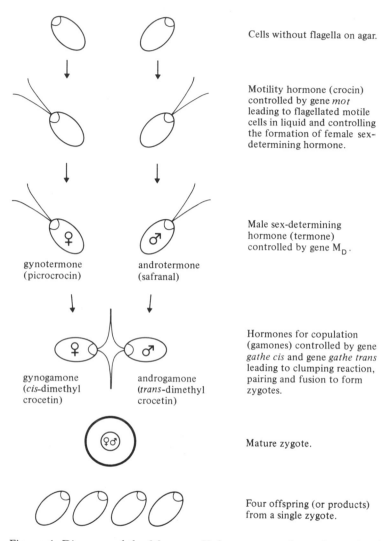

Cells without flagella on agar.

Motility hormone (crocin) controlled by gene *mot* leading to flagellated motile cells in liquid and controlling the formation of female sex-determining hormone.

Male sex-determining hormone (termone) controlled by gene M_D.

gynotermone
(picrocrocin)

androtermone
(safranal)

Hormones for copulation (gamones) controlled by gene *gathe cis* and gene *gathe trans* leading to clumping reaction, pairing and fusion to form zygotes.

gynogamone
(*cis*-dimethyl crocetin)

androgamone
(*trans*-dimethyl crocetin)

Mature zygote.

Four offspring (or products) from a single zygote.

Figure 4. Diagram of the Moewus–Kuhn account of sex determination in *Chlamydomonas eugametos*. (From J. Sapp, What Counts as Evidence, or Who Was Franz Moewus and Why Was Everybody Saying Such Terrible Things About Him? *History and Philosophy of the Life Sciences 9*: 277–308, 1987; used with permission.)

Each one of these stages in the degradation of protocrocin was held to be brought about by a different gene, which Moewus identified (Figure 4). In this scheme, genes controlled chemical reactions by way of enzymes. Moewus provided evidence for the existence of many mutations (64) of the gene he called *mot*, which exhibited the same characteristic – the inability to produce both the motility hormone and the female-determining hormone (Moewus, 1940b). Moewus's procedures were similar to those developed later by Beadle and Tatum (1941). It will be recalled that Beadle and Tatum detected the effects of mutations in *Neurospora* by their ability to grow on a well-defined chemical medium. By ascertaining what substances needed to be added to the medium in order for mutant cells to grow, they determined which enzyme was missing, and were able to identify the mutant gene responsible for controlling the enzyme in question. Similarly, Moewus detected the effects of mutations by ascertaining which hormones had to be added to the medium to render the cells capable of copulating. Moewus induced mutations through various means – cold and heat shocks, chemicals, and gamma rays. On the basis of these combined biochemical, genetic, and mutagenic techniques, he held that a gene (M_D) controlled the production of an enzyme having the specific action of cleaving picrocrocin and thereby converting this female-determining "termone" into safranal, the male-determining one (see Figure 4). Gene M determined male valence, and Moewus claimed to have induced 43 mutations at this locus (Moewus, 1940b).

As mentioned above, Kuhn and Moewus held that *Chlamydomonas eugametos* normally produced both cis and trans isomers of dimethylcrocetin, the hormone responsible for permitting the two cells to copulate. The synthesis of the two isomers was supposed to be carried out by means of two enzymes (or enzyme systems) with different conditions for activity. The specificity or activity of these two enzymes would be determined by two *gathe* genes. The production of the cis isomer, they suggested, was under the control of the gene *gathe*[cis]; and production of the trans isomer was under the control of the gene *gathe*[trans]. Moewus provided evidence for the existence of 17 mutations affecting the cis enzyme and 18 affecting the trans enzyme. He brought this work together in four remarkable papers (Moewus, 1940a–d). In effect, Moewus described, for the first time in the history of biology, a complete sequence of the life history of an organism in terms of genes and molecules.

4. Too good to be true, and too true to be dismissed

> In a series of papers, Dr. Franz Moewus, working first at the Kaiser Wilhelm Institut für Biologie in Berlin, and later at Erlangen and Heidelberg, has published a number of very remarkable results on relative sexuality in unicellular algae. The qualitative results are in agreement with those of other workers of the same school. We wish, however, to direct attention to a feature of his quantitative data which casts grave doubt on their reliability. (Philip and Haldane, 1939: 334)

> Yet, I seem vaguely to recollect that a similar study of Mendel's data was published long ago and led to similar implications. Moreover, I have always believed that Mendel must have run into cases of linkage in his experiments and simply chose not to mention them because they seemed to conflict with his evidence of independent assortment. Considerations of this kind make me dread the danger of throwing out the baby with the bath water. (T. M. Sonneborn to Sewall Wright, February 19, 1944)

In Germany, during World War II, many scientific laboratories were closed, journals were discontinued, and scientists who were members of Jewish families were forced to leave their country, if they could. Some committed suicide, others were tortured and died in prison camps. Many of the non-Jewish scientists who remained were drafted, and communication among the rest was minimal. However, in the United States, during the war, many scientists remained far removed from events in Europe. Many continued to work on fundamental problems, often combining their work with the war effort whenever possible. Moewus's work attracted considerable attention in the United States during the war and was frequently reviewed from various perspectives. Plant physiologists addressed Moewus's work in terms of his physiological results as they pertained to hormonal regulators of sexuality. Microbiologists reviewed his work with respect to the relations among "growth factors," vitamins, and enzymes. The few geneticists who were working on microorganisms before the war celebrated Moewus's demonstration of Mendelian principles in these organisms and his biochemical genetic analysis of sex determination.

However, no one accepted Moewus's physiological and genetic

claims outright. Instead, they received them with a great deal of skepticism and studied them with circumspection. In fact, the first time Moewus's work was reviewed outside Germany, it was attacked. Indeed, there were charges launched against Moewus's objectivity and the reliability of the data reported in several of his papers. The criticisms of the reliability of Moewus's work led investigators into a deep discussion of scientific procedures in general and of the difficulties in establishing objective criteria for verifying scientific results. These discussions centering around Moewus's work would continue for two decades.

Statistical regression

The first widely publicized criticism of Moewus's work appeared in a short note entitled "Relative Sexuality in Unicellular Algae" published in the central British journal *Nature* at the outset of World War II. It was based on a statistical critique of two of Moewus's early papers on the genetics of sex determination (Moewus 1935b, 1938b), and it was set forth by Ursula Philip and the celebrated geneticist J. B. S. Haldane, working at the Department of Biometry, University College, London (Philip and Haldane, 1939). Haldane is best known today for his statistical work which helped to bring Mendelian genetics and natural selection together to form the so-called evolutionary synthesis of the 1930s and 1940s.

During the mid-1930s, Haldane's writings extended over a broad range of topics, including quantum mechanics as a basis for philosophy; anthropology and human biology and politics; and various problems of mathematical biology (see Pirie, 1966). He was also highly celebrated for his theorizing on how the gene might control biochemical processes. Haldane had been trained as a biochemist before turning to genetics in the 1930s. In 1932, he had left Cambridge University to develop genetics at the John Innes Horticultural Institution. William Bateson had played a leading role in establishing genetics work at John Innes, and, as mentioned in Chapter 1, Bateson's student Muriel Wheldale Onslow had begun to do biochemical genetics studies of plant pigments in flowering plants at John Innes before accepting a position at University College, Cambridge. Haldane had accepted a position at John Innes in 1932, with the intention of developing biochemical genetics. However, institutional difficulties intervened, and his intention was not realized. In 1936, he resigned from the John Innes Institution to take the Weldon Chair of Biometry at University College, London (see Lewis, 1983: 121). His research interests shifted progressively from the biochemical genetics of plants to formal genetics, the study of linkage and sex-linked inheritance in the human species. As Haldane (1954: 53) later lamented:

> In 1932, I left the Department of Biochemistry at Cambridge, with a promise that I should be given facilities for the study of biochemical genetics at the John Innes Horticultural Institution. I had hoped, in particular, first to investigate the enzymes concerned in pigment production, and later those concerned in carbohydrate metabolism. For various reasons this promise was not kept, and I found that I was cut off from biochemical research. No one has, in fact, attacked the problems in question, and it was left to Beadle's school to do similar work on simpler organisms.

Although Haldane did not have the opportunity to do biochemical genetics work when he wanted to, biochemical genetics had a high profile at the John Innes Horticultural Institute. In fact, Haldane (1937: 3–4) could write that the "most striking success in the analysis of gene action has been that of my colleague, Miss Scott-Moncrieff, who has given a biochemical account of the action of thirty-five genes concerned in the production of anthocyanin pigments in fourteen species or genera of plants." Moreover, Haldane's own theorizing on the nature of gene action during the 1920s became well-known during the 1940s and 1950s. As a Marxist, Haldane was also persistently concerned with the relations between science and politics. With the rise of nazism, he worked hard for the British Academic Assistance Council which enabled Jewish scientists to take up refuge in England. In 1933, he had already begun to take Jewish refugee scientists into his laboratory. Ursula Philip was among them (see Clark, 1968).

Philip and Haldane's statistical criticisms of Moewus's work were severe, and their message was clear: Some of the genetic data supplied by Moewus pertaining to crossing-over and segregation of sex-determining genes in algae showed so little variation from theoretical expectations that it was unreasonable to believe that they could have been obtained by chance. With regard to Moewus's (1938b) experimental data on the inheritance of sex in *Chlamydomonas eugametos*, Philip and Haldane claimed that "only once in about 10^{11} trials would so good a fit be obtained by chance." They made similar remarks based on statistical calculations of Moewus's (1935b) segregation data pertaining to sex inheritance in another genus: *Protosiphon botryoides*. They calculated that the probability that "so good" a general agreement between theory and observation could be obtained by chance was once in 3.5×10^{22} trials. Conveying the meaning of their statistical formulation into a sensational analogy, they wrote that "if every member of the human race conducted a set of experiments of this type daily, they might reasonably hope for such a success once in fifty thousand million years" (1939: 334).

They suggested two possibilities to account for Moewus's data. One was that a "wholly new type of biological regulation has been unwittingly discovered, by which the frequency of segregation is controlled with extreme sharpness, and that this applies to three segregations in two species." The other was that "the author has consciously or unconsciously adjusted his observations to fit his theory." They concluded (1939: 334):

> As this work is of fundamental importance for the theory of sex-determination, it is desirable that these experiments should be repeated by an independent observer. Until this is done, those parts of Dr. Moewus' genetical theory which are based on these numbers cannot, we think, be accepted.

Although Philip and Haldane's critique became one of the most widely known statistical criticisms launched against Moewus's work, it was not the only one of its kind. During World War II, two German investigators also launched statistical critiques. Klaus Pätau (1941) followed Philip and Haldane's lead and extended the statistical attack to some other genetic results of Moewus published in 1940 (see Moewus, 1940d). He claimed similar inexplicable statistical correlations. However, the subject of primary importance to Pätau was not solely the significance of Moewus's work for genetics or for a theory of sexuality. Moewus's credibility clearly superseded the importance of his work. Pätau went one step further than Philip and Haldane and attempted to turn this question into a problem for statistics proper. Thus, unlike Philip and Haldane, who suggested that someone should repeat Moewus's experiments, Pätau (1941: 319) concluded with an urgent request for statistical checks on the statistical critiques: "So here we can conclude that three papers about decidedly different biological manifestations revealed the same statistical distributions. A confirmation of these findings by a third party is imperative."

Pätau's call for a third party was answered the next year by Wilhelm Ludwig (1942). Statistics was not well known among biologists of the 1930s. Ludwig, a zoologist at the University of Heidelberg, turned to statistical studies while involved in so-called war projects in Germany. Ludwig checked Pätau's statistical work and argued that Pätau had made an error in calculation. Pätau had calculated that the probability of obtaining such good agreement with the expected results in question was less than 10^{-9}, whereas according to Ludwig, the probability was less than 10^{-22}. However, this criticism was answered by Ludwig himself. Ludwig visited Kuhn's laboratory in Heidelberg in 1942, and he and Moewus poured over Moewus's manuscripts and papers, trying to find out why the rates were so good. Based on examination of Moewus's notes, Ludwig found that Moewus had not published all of his data. He concluded that the numerical results Moewus used to support his conclusions were actually twice the size of the published results. Nonetheless, according to Ludwig, the probability of obtaining such good agreement with the theoretical was still only .001.

But the close conformity between data and theoretical expectations in Moewus's published reports did not result simply from Moewus's having not published all of his data. The data were also shaped by the procedures Moewus used to select zygotes for genetic analysis. As discussed in Chapter 2, when scientists analyzed the criticisms of Beadle's one-gene–one-enzyme theory, their discussions often focused on technique. It will be recalled that Max Delbrück pointed out bias in the results of the *Neurospora* school in terms of bias in the procedures they used in selecting mutations for biochemical genetic investigation. Moewus's results were biased in a similar way. Mrs. Kobb recalls that Moewus went through his notes and found a little note saying that he had picked for germination zygotes that were all in one stage of development. He must have done some preselecting. It was not done at random.

Moewus himself replied to Philip and Haldane's criticisms in two published papers. It will be recalled that Philip and Haldane suggested that Moewus either adjusted his data to suit his theory or had discovered a new type of biological regulation. Moewus (1940c) implied that a new type of biological regulation might indeed be the case at least for some of the data examined by Philip and Haldane. He addressed the statistical criticisms at length in 1943. He argued that most of the data examined by Philip and Haldane were not what they assumed them to be. He held that they were not segregation ratios but comparisons of physiological properties of two types of cultures and that he had tried to select for each comparison exactly 15 times. Despite Moewus's attempts to account for the statistical criticisms, and dispel the rumors about the reliability of his work, as we shall see, his own replies had little impact, at least on biologists outside of Germany.

Biochemical transformation of carotene

Still other criticisms of other aspects of Moewus's work appeared. These concerned some biochemical descriptions pertaining to the conversion by visible light of 1% *cis*- to *trans*-dimethylcrocetin per minute. Moewus claimed this step to be responsible for the organism's ability to copulate. In 1940, Kenneth Thimann, a leading plant physiologist at Harvard University, addressed this problem in relation to the physiological aspects of Moewus's work. As will be discussed later, the linear transformation was held by influential American microbial geneticists to be impossible on physical grounds.

When reviewing Moewus's work on the biochemical transformations of the carotene molecule, Thimann was struck by the theoretical novelty – that one molecule could have such diverse physiological effects. The pluralistic and diverse effects of one molecule on a cell was a feature of Moewus's work that was novel and striking to many biologists. According to Moewus, as Thimann (1940: 32) pointed out, *Chlamydomonas* made very complete use of the carotene molecule. Intact, it served as a plastid pigment; split, it produced one substance, which in turn served to determine sex type, and another substance, crocetin, which in one specific form caused copulation and in another caused motility. Thimann (1940: 32) was equally impressed, however, by the "elegance" of the numerical results concerning the chemical transformations:

> In each minute exactly 1% is converted; right up to the end. A more striking convenience in numerical results could hardly be imagined . . .
> The data are remarkably elegant and consistent. This will have been noticed above in the *cis–trans* ratios for the two groups of exactly equidistant sexes and the linear conversion at just 1% per minute. There is scarcely a single discordant note in the tables; as irradiation progresses activity disappears completely for the next 2 to 4 minutes later in every instance.

After noting the critique of Philip and Haldane, Thimann (1940: 32) concluded:

> As to the elegance of the physiological results, had Lewis Carroll been interested in the biology of the carotenoids, he could not have written a better "*Chlamydomonas* in Crocin-land". Confirmation of the experiments is much to be desired.

Although many investigators seemed to be convinced that Moewus, either consciously or unconsciously, did adjust his data to some extent, this was not enough to discount his claims. Indeed, in many respects Moewus's work seemed to be on firm ground. Among physiologists, Moewus's work fitted well with the great interest of the times in biosynthetic pathways; there was great interest in the role of steroids in the development of sexuality. To suggest that carotenoids, simple linear steroids, were involved was very intriguing from this point of view.

Among plant physiologists, Moewus's work seemed to be part of the auxin story which came into prominence in the early 1950s. Auxins were generally considered to be growth regulators (hormones) found in plants.

Plant physiologists such as Thimann at Harvard University carried out extensive work on auxins from the point of view of the relation between their structure and function. Others, such as A. E. Murneek at the University of Missouri, had been working on the physiology and reproduction of horticultural plants since the mid-1920s. During the late 1940s and early 1950s, Murneek carried out studies on certain aspects of the function of plant hormones and the practical application of so-called synthetic growth regulators in reference to the reproduction of horticultural crops. To plant physiologists like Murneek, the physiology of sexual reproduction had been a neglected field; Moewus's work offered great possibilities. The plausibility of Moewus's work could be evaluated in terms of its fit with statistical theory and contemporary physiological theory. But it also could be evaluated in terms of his social relations in the field. In light of these last two issues, Murneek (1941) defended its general plausibility, and he encouraged work on similar lines.

Moewus's immediate professional associations tended to strengthen, or lend credibility to, the truth of his claims. Moewus's work was certified by Max Hartmann, celebrated as the leading protozoologist in Germany, and by Richard Kuhn, a Nobel Prize winner in biochemistry, who had worked in collaboration with him. The role played by Kuhn in helping to dissuade biologists from dismissing Moewus's claims is well summarized in the following comments of Murneek (1941: 617–618).

> Because of the great complexity but definiteness and "elegance" of the results, Moewus and Kuhn's investigations have been subjected to severe criticism, especially as regards the statistical reliability of the data. The one by Philip and Haldane (1939) does not refer directly to the physiological side but more specifically to Moewus' analyses of genetic behavior of various races of this algae. Thimann (1940), on the other hand, criticizes chiefly the physiological studies. Considering Kuhn's world-wide reputation as a pigment chemist, it would seem to be advisable not to dismiss these spectacular investigations as unbelievable but study them with circumspection. Verification, of course, would be highly desirable. Moewus may belong to the category of scientists who like to make their studies "beautifully complete." But even if a small part of the results be verified, it would be a notable advance in our understanding of the physiology of sexual reproduction in plants.

Those who were concerned with the biochemistry of microorganisms, such as the celebrated microbiologist C. B. van Niel at Stanford University, clearly recognized the great theoretical contributions of Moewus. To microbiologists, like plant physiologists and biochemists, the work of Moewus, Kuhn, and their collaborators clearly indicated a relationship among "growth factors," vitamins, and enzymes. As will be discussed later, several biologists, including Lewin and Sager who were to center their research on *Chlamydomonas,* took a summer course on the biochemistry of microorganisms offered by van Niel at Stanford University. In his review of biochemical studies of microorganisms, van Niel (1940: 113) referred to

> the spectacular studies of Moewus et al. (Moewus 1938a,b; Kuhn, Moewus, and Jerchel 1938), from which it appears that only one molecule of crocin per cell might induce the activity of flagellae in *Chlamydomonas eugametos,* and that the methyl esters of *cis-* and *trans-*crocetin in somewhat larger quantity and in different proportions cause either the male or the female gametes to become capable of conjugation. . . . These illustrations suggest future discoveries of as many more enzyme systems as at present known.

Early genetic apologists

Van Neil did not refer to any criticisms of Moewus's work such as those launched by Philip and Haldane. However, the celebrated zoologist H. S. Jennings, at Johns Hopkins University, did when he reviewed Moewus's work in a review of "inheritance in protozoa" in 1941. Jennings wrote the first exhaustive review of Moewus's work from the vantage point of genetics research in America. He saw in Moewus's work an unprecedented use of microorganisms for genetics. In his view, Moewus's work seemed to swing wide open the doors to microbial genetic technology by showing that microorganisms possessed genes that were inherited in the classical Mendelian way. After mentioning Philip and Haldane's critique, Jennings (1941: 750) wrote:

> Aside from this difficulty, the work of Moewus has placed the genetics of protozoa on a new footing. It has brought the phenomena of inheritance in these organisms into the same system that is manifested in the Mendelian inheritance of higher organisms. It has brought to light in the flagellated protozoa instances of most of the phenomena in such inheritance, as before known only in multicellular organisms.

Indeed, there were some influential apologists for Moewus's work among American biologists. The most important of these was undoubtedly T. M. Sonneborn, a former student of Jennings. Between 1941 and 1951, Sonneborn popularized and skillfully defended the originality and significance of Moewus's technical procedures and concepts. Although Sonneborn was one of the first American geneticists to criticize Moewus's work publicly, he would not allow his criticisms, nor those of others, to close his eyes to what still seemed to him to be Moewus's important accomplishments.

Any complete account of Sonneborn's support for Moewus would first have to take into consideration that he, like Moewus, began his investigations of the genetics of microorganisms long before other geneticists turned to microbial genetic technology. Sonneborn was well aware of the technical obstacles that had to be overcome in order to utilize microorganisms for genetic analysis. After learning how to control mating in *Paramecium* (see Figure 2e), a task that took seven years of systematic investigation, Sonneborn (1937) provided the first demonstration in the United States that, like higher organisms, unicellular organisms possessed genes that behaved in the classical Mendelian way. Sonneborn received great acclaim for his early work and quickly rose to prominence in the American biological community as one of the few authorities on the genetics of microorganisms (see Sapp, 1987a).

However, this was not the whole story. Sonneborn investigated various novel phenomena involving complex relationships among nucleus, cytoplasm, and environment which could not be easily embraced within the confines of Mendelism and contemporary theories of the nature of the gene. During the 1940s and 1950s, Sonneborn built up a major program of *Paramecium* genetics at Indiana University. He was well regarded as a brilliant and ingenious experimentalist, someone who carefully scrutinized genetic procedures and experimental processes, and a major critic of Mendelian genetic orthodoxy. Although Sonneborn played an important role in the development of the genetics of microorganisms, *Paramecium* presented problems for obtaining detailed biochemical descriptions of gene action. As discussed in Chapters 2 and 3, the biochemical genetics studies that emerged into prominence following World War II relied on microorganisms that could be grown on a defined and preferably synthetic culture medium. However, in nature, *Paramecium* feeds on bacteria; a synthetic growth media could not be constructed. Thus, *Paramecium* could not be fully utilized in the kind of biochemical genetics done on *Neurospora*, yeast, and algae. Nonetheless, it was highly valuable for investigating the role of the cytoplasm in heredity. Sonneborn's experimental work and theorizing on nucleocytoplasmic relations had a major impact on theories of gene action during the 1940s and 1950s (see Sapp, 1987a).

As one of the few biologists who worked on the genetics of microorganisms prior to World War II, Sonneborn had kept a close eye on the work published by protozoologists in Germany. Indeed, Jennings's and Sonneborn's early work on microorganisms was highly informed by genetic work on Protozoa in Germany during the 1920s and 1930s, especially that concerned with the inheritance of acquired characteristics *(Dauermodifikationen)*, carried out in the laboratory of Max Hartmann. In fact, when Sonneborn completed his doctoral work at Johns Hopkins University, under the direction of Jennings, he had considered going to Germany to work in Hartmann's laboratory (Sonneborn, unpublished autobiography, 1978). Since the early 1930s, when he was carrying out his pathbreaking work on the genetics of *Paramecium*, Sonneborn had also followed, with interest and care, Moewus's remarkable publications on *Chlamydomonas* and other algae. Sonneborn was very interested in problems of sexuality in microorganisms. After he learned to crossbreed different strains of *Paramecium*, he turned to a genetic analysis of the complexities of mating-type inheritance itself (Sonneborn, 1937). The determination of mating type in *Paramecium* would baffle researchers for many years (see Nanney, 1954, 1958).

In 1941, Sonneborn published an extensive review of Moewus's work on sex determination in *Chlamydomonas* in relation to his own research and that of his students on mating-type determination in *Paramecium*. Sonneborn could not accept the concept of relative sexuality as developed by Hartmann, which he claimed, appealed to "abstract, ill-defined, and confusing concepts as fundamental maleness and femaleness" (Sonneborn, 1941: 705). However, he found it easy to distinguish Moewus's *results* from the *interpretations* he gave to them. He rejected the idea that only two sexes existed, and reinterpreted Moewus's experimental results in terms of the existence of multiple sex systems.

Sonneborn also recognized that there were "certain difficulties in Moewus's observations that raise serious questions concerning the reliability and accuracy of his reports" (Sonneborn, 1941: 686). In addition to those already reported, he found "irreconcilable contradictions" in Moewus's reported observations on the percentage of the sex stuffs – cis- or *trans*-dimethylcrocetin – that was held to change the organism's sex. Nonetheless, he did not reject Moewus's published results as being essentially unsound. On the contrary, he held them to be of great significance for understanding many problems in genetics. In view of his criticisms, those of Philip and Haldane (1939), and the failure of Pringsheim and Ondratschek (1939) to confirm parts of Moewus's work and "the great theoretical importance of the work," Sonneborn concluded that independent repetition of the work was urgently needed.

The same year, at a Cold Spring Harbor Symposium on Long Island, New York, Sonneborn (1942) gave another exhaustive review of the work of Moewus, Kuhn, et al., in a paper on "Sex Hormones in Unicellular Organisms." Again, he applauded the novelty and importance of the Moewus–Kuhn work on sex hormones. Sonneborn (1942: 111) began his account by mentioning some aspects of the social context of the work on sex hormones in *Chlamydomonas*. "The early work," he pointed out, "was done by Moewus alone,

> but the biochemical aspects of the problems became so alluring that R. Kuhn of Heidelberg, the distinguished student of carotenoids and winner of a Nobel Prize, became a leading partner in the investigations appearing from 1938 on. As a result of the war conditions, we can at present know only how matters stood at the end of 1940. Important advances have doubtless been published since the 18 months that scientific communication with Germany has been cut off.

After reviewing the work of Moewus, Kuhn, and their collaborators, Sonneborn (1942: 117) remarked, "Few known cases equal this as a point of contact between genetics, development and hormones." Indeed, this was an understatement, for it is safe to say that no other work by 1942 equaled the *Chlamydomonas* work in this regard. After reviewing the chain of reactions "partly known and partly assumed" to lead to the production of the hormones and its genic control, Sonneborn (1942: 119) wrote: "The strict economy shown in the derivation of these hormones from protocrocin is remarkable and the combination of the biochemical with the genetic analysis makes this one of the most perfect such studies in the literature."

Although Sonneborn celebrated the Moewus–Kuhn work, he was also critical of it. His review included a long section entitled "Comments and Critique" that began with the following remark: "Although the accomplishments of Moewus and Kuhn are already magnificent, there remain several matters of great importance urgently requiring further study" (Sonneborn, 1942: 119).

Sonneborn wanted only to build on and refine Moewus's work, not destroy it. He suggested various technical problems that needed further investigation. For example, none of the hormones controlling sexuality had been isolated in pure crystalline form from the filtrates themselves; they had been identified only by "indirect methods." Sonneborn reasoned that additional physiological, chemical, and genetic explanations were required to account for some problems of sexual behavior observed by Moewus. He also offered alternative theories for the activity of the gene *mot*, which controlled the motility hormone, crocin, as well as for the genes *F* and *M*. Sonneborn also highlighted some contradictory statements in Moewus's papers relating to the rate at which *cis*-dimethylcrocetin (female gamone) is transformed into the trans ester (male gamone) when irradiated. "These numerical discrepancies are disturbing," Sonneborn (1942: 122) wrote, "in relation to the very precise and but slightly variable numerical relations found elsewhere in their papers."

Finally Sonneborn (1942: 122) came to the question of the reliability of Moewus's data: "This critique would be incomplete without calling attention to the unpleasant but possibly important question as to whether Moewus reports his data accurately." Sonneborn began to weigh carefully the arguments for and against the reliability of Moewus's work. First, he examined Philip and Haldane's criticism that some of Moewus's experimental data were too close to theoretical expectations, and their suggestion that this implied either a conscious or unconscious adjustment of observations to fit a theory or a new type of biological regulation. On the other hand, he pointed out that Moewus (1940c) had replied to this criticism and implied that an alternative of a new type of biological regulation was indeed the case, at least for a part of the data examined by Philip and Haldane. Moewus (1940c) stated that this uniformity was peculiar to the crossover values observed between genes *F* and *M* and was not found in crossovers between other loci. Again, Sonneborn suggested, "Only independent confirmation can resolve these doubts."

However, up to that time no one had published any attempts to repeat Moewus's work on *Chlamydomonas*. Moewus's competitors in Prague, E. G. Pringsheim and K. Ondratschek, Sonneborn (1942: 122) pointed out, "did try their hand at certain technical matters (conditions for copulation; analysis of sex and copulation in synoecious clones) on another form, *Polytoma uvella*, investigated by Moewus." They failed to confirm the existence of sex substances affecting sexuality. However, Sonneborn noted that Moewus (1940a) himself ascribed this result to their failure to employ suitable methods. Indeed, just because Pringsheim and Ondratschek could not replicate Moewus's results did not mean they had disconfirmed Moewus's work. The difficulties scientists had in distinguishing a "failed replication" from a "disconfirmation" of Moewus's work will be dealt with at length later when discussing some systematic attempts to repeat crucial aspects of Moewus's experiments.

In the meantime, other indications of the reliability of Moewus's work could be considered. In a general way, Sonneborn argued, the results on *Chlamydomonas* were supported by "comparable observations" of the effects of "sex substances" on other types of algae. Moewus himself had carried out studies on several genera of algae. Similar effects of hormones affecting sexual behavior in algae had been also independently reported by other investigators in Hartmann's laboratory (see, e.g., Jollos, 1926). "But," Sonneborn argued, "in no other studies has the analysis of those phenomena been carried out in anything like so full a manner as in the work of Moewus and Kuhn on *Chlamydomonas:* in none of them have any hormones been identified" (Sonneborn, 1942: 123).

Sonneborn further argued that the possible general significance of Moewus's work was suggested by the widespread occurrence of carotenoids among plants. Moewus (1940e) reported their existence in the pollen and stigmas of many genera of flowering plants. He had "even shown that these contain crocin by means of the delicate *Chlamydomonas* motility test (for which only one molecule of crocin is required per cell): extracts of these organs will induce *Chlamydomonas* to become motile in the dark under anaerobic conditions" (Sonneborn, 1942: 123). Sonneborn concluded his paper by pointing out that his work on *Paramecium* (Sonneborn, 1937) also demonstrated group formation when cultures of diverse mating types were mixed. Moreover, work done in his laboratory also suggested the possible existence of hormones that have an effect on agglutination of group formation – comparable to substances functioning as hormones in *Chlamydomonas* (Sonneborn, 1942: 123).

Mendel and Moewus

Sonneborn's review of the papers of Moewus, Kuhn, and their collaborators did not cause a great deal of excitement at Cold Spring Harbor in 1941. The big rush to the study of microorganisms and to combining biochemical and genetic analysis had not yet occurred, and few researchers recognized (and some did not want to recognize) the great potential of microbial genetic research. The Cold Spring Harbor conference of 1941 focused on "The Relations of Hormones to Development."

The ultimate objective of workers in this field was "growth control." Once normal growth was understood, workers claimed, "it should be possible to control abnormal growth." However, studies of the relation between genes and the control of growth were largely omitted from the conference papers. With the exception of Sonneborn's paper and one other (by Boris Ephrussi, on eye differentiation in *Drosophila*), there was no discussion of the role of genes. The remaining 16 papers dealt only with the chemistry and effects of hormones on the growth and development of higher plants and animals. They did not mention the role of genes in these processes. As will be discussed in Chapter 8, a decade later, when Sonneborn again addressed the question of the importance of Moewus's work, at another Cold Spring Harbor Symposium, the response was staggering.

In the meantime, the few geneticists who were working on biochemical aspects of genetics in the United States, such as Beadle and his co-workers, had become involved in war-related projects (see Kay, 1988). In 1940, several leaders of the American scientific establishment had organized a National Research Committee in order to coordinate the nation's scientific research for the interests of war. By 1941, an Office of Scientific Research and Development (OSRD) was created with powers to initiate and coordinate scientific research in academic institutions, with industry and the military (see Greenberg, 1967). Many scientists working in the physical sciences organized their war-related projects under the auspices of the OSRD. War-related projects led by the OSRD's Committee on Medical Research in Pharmacology and Biochemistry, such as the production of penicillin, were coordinated with the private sector, including the pharmaceutical firms of Merck & Company, E. R. Squibb & Sons, Sharp & Dohme, etc.

The biochemical genetic work of Beadle and his group, which focused on the nutritional requirements of microorganisms, gained significance in this context (see Kay, 1988). The *Neurospora* work pertaining to mutations affecting vitamins and amino acid synthesis was important not only for resolving fundamental problems of the nature of the gene and gene action, but also for the food and drug industries who were concerned with manufacturing vitamins and amino acids. With the projected meat shortages, Beadle himself pointed out the value of *Neurospora* work for assessing the vitamin content of dehydrated food products (Kay, 1988: 18). Beadle and his co-workers were not interested in developing or evaluating Moewus's work. During the early 1940s, Beadle and his group were just beginning to develop and refine their own techniques, and Beadle was involved in negotiations with pharmaceutical companies and philanthropic institutions (see Chapter 9).

Beadle mentioned the work of Moewus only once in publication, in 1945 (see Chapter 5). Although Beadle and Tatum (1941) never referred to Moewus's work in their first paper on *Neurospora*, they had known about his work. Moewus's wife, Liselotte, recalls (interview, May 26, 1987) that before the outbreak of war, Moewus had written to Beadle and had sent his recent papers on the genetics and biochemistry of sexuality in alga. Beadle wrote back, congratulating him on his brilliant work. Beadle at that time was working on a higher organism, *Drosophila*; he had not yet appreciated the great value of microorganisms for investigating gene action. As discussed in Chapter 3, Beadle later claimed that the idea of switching to microorganisms occurred during a lecture given by Tatum on comparative bichemistry in bacteria. He has never claimed that Moewus's work on *Chlamydomonas* influenced him to work on microorganisms.

It should be mentioned here that multiple discovery – that is, discovery by two or more individuals working separately – occurs frequently in science and often gives rise to bitter priority disputes (see Merton, 1957, 1961). Indeed, historians have come to recognize that although some discoveries take one completely by surprise, there are many more which are more or less expected, and toward which several investigators are working simultaneously. Multiple discovery falls into the latter class, though often the disputants hotly contest that their prize discovery was expected, and insist that their competitor had stolen their idea. To the historian, however, the analysis of these disputes are more important than trying to settle them. Moewus himself never publicly claimed to have influenced Beadle to use microorganisms for the study of genic action. Although he and Kuhn were well aware that they had begun to develop a whole new research domain, Moewus never publicly claimed priority for first demonstrating how to use microorganisms for biochemical genetics. This is not to say, however, that there was no priority dispute concerning Beadle and Moewus. As will be discussed later (see Chapter 8), some of Moewus's allies, most prominently Sonneborn, argued forcefully that Moewus's work deserved priority recognition over the work of Beadle and his group. However, both Moewus and Beadle remained aloof from this dispute.

In the early 1940s, when Sonneborn was discussing and evaluating the validity of Moewus's work, Beadle and his co-workers were engaged in their own controversy. Beadle's group claimed to have discovered a previously unknown amino acid which they called "neurosporin" (see Kay, 1988: 12). Between 1942 and 1943, the neurosporin story caused a great deal of excitment in Beadle's laboratory. According to Beadle, "Tatum and Bonner were burning the night lights trying to get the structure established and a synthesis worked out" (quoted in Kay, 1988: 21). However, according to Kay (1988), by 1943 Beadle retracted the claim. The amino acid, neurosporin, did not exist! It turned out to be a substance made up of several amino acids and of marginal significance to biology. The discovery of neurosporin was a mistake; no charge that Beadle or his co-workers had faked data was ever made.

Beadle was not concerned with problems of sexuality and for whatever reasons did not pay a great deal of attention to the work of Moewus. Indeed few geneticists in America showed any interest in Sonneborn's summary and evaluation of the work of Moewus, Kuhn, and their collaborators. Sewall Wright was an exception. Having widely promoted the study of genic action and having organized lectures and seminars on recent developments in the domain, he was very interested in Sonneborn's useful summary and evaluation of Moewus's work. Wright is best known today as one of the central architects of the evolutionary synthesis of the 1930s and 1940s which brought Mendelian genetics and Darwinism together into a unified perspective. Like Haldane, he is widely applauded by scientists and historians for his statistical work in population genetics and for his theorizing on evolutionary mechanisms generally (see Provine, 1986). But Wright was also persistently concerned with the physiological aspects of the gene; since the mid-1920s, he worked, at the University of Chicago, largely on the inheritance of coat color in guinea pigs (see Provine, 1986). Wright (1941; 1945) wrote some of the earliest extensive reviews of physiological genetics, alluding to the diverse problems in understanding genetic control of physiological processes.

Like many other American geneticists, during the war, Wright combined his work on genetics with war projects. In January 1944, he was busy doing "a statistical analysis of data in some war projects and lecturing to ASTP [Army Special Training Program] premedics on what the army should know about annelids, evolution, etc." (Wright to Sonneborn, January 29, 1944). At the same time, he was organizing the first major symposium on gene action for the American Zoological Society. In particular, both Sonneborn and Wright were theorizing on the nature of cytoplasmic genetic factors, or plasmagenes, and on their role in cell physiology and cellular differentiation (see Sapp, 1987a). Wright himself was planning to give a theoretical paper on "Gene and Cytoplasm in the Control of Cellular Physiology" (Wright to Sonneborn, May 4, 1944). Sonneborn was billed as a major speaker who would be lecturing on his experimental work on nucleocytoplasmic interactions in *Paramecium*.

Like Sonneborn, Wright began to integrate and discuss Moewus's biochemical genetic work in his course on physiological genetics at the University of Chicago. In his view, it seemed unlikely that Moewus's work was completely fabricated. It was too extensive, and it had been witnessed by others. The question then centered on whether there had been some polishing of the data. In a long letter to Sonneborn, Wright (January 29, 1944) gave his opinion about the validity of some of Moewus's genetic reports, and offered a statistical critique of various other aspects of Moewus's numerical results, left untreated by Philip and Haldane and others:

> Another point that I should discuss soon in my course is Moewus' work, of which you have written such a useful review. I am in great doubt as to what to make of it and I gather that you are not totally certain. If sound, it is of the greatest importance in many respects – (1) a clear demonstration that crossing over can occur in the 2-strand stage (ruling out both Billing's and Darlington's theories at least in *Chlamydomonas eugametos* though not necessarily in Pascher's species), (2) the extraordinary infrequency of double crossovers, (3) as a case in which all differences among 5 species and many *natural* races as found in nature are simple Mendelian (instead of multiple factor) including such characters as size, form, reactions to differences in pH, temperatures . . ., osmotic situation, Ca concentration, desiccation etc.) – all differences in the same character being (with some qualification) due to a single series of alleles, (4) inheritance of *absolute* length but *relative* breadth (complicated in the latter by surprising interaction with the form (cefo) series, (5) the remarkable sex-determining scheme with its sharp specificity for each of various proportions of known substances, lethal effects of certain combinations and other complications which you have discussed and finally, (6) the ratios, enormously more consistent with each other in separate experiments than expected, allowing for accidents of sampling, as pointed out by Philip and Haldane. I tested all of the ratios in the long 1940 paper, written after their criticism to see whether his statement that this peculiar result applies only to the group of closely linked genes relating to sex [is true].
>
> It applies to everything, although not to as extreme an extent, per degree of freedom, as in the earlier figures. For example, Moewus gives a frequency distribution of 512 types of zygotes (each producing 2 each of 2 complementary genes in most cases), expected in equal numbers among some 100,000 zygotes from a 10-factor cross (indicators of the 10 chromosomes). The mean (199.75) and standard deviation (14.14) are stated correctly, but the author notes as evidence of agreement with statistical the-

ory the fact that no frequencies exceed twice the standard deviation. . . . But there should be nearly 5 percent (23 cases in 512), and in fact the standard deviation of his distribution (not reported) turns out to be only 9.0, more than 11 times its standard error too small, with a probab[ility] of a "better" result by repetition of about 10^{-30}. The same figure is obtained from X^2 ($X^2 = 208$; 511 degrees of freedom).

Summarizing all but a few sporadic ratios (with Prob[ability =] .87):

	n	X^2	Probability
10-Factor cross	511	208	$1-10^{-30.4}$
Cases of random assort.	355	200	$1-10^{-10.9}$
Cases of linkage	48	26	.994
Total	914	434	$1-10^{-39.9}$
Philip and Haldane (not included above)	61	8	4×10^{-22}
	975	442	$1-10^{-46.8}$

There is clearly something wrong. A result closer to expectation than reasonable is no doubt often due to conscious or unconscious classification of doubtful cases to favor the expected ratio. This should not be the case here since there is exact[ly] 2:2 segregation among the genes from a zygote and each gene gives rise to a clone. The character differences are with few exceptions described as so clear-cut that there should be little doubt even about individuals and none about clones. The most favorable hypothesis is that the counting was done by a group of assistants, some of whom were either so inaccurate that their work had to be rejected, leading to selection against extreme real deviations from expectation or that some were dishonest and reported the exact or nearly exact results expected. At the other extreme is the hypothesis that the entire array of results is manufactured, which seems hardly probable or possible of work done on such a scale, with progress necessarily known in a general way by many persons. Yet it is hard to decide what weight to give the various results. What do you think?

Wright's statistical criticisms of Moewus's work were never published. But they did circulate among some of Sonneborn's students during the late 1940s and 1950s (see Chapter 7). Sonneborn (February 19, 1944) wrote back to Wright, pointing out their similar views regarding Moewus's work. At the same time, he offered an extremely lucid and sophisticated reconstruction of the thought processes underlying Moewus's experiments; he pointed out the dangers of dismissing Moewus's claims too quickly, and he emphasized the difficulties of having Moewus's experiments repeated and evaluated by an independent and objective observer:

Concerning the work of Moewus, my reaction is much the same as yours. I can see many objections, difficulties and indications of dishonesty, yet I am unable to conclude that the whole story is a mere fabrication. The calculations that you have made, like those of Philip and Haldane, seem convincing evidence of dishonesty. Yet I seem vaguely to recollect that a similar study of Mendel's data was published long ago and led to similar implications. Moreover, I have always believed that Mendel must have run into cases of linkage in his experiments and simply chose not to mention them because they seemed to conflict with his evidence of independent assortment. Considerations of this kind make me dread the danger of throwing out the baby with the bath water. The many genetic peculiarities that you list are certainly disturbing. Some of them might well be due to a combination of ignorance and dishonesty, for example, the 2-strand crossing-over. The first paper in which evidence for this appeared was presented in such a way as to show clearly that Moewus was not aware of either the significance or peculiarity of his results. Only later, presumably after the matter had become called to his attention by others, did he explicitly discuss the matter; and, when he did so, he took pains to add that crossing over occurred also in the orthodox 4-strand stage under certain conditions. Some of the other difficulties you mention, such as the lack of multiple factor determination for certain differences between species and natural races, I had not before considered. I wonder whether this could not be due in part to the combination of haploidy with vegetative reproduction and in part to selection of only "favorable" characters for study. The number of "species" reported in *Chlamydomonas* is enormous: practically every biotype with a distinct complex of characters is called a species. Haploidy, assisted by vegetative reproduction, of course renders each genic combination a pure breeding unit. Would not this render single gene differences between races and "species" more likely to occur than in diploids lacking vegetative reproduction? In the present case, however, I should think that selection of material is the more important factor. Moewus was not so much interested in describing fully the genetic basis of differences between races and "species" as in finding a large number of genes to work with. For this reason, I should guess that any complex case he ran into he would put aside as unsuited to his purposes and would report only the simply analyzed and striking differences. This would then result in leading to the situation you have remarked as extraordinary. I fear there is little more that I could add to help reach a satisfactory conclusion about Moewus. As many of us who have been perplexed about this work have agreed, the only really helpful thing to do would be to have the work repeated by some one who is both competent and above suspicion; but that is a task for which few are prepared, unless the materials used by Moewus and Kuhn were supplied. And, even granting the great importance of the issue, most of us are too deeply involved in our own projects to drop them and take up such an exhausting and prolonged task as this would be.

Sonneborn's letter raises a number of issues that will be discussed throughout this book – bias in selection of data to fit theory, conscious bias (dishonesty), ignoring messy results, the difficulty of having the work repeated by an independent observer "who is both competent and above suspicion," etc. In fact, many of the issues raised by Sonneborn can be highlighted in the case of Mendel to which he refers. Indeed, Wright's statistical critique and that of Philip and Haldane were of the same type used by R. A. Fisher in 1936 in an attempt to reconstruct Mendel's methods. As in the case of Moewus, Fisher claimed that Mendel could not have obtained all the experimental data that he gave to justify the laws that would carry his name. A brief glance at the scientific discussion surrounding Mendel's results will help us to understand many of the complexities and uncertainties underlying scientists' attempts to evaluate knowledge claims and to elucidate the thought processes of their colleagues through reading their published papers.

5. Mendel revisited: Are all genetic reports faked?

> Sewall Wright once pointed out to me that many reported data in genetics are "too good" to be random samples. Most of us are subject to the same prejudices as are here assumed to have influenced Mendel.
> (Beadle, 1967: 338)

Since Fisher wrote his paper, "Has Mendel Been Re-discovered?" (1936), a great deal of attention has been given to the question of whether or not Mendel deliberately fudged his data. The debates focus primarily on the moral element – whether Mendel was "honest" or "dishonest" (e.g., see Dunn, 1965; Iltis, 1966; Wright, 1966; Beadle, 1967; Campbell, 1976; Pilgrim, 1984). At first glance, such a debate might seem trivial. Who really cares if Mendel fudged some data? After all, he was right. However, once we consider the role of founding fathers in defining groups and the rhetorical roles they play in science, the purity of motives of the celebrated originator becomes extremely important.

The search for purity of motives in "founding fathers" is pervasive in the history of science. The reconstruction of the thought process of a creative genius is central to the "Darwin industry" (see Shapin and Barnes, 1979). What is at stake in this controversy is whether or not Darwin was in any way part of or responsible for the political and ideological uses of his theory. It is well known today that "evolution had been invoked to support all sorts of political and ideological positions from the most reactionary to the most progressive" (Young, 1971: 185). Several writers have charged that Darwin was influenced by the notorious socioeconomic views of Thomas Malthus. Others argue that he was as much a Social Darwinist as his contemporaries who appealed to "nature" to legitimize their political views (Moore, 1986). As Shapin and Barnes (1979: 127) have pointed out, "Darwin's defence" has rested upon three assertions:

> The first is that of internal purity: Darwin's *intentions* and *motives* in writing the *Origin* were above reproach, and his personal beliefs in 1859 were innocent of "ideological" taint. The second is purity of ancestry: influences upon the *Origin* were entirely wholesome and reputable, nothing "ideological" was gleaned from Malthus. The third assertion is purity of germ-

plasm: nothing outward could *properly* be deduced from the theory in the *Origin;* truth does not blend with error; insofar as truth was used to justify Social Darwinism, it was misused.

Shapin and Barnes (1979: 133) concluded that "Darwin's defense is far better staffed and funded than its opposition," but the more interesting question for us is, Why has the trial been conducted at all? Shapin and Barnes (1979: 134) can suggest only an anthropological explanation:

> The scientific discipline of evolutionary biology had its font and origin in the person of Charles Darwin and in the text of 1859. Darwin is a sacred totem by virtue of his "foundership" of modern biology: science is sacred, so must Darwin and his Book be sacred; both must be protected from contamination by the profane. As the author of the *Origin* he must himself be pure; his thought must be unmingled with wordly pollutions and incapable of satisfactorily blending or combining with the suspect formulations of social Darwinism. Thus, "influences" from the "profane" Malthus can only be the spiritual emanations of mathematics and genuine science, or nonessential stimuli or manners of speech. And implications for social Darwinism can only be misunderstandings.

Although this is a wildly speculative suggestion, it does have a great deal of merit in helping us to understand why the motives of so many "founding fathers" have been put to scrutiny. It helps us to explain why Darwin's belief in the inheritance of acquired characteristics is downplayed by scientists and historians, why the later editions of the *Origin,* in which Darwin becomes more and more "Lamarckian," are considered as having less integrity than the first edition. Darwin scholars habitually apologize for the later editions; they claim that the views Darwin expressed in them were not the "pure" Darwin; they reflected only his bowing to contemporary criticisms. It helps us to understand why Newton scholars focus so much attention on reconstructing Newton's thought, and why they habitually overlook his metaphysical alchemical views; for Newton himself was no Newtonian. And, as mentioned in Chapter 2, Olby (1979), Brannigan (1981), and Callender (1988) have argued that Mendel was no Mendelian. An anthropological perspective may help us to understand why those critical of psychoanalysis attempt to discredit it by examining Freud's motives. In each case, the purity of motives of the "found ıg father" becomes a major issue.

It is not surprising that the interest in whether or not Mendel fudged his data was first brought to great public attention in 1965 at centennial symposia of the genetics clan (see Dunn, 1965: 12; Iltis, 1966: 209; Wright, 1966: 173–175; Beadle, 1967: 337–338). In view of Mendel's stature in genetic culture, and the defense accorded him by geneticists who attempt to account for his data, it remains a mystery why R. A. Fisher, a Mendelian himself, would make such a charge in the first place. After all, fraud charges are often made to discredit an individual and/or competing theory. Indeed, no attention has been given to the historical context in which Fisher wrote his paper of 1936: Why was Fisher concerned with reconstructing the thought processes of Mendel? Many of the issues raised in Fisher's paper are of the greatest importance in appreciating the attitude of geneticists toward Moewus's work. However, before addressing these issues, it is necessary to set Fisher's paper in its own historical context.

Statistics versus experimentation

Fisher's critique of Mendel's work came at a time when statistical studies of populations were becoming intimately allied with Mendelian genetic principles. In fact, it represented an attempt to understand the famous dispute between the biometricians (statisticians) and the Mendelians (experimentalists). Conflicts between statisticians and Mendelian geneticists emerged at the turn of the century, soon after Mendel's laws were "rediscovered" (see Provine, 1971; Kevles, 1980; MacKenzie, 1981). Essentially, the theoretical element of the dispute revolved around the question of whether or not evolution was continuous or discontinuous – whether or not new species emerged slowly by natural selection, or quickly through large mutations with natural selection playing only a negative role in selecting out those new species or mutants that cannot survive. The biometricians, who were led by Karl Pearson and W. F. L. Weldon in England, saw themselves as Darwinians. They supported continuous evolution. The Mendelians, led by William Bateson, were non-Darwinians and supported discontinuous evolution. The dispute raged on in private correspondence and in published journals throughout the first decade of the century.

Bateson found it "impossible to believe" that biometricians had "made an honest attempt to face the facts." He doubted that they were "acting in good faith as genuine seekers of the truth" (quoted in Kevles, 1980: 442). The leading biometrian, Weldon (1901), for his part, attempted to test the validity of Mendelism by subjecting Mendel's results to statistical tests. He did not claim that Mendel's results were statistically too good to be true, but doubted the possibility of reproducing Mendel's results with further pea experiments. Weldon concluded his critique with the remarks that Mendel was "either a black liar or a wonderful man." He remarked to Pearson, in 1901, "If only one could know whether the whole thing is not a damned lie!" (quoted in Kevles, 1980: 445).

But there was more to the debate than a theoretical discussion about evolution. Both Mendelians and biometricians were struggling to dominate the field; both based their work on different methods as well as different theories. Methodological issues became principal stakes in this controversy. Which methods, those of the experimentalist or those of the statistician, were most appropriate for biological – that is, evolutionary – problems? Geneticists, using experimentation as their polemical tool, attempted to exclude biometricians from the field by denying the legitimacy of purely statistical approaches to heredity and evolution. The views of Wilhelm Johannsen (who had provided the central terms of genetics: genotype, phenotype, and the gene) were highly representative of those experimentalists who supported discontinuous evolution:

> Certainly, medical and biological statisticians have in modern times been able to make elaborate statements of great interest for insurance purposes, for the "eugenics movement" and so on. But no profound insight into the biological problem of heredity can be gained on this basis. (Johannsen, 1911: 130)

Thus the non-Darwinian geneticists attempted to exclude Darwinian statisticians from the field. In fact, the dispute between the biometricians and Mendelians had come to a head in England by 1905. Bateson was judged to be the victor (see Provine, 1971). However, by the 1920s and 1930s, statisticians began to reestablish their authority in the field. They gained their legitimacy primarily from the statistical studies of populations led by the contributions of Fisher, Haldane, and Wright. As mentioned earlier, they were central architects of what Julian Huxley in 1942 called the "modern synthesis." The evolutionary synthesis of the 1930s and 1940s was based upon Mendelian gene recombination, mutation, and Darwinian selection theory. Evolution according to this theory was continuous after all; Bateson and the first generation of geneticists were judged to be wrong in allying Mendelism with non-Darwinian views of discontinuous evolution.

Fisher's paper of 1936 fits squarely within this theoretical shift. One can understand it as an attempt to put the last nail in the coffin of the controversy. The core of Fisher's paper is an interpretation of Mendel's motives and theoretical views. The principal stake in Fisher's paper is not whether or not some of Mendel's data were faked. It is rather an historical dispute over Mendel's attitude toward Darwinian natural selection. Bateson had characterized Mendel as a non-Darwinian ally in his struggle against Darwinian biometricians. Fisher, on the other hand, attempted to redefine the "founding father," Mendel, as a good Darwinian. The main thrust of his criticisms was launched against Bateson, not Mendel. Both Bateson and Fisher superimposed the context in which they found their own work onto that of Mendel and his times.

Fisher used Bateson as a scapegoat for the heated controversy between Darwinians and Mendelians. He charged that Bateson had deliberately intended to deceive scientists by allying Mendel with non-Darwinian views and by fabricating and distorting history to suit his interest. As mentioned, Bateson and the first generation of Mendelians were engaged in a struggle with nonexperimentalists. In his book, *Mendel's Principles of Heredity*, Bateson claimed that like himself, Mendel had worked in virtual conflict with nonexperimentalists and Darwinians and that this was partly responsible for his "neglect" for 35 years. Thus Bateson (1909: 2) wrote:

> While the experimental study of the species problem was in full activity, the Darwinian writings appeared. Evolution, from being an unsupported hypothesis, was at length shown to be so plainly deducible from ordinary experience that the reality of the process was no longer doubtful. With the triumph of the evolutionary idea, curiosity as to the significance of specific differences was satisfied. The *Origin* was published in 1859. During the following decade, while the new views were on trial, the experimental breeders continued their work, but before 1870 the field was practically abandoned.

Again Bateson (1909: 311) claimed:

> With the views of Darwin which were at that time coming into prominence Mendel did not find himself in full agreement, and he embarked on his experiments with peas, which as we know he continued for eight years.

"Had Mendel's work come into the hands of Darwin," Bateson (1909: 316) declared, "it is not too much to say that the history of the development of evolutionary philosophy would have been very different from that which we have witnessed."

Fisher strongly opposed Bateson's interpretation of Mendel's experimental program, which he claimed was self-interested and held no truth value. Bateson's views were only those of a "zealous partisan":

> It cannot be denied that Bateson's interest in the rediscovery was that of a zealous partisan. We must ascribe to him two elements in the legend which seem to have no other foundation: (1) The belief that Darwin's influ-

ence was responsible for the neglect of Mendel's work, and of all experimentation with similar aims; and (2) the belief that Mendel was hostile to
Darwin's theories, and fancied that his work controverted them. (Fisher,
1936: 116)

Fisher (1936: 117) argued:

> Bateson's eagerness to exploit Mendel's discovery in his feud with the
> theory of Natural Selection shows itself again in his misrepresentation of
> Mendel's own views. Although he was in fact not among those responsi-
> ble for the rediscovery, his advocacy created so strong an impression that
> he is still sometimes so credited.

Fisher, who "knew" that *Mendelism* was not opposed to natural selection, believed that *Mendel* also knew that his work was allied with Darwinism. Those who believed that natural selection was the principal
driving force in evolution could share both Mendel and Darwin as common intellectual ancestors. When reconstructing Mendel's thought process, Fisher claimed that Mendel's experimental program could be made
intelligible only on the basis that Mendel worked squarely within a Darwinian framework. For example, in Mendel's day, most hybridists
crossed different species. They believed that species did not evolve and
that they possessed essential qualities, specific natures, or "essences."
They were concerned with crosses between species to investigate the
ways in which the forms of the hybrid reflected the parental "essences."
Mendel's approach, Fisher reasoned, conflicted with this: He crossed
closely allied varieties, not different species.

In Fisher's view, this suggested not only that Mendel was an evolutionist, but also that his work was actually shaped by Darwinian theory.
Thus, Fisher (1936: 117) wrote: "It's a consequence of Darwin's doctrine,
that the nature of hereditary differences between species can be eluci-
dated by studying heredity in crosses within species." The issue of
whether the characters responsible for differences between species
could be detected by crossing individuals within a species was a highly
contentious one during the first half of the twentieth century. Many
biologists, opposing the all-exclusive role of natural selection and genes
in evolution, argued that Mendelian genetics applied only to trivial characteristics, such as eye color, hair color, tail length, etc., which did not
account for species differences. Moreover, they maintained that "fundamental" characteristics of the organism, which distinguished higher tax-
onomic groups (macroevolution), lay beyond the Mendelian chromosome theory and Darwinian selection theory (see Sapp, 1987a). Fisher
(1936: 118), on the other hand, suggested that Mendel himself would

have opposed such views: "Had he [Mendel] considered that his results were in any degree antagonistic to the theory of selection, it would have been easy for him to say this also."

It was in the course of delineating Mendel as a good Darwinian that Fisher made the claim that Mendel's results were too good to be true and calculated that in the overall results one would expect a fit as good as Mendel reported only once in 30,000 repetitions. However, this charge was not meant to discredit Mendel; it was meant to celebrate his power of abstract reasoning. Fisher (1936: 123) claimed that Mendel had his laws in mind before he did his experiments:

> In 1930, as a result of a study of the development of Darwin's ideas, I pointed out that the modern genetic system, apart from such special features as dominance and linkage, could have been inferred by any abstract thinker in the middle of the nineteenth century if he were led to postulate that inheritance was particulate, that the germinal material was structural, and that the contributions of the two parents were equivalent. I had no idea that Mendel had arrived at his discovery in this way. From an examination of Mendel's work it now appears not improbable that he did so and that his ready assumption of the equivalence of the gametes was a potent factor in leading him to his theory. In this way his experimental programme becomes intelligible as a carefully planned demonstration of his conclusions.

In Fisher's account, the claim that Mendel's data were too good to be true provided testimony to his claim that Mendel had his ideas in mind before doing his experiments. Mendel was a thinker, not just a tinker. In Fisher's view, Mendel's whole research program, his experimental design, and results could be made intelligible on the assumption that he had supported Darwinian evolution in the back of his mind.

Then did Mendel "cook" his results to suit his theory? Fisher entertained three possibilities to account for Mendel's results: (1) that Mendel was lucky; (2) that he unconsciously biased the results, and (3) that he consciously biased the results in favor of his theory. Fisher ruled out the first two possibilities as providing inadequate accounts and instead proposed a conscious bias of "fudging the data." However, he did not place the responsibility squarely on Mendel. Instead of questioning Mendel's integrity as an honest monk and scientist, he suggested that "Mendel was deceived by an assistant who knew too well what was expected" (Fisher, 1936: 132).

If this was the correct interpretation of Mendel's results, then how did it go undetected by geneticists for so long? Fisher (1936: 137) concluded his attempts to recast Mendel as a good Darwinian by raising two issues in this regard. First, he claimed that Mendel's opinions had been misrepresented because his work was not examined with sufficient care. Writers relied on accounts of others. But, as Fisher remarked, "There is no substitute for a careful, or even meticulous, examination of all original papers purporting to establish new facts." Second, Fisher suggested that biologists before him had imposed their own meanings on the work of Mendel. Their interpretations were influenced by the theory of their times:

> Each generation, perhaps, found in Mendel's paper only what it expected to find; in the first period a repetition of the hybridization results commonly reported, in the second a discovery in inheritance supposedly difficult to reconcile with continuous evolution. Each generation, therefore, ignored what did not confirm its own expectations. (Fisher, 1936: 137)

Indeed, the reading of scientific papers, like the construction of original scientific data, is not a straightforward affair. Meaning is not embedded in raw observations. It is bestowed upon the data by the intentions of the observer. Meaning is often not embedded in scientific papers reporting original data either. It is often superimposed onto such papers. As Fisher suggests, when one reads a scientific paper, one does so with theoretical expectations in mind. In his view, the biases of others were obstacles to the recognition of Mendel's discovery. It is striking that Fisher excluded his own interpretation from any biases. It would not be difficult to show some of the ways in which Fisher himself shaped the evidence from Mendel's paper in order to impose a Darwinian framework on it (see Sapp, 1989). It is enough to mention here that perhaps the most detailed attempt to reconstruct Mendel's experimental program has been conducted by Callender (1988), who argues forcefully that Mendel, far from being a Darwinian, "was an opponent of the fundamental principle of evolution itself" (Callender, 1988: 72). We will return momentarily to the problem of reading scientific texts.

Fisher's analysis of Mendel's data raised two additional issues for methodological reflection: observer bias in analyzing experimental results, and whether or not the validity of experimental results could be tested by statistical means. Indeed, although Fisher's paper represented an attempt to close the dispute between Mendelism and Darwinism once and for all, at the methodological level the conflict between statistical and experimental modes of reasoning continued. Geneticists who subsequently addressed Fisher's claims found it necessary to consider these methodological differences when attempting to assess the strength and meaning of statistical critiques of experimental results.

Some of the difficulties to be encountered are well illustrated by a critique of Fisher's paper written by George Beadle in the proceedings of the "Mendel Centennial Symposium" sponsored by the Genetics Society of America in 1965. Beadle charged that Fisher's reconstruction of Mendel's methods was incomplete, and he explored the phenomenon of unconscious bias to account for Mendel's results. He claimed that Fisher had considered one kind of bias only, due to "misclassification" of some hereditary variations – for example, a shriveled round pea scored as "unwrinkled." Beadle remarked, "As every experimenter in genetics knows, some classifications are difficult and may easily be unconsciously biased in favor of a preconceived hypothesis" (Beadle, 1967: 338). Beadle himself was personally very sensitive to this source of error, for as he recalled:

> I once discovered a loose genetic linkage in maize between floury endosperm and a second endosperm character known to be on chromosome 9, a linkage that I subsequently concluded was the result of my "wanting" to find it. The floury character is often difficult to score, and I believe I unconsciously put the doubtful ones in the piles that would suggest linkage.

However, Beadle was careful to protect his own credibility and added: "Fortunately, I recognized the possibility of this kind of error in time to withdraw a manuscript I had submitted for publication" (1967: 338).

Observer bias in selecting and sorting data is indeed a serious obstacle for those who want to claim objective status for their results by invoking claims of naive empiricism. However, observer bias in selecting data in genetic analysis is only one difficulty. Beadle discussed a second problem resulting from the theory-ladenness of observations: how much data to include in a scientific paper, and how an experimenter knows when the experiment is over. He suggested that it was entirely possible that Mendel had stopped counting when he had obtained results close to expectation. This possibility was also suggested by Dunn (1965) and Olby (1966). Beadle (1967: 338) explained:

> As he [Fisher] points out, Mendel clearly had his hypothesis in mind before completing all his work and therefore rejected certain numerical ratios. It is also clear, as Fisher deduces, that Mendel did not classify all the pea plants and seeds he grew. Presumably he classified enough to convince himself that the result was as expected. It is perfectly natural under these circumstances to keep running totals as counts are made. If, then, one stops when the ratio "looks good," statistically the result will be bi-

ased in favor of the hypothesis. A seemingly "bad" fit may be perfectly plausible statistically, but one may not think so and add more data to see if it improves, thereby raising interesting questions, some mathematical and some psychological.

What is of concern to us is *not* whether Beadle's remarks actually account for Mendel's particular results, but the methodological issues they raise. The last two sentences are very significant: The data that look "good" to the experimentalist look "bad" for the statistician, and vice versa. There seems to be a methodological incommensurability concerning the nature of statistical and experimental modes of reasoning. This might be called the "experimentalist–statistician paradox." The idea is that, from a statistical point of view, the geneticist should not provide "so many data" that his or her results come too close to the theoretical expectations, for the closer they come to the "truth" the less true they will be. This is a strange paradox indeed, and it is based on faulty reasoning. The principle that data are considered to be less true as they reach theoretical expectations is based on the idea that geneticists *should* be studying a random sample. It assumes that experiments should be carried out independently of the law or theory that the observer is using for explanation. In other words, it appeals to naive empiricism and ignores the theory-ladenness of observations. The theory itself informs the experimenter about what kind of experiment to perform, what kind of phenomena to examine, and how results are to be understood; it also tells the experimenter when the experiment is over. This last issue is at the heart of Beadle's suggestion that Mendel simply stopped counting when he obtained the results expected.

What "liberties" scientists are allowed in selecting positive data and omitting conflicting or "messy" data from their reports is not defined by any timeless method. It is a matter of negotiation. It is learned, acquired socially; scientists make judgments about what fellow scientists might expect in order to be convincing. What counts as good evidence may be more or less well-defined after a new discipline or specialty is formed; however, at revolutionary stages in science, when new theories and techniques are being put forward, when standards have yet to be negotiated, scientists are less certain as to what others may require of them to be deemed competent and convincing. As we shall see later in discussions of Moewus's work, the role played by theory in the experimental process and the biases of the observer present a number of difficulties when experimenters attempt to use replication as a test of the validity of scientific knowledge claims.

The rhetorical nature of the scientific paper

Before returning to the Moewus saga proper, it is necessary to examine still another aspect of the experimental process – the extent to which published experimental reports can be taken as literal accounts of how scientists generate and interpret their data. This was an issue of concern in the Moewus controversy and in the attempts to elucidate his motives and methods and assess the liberties he may have taken. It was also raised by Bateson and Fisher when attempting to understand Mendel's conduct. Contrary to what is generally believed, Fisher was not the first to question the authenticity of Mendel's reported experimental results. Although it has been ignored by commentators who have examined Fisher's statistical criticisms, Bateson's comments raised the possibility that all of Mendel's "experiments" were fictitious. He suggested that Mendel could not have had the varieties of plants he described.

Bateson (1909) questioned the authenticity of Mendel's celebrated experiments in a footnote to a passage in the translation of Mendel's experiments which Bateson used in his book, *Mendel's Principles of Heredity*. After describing his first seven experiments, Mendel opened his subsequent section with the following claim: "In the experiments described above, plants were used which differed only in one essential character." Bateson commented:

> This statement of Mendel's in the light of present knowledge is open to some misconception. Though his work makes it evident that such varieties may exist, it is very unlikely that Mendel could have had seven pairs of varieties such that the members of each pair differed from each other in *only* one considerable character. (Bateson, 1909: 350)

Fisher fully realized the weight of this criticism. One would expect that some or all of the crosses would have involved more than one contrasting pair of characters. Fisher believed that Mendel meant his reports to be taken literally. In response to Bateson's remarks, Fisher offered two possibilities to account for Mendel's statement. Both involved how Mendel wrote up his reports and what he regarded as an "experiment." The first possibility was that "he might, for each cross, have chosen arbitrarily one factor, for which that particular cross was regarded as an experiment, and ignored the other factors" (Fisher, 1936: 119). Although this course might seem to be wasteful of data, Fisher claimed that Mendel, in fact, "left uncounted, or at least unpublished, far more material than appears in his paper." In other words, he published only enough data that he believed would be sufficient to convince readers of his theory.

The second possibility was that "he might have scored each progeny in all the factors segregating, assembled the data for each factor from the different crosses in which it was involved, and reported the results for each factor as a single experiment" (Fisher, 1936: 119). This course of action, Fisher claimed, was what most geneticists would take, unless they were discussing either linkage or multifactoral interaction.

On the other hand, Bateson's intimation that Mendel's data were wholly fabricated and that his "experiments" were fictitious remained a possibility. As Fisher noted, Mendel did not give summaries of the aggregate frequencies from different experiments. This conduct would be easily intelligible if the "experiments" reported in the paper were fictitious, being in reality themselves such summaries. This kind of over-simplification is often used when teachers illustrate principles to students in a lecture. Fisher (1936: 119) continued:

> Mendel's paper is, as has been frequently noted, a model in respect of the order and lucidity with which the successive relevant facts are presented, and such orderly presentation would be much facilitated had the author felt himself at liberty to ignore the particular crosses and years to which the plants contributing to any special result might belong. Mendel was an experienced and successful teacher, and might have adopted a style of presentation suitable for the lecture-room without feeling under any obligation to complicate his story by unessential details. The style of presentation with its conventional simplifications, represents, as is well known, a tradition far more ancient among scientific writers than the more literary narratives in which experiments are now habitually presented. Models of the former would certainly be more readily accessible to Mendel than of the latter.

It is difficult to know exactly what Fisher meant by the tradition of "ancient" scientific writers. However, one can easily challenge any sharp distinction between what Fisher calls the "simplifications" of "ancient" scientific writers, and the "more literary" accounts of modern scientists. In effect, this has been done to some degree by the Nobel Laureate, Peter Medawar, who in 1963 posed the question, "Is the Scientific Paper a Fraud?" In raising this question, Medawar did not mean that the scientific paper misrepresents "facts," nor that the interpretations found in a scientific paper "are wrong or deliberately mistaken." What he meant was that "the scientific paper may be a fraud because it misrepresents the thought process that accompanied or gave rise to the work that is described in the paper" (Medawar, 1963: 377).

Medawar (1963: 377) was perhaps the first to emphasize that "the scientific paper in its orthodox form *does* embody a totally mistaken conception, even a travesty, of the nature of scientific thought." The struc-

ture of the "orthodox scientific paper" itself, Medawar argued, is telling
in this regard. He described the structure of the typical scientific paper
in the biological sciences as follows:

> First, there's a section called the 'introduction' in which you merely de-
> scribe the general field in which your scientific talents are going to be
> exercised, followed by a section called 'previous work' in which you con-
> cede, more or less graciously, that others have dimly groped towards the
> fundamental truths that you are now about to expound. Then a section
> on 'methods' – that's O.K. Then comes the section called 'results'. The
> section called 'results' consists of a stream of factual information in which
> it's considered extremely bad form to discuss the significance of the results
> you're getting. You have to pretend that your mind is, so to speak, a virgin
> receptacle, an empty vessel, for information which floods into it from the
> external world for no reason which you yourself have revealed. You re-
> serve all appraisal of the scientific evidence until the 'discussion' section,
> and in the discussion you adopt the ludicrous pretence of asking yourself
> if the information you've collected actually means anything; of asking
> yourself if any general truths are going to emerge from the contemplation
> of all the evidence you brandished in the section called 'results'. (Meda-
> war, 1963: 377)

The above description is somewhat of an exaggeration, for certainly
many scientific papers do not follow this structure. But we can agree
with Medawar that there is "more than a mere element of truth in it."
"The conception underlying this style of scientific writing is that scientific discovery is an *inductive* process" (Medawar, 1963: 377). In its crud-
est form, induction implies that scientific discovery, or the formulation
of scientific theory, begins with the "neutral" evidence of the senses.
The scientific paper gives the illusion that discovery begins with simple
unbiased, unprejudiced, naive, and innocent observation. Out of this
unbridled evidence and tabulation of facts, orderly generalizations
emerge, crystallize, or at least gel.

Yet scientists know full well that theories do not emerge and gel in
this way. They know what meaning to place on their results before they
conduct their experiments. Indeed, as mentioned above, it is their anticipation of results that informs them of what experiments to perform,
what phenomena to examine, and what data to report. Medawar traces
the inductive structure often framing modern scientific papers to the
nineteenth-century writing of the philosopher John Stuart Mill. However, it would be naive to believe that scientists are the dupes of philosophers. It is also wrong to suggest that "the scientific paper" is a fraud.
"The scientific paper" is not a fraud; it is rhetoric. The structure of the

narrative of the scientific paper plays an important persuasive role in science. First, it is important to remember that scientific work is steeped in the biases of two cultures: the larger culture in which science is allowed to persist, and the scientific culture itself. Scientists' belief in theories, in experiments, and in observations they make and report are often influenced by forces arising from both. In the larger culture, scientists have maintained their legitimacy in part by appealing to their "objectivity." The literary style of the scientific paper is designed to protect scientists' interests as purveyors of truth and also to maintain public support.

The structure of the scientific paper plays a similar rhetorical role within the scientific culture. Shapin (1984) presents a detailed study of an early attempt to establish conventions for writing scientific papers. He shows that the seventeenth-century experimentalist Robert Boyle set out rules to distinguish authenticated scientific knowledge from mere belief. This was done, in part, by what Shapin (1984: 484) calls "the literary technology of virtual witnessing." This *"literary technology* by means of which the phenomena produced . . . were made known to those who were not direct witnesses" involved providing protocols for experiments, recounting unsuccessful experiments, and displaying humility so as not to look self-interested and untrustworthy, citing other writers not as judges but as witnesses to attest matters of fact, etc.

Boyle's literary techniques would give a veneer of objectivity and "matter-of-factness" to published scientific claims. But this often obscures the intentions of the author and the process by which results are produced. Therefore, the rhetoric and conventions often embodied in scientific papers make it more difficult for scientists to evaluate critically their competitors' knowledge claims. When commenting on Mendel's paper, one writer remarked: "All geneticists admitted that it was written so perfectly that we could not – not even at present – put it down more properly" (Nemec, 1965: 13). Yet, it was this very "perfection" that has made Mendel's conduct so difficult to ascertain. As in the case of Mendel, interpretations of a scientist's intentions and procedures based on readings of published papers often involve considerable speculation and conjecture.

But there are other problems. Often, the "methods" section is imprecise. Medawar's analysis of the scientific paper is incomplete in this regard, for the "methods" section is not always "O.K." Seemingly trivial, but yet vital information concerning procedures are often left out of the abbreviated "methods" sections of scientific papers. Moreover, much of modern science involves special technical skills, what Michael Polanyi (1958, 1967) called "tacit knowledge," and "craftsman's work" (Ravetz, 1971: 75–76). These personal skills are difficult to articulate – like riding a bicycle. H. M. Collins (1974: 176) wrote of experiments in the transversely excited atmospheric (TEA) laser field: "To date, no one to whom I have spoken has succeeded in building a TEA-laser using written sources (including pre-prints and internal reports) as the sole source of information, although several unsuccessful attempts have been made, and there is now a considerable literature on the subject."

As will be discussed in more detail later, the fact that science involves elements of tacit knowledge or "craftsman's work" poses another severe constraint on using replication as a definitive test of the validity of a colleague's knowledge claim. As Rosenthal (1966) and Collins (1985) have emphasized, it is often difficult to distinguish a "disconfirmation" through replication from a "failed replication." As we shall see, when attempts to repeat Moewus's experiments were made, Moewus frequently accounted for the difficulties others had in confirming his results in terms of their lack of personally acquired skills.

In the meantime, as we have seen, Moewus's work was assessed in terms of its fit with contemporary physiological and genetic theory, and his social relations in the field (the validity of his work seemed to have been certified by several virtuous witnesses, some of whom, such as Hartmann and Kuhn, were recognized scientific leaders). Statistical criticisms of Moewus's work had also been used as an indicator of the reliability of his results. These were the most serious criticisms of his work discussed thus far. However, statistical criticisms in themselves were weak and could easily be trivialized inasmuch as they ignored various aspects of the experimental process: the conscious and unconscious biases of geneticists in selecting certain phenomena to investigate and certain data to report. One could not dismiss Moewus or judge the validity of his claims in genetics on statistical reasoning alone. Genetics was not based on a foundation of statistical stones. With regard to statistical criticisms, most geneticists lived in glass houses. As Sewall Wright (1966: 173–174) remarked, regarding Fisher's appraisal of Mendel:

> I do not think that Fisher allows enough for the cumulative effect on X^2 of a slight subconscious tendency to favor the expected result in making tallies. Mendel was the first to count segregants at all. It is rather too much to expect that he would be aware of the precautions now known to be

necessary for completely objective data. . . . Checking of counts that one does not like, but not of others, can lead to systematic bias toward agreement. I doubt whether there are many geneticists even now whose data, if extensive, would stand up wholly satisfactorily under the X^2 test.

Indeed, Mendel's integrity has been defended well. The claim that there was no deliberate effort at falsification in Mendel's work has a great deal of support. However, if many reported data in genetics are "too good" to be random samples, then we may well ask, Why were geneticists so quick to attack Moewus's integrity and so critical of his work? Perhaps geneticists of the 1940s were not always aware of the limitations of statistical critiques and not reflective about the biases in their own data. However, this was not all there was to it. In the following chapter, we shall see that to answer this question fully, one has to take into consideration other factors in addition to these methodological issues.

6. The context of discovery and the failure of justification

I think you can't ignore the political context. If Moewus had not been a Nazi then everything might have been looked at differently. . . . That was everybody's attribution about him.
(Joshua Lederberg, interview, May 26, 1986)

The biochemical genetics of microorganisms was barely a field of activity prior to World War II. Moewus's work had been reviewed largely by plant physiologists and biochemists and in terms of his contributions to a physiological understanding of sexuality. Only a few geneticists who had worked on microorganisms, such as Jennings and Sonneborn, celebrated Moewus's breakthrough of using microorganisms for genetic analysis. The value of using microorganisms for genetic analysis had not yet been fully recognized by geneticists.

After World War II, the situation changed dramatically. Beadle and Tatum's work on biochemical genetics of *Neurospora,* begun in 1941 at Stanford University, had been transformed into a major research program. In 1945, Beadle and his group had moved to the California Institute of Technology, where a great deal of emphasis was placed on chemical aspects of genetics. Beadle's group developed in cooperation with the Chemistry Department there, especially with research led by Linus Pauling who worked on the structure of proteins. Yeast genetics began to emerge into prominence under the leadership of Carl Lindegren and Sol Spiegelman in the United States, and Boris Ephrussi in France. In 1946, Joshua Lederberg and Edward Tatum had learned to use bacteria for genetic work, and Lederberg began to establish a major bacterial genetics research program at the University of Wisconsin. In France, researchers at the Institut Pasteur, led by André Lwoff, began to focus on bacteria and its viruses. The message was quickly spreading throughout biological research communities that microorganisms had sex and could be crossbred, and that their genetic study was extremely valuable to working on fundamental problems of the nature of the gene and genic action.

In the decade following World War II, Moewus's work provoked phenomenal controversy in microbial genetic communities throughout the

120

Western world. Moewus's work on sexuality in algae now seemed to be leading the new revolutionary domain of biochemical genetics. Not only did Moewus anticipate geneticists generally by using microorganisms for biochemical genetics, but *Chlamydomonas* itself also proved to be an exceptionally useful genetic tool. In fact, the decade following World War II witnessed a burst of genetic studies on *Chlamydomonas* (see Gowans, 1976). At Stanford, the leading algologist, G. M. Smith, took up the study of sexuality in *Chlamydomonas* in an attempt to confirm and extend Moewus's work. Ralph Lewin at Yale and also G. S. Gowans at the University of Missouri turned to genetic investigations in *Chlamydomonas*. Ruth Sager, at the Rockefeller Institute for Medical Research in New York, began her work on genetic analysis of chloroplasts, which eventually led to standard genetic procedures for investigating cytoplasmic organelles (see Sager, 1972; Sapp, 1987a).

Undoubtedly, Moewus's work seemed to offer insightful ideas and promising techniques and problems. However, Moewus did not always receive credit, especially among geneticists, for what some considered to be his pathbreaking investigations. The early critiques of Moewus's published results by Philip and Haldane, Sonneborn, and others had not been forgotten. They still cast doubt on the reliability of his results. By the early 1950s, Moewus's work on sexuality and the precise mechanisms of biochemical genetic control of sexual processes in *Chlamydomonas* was frequently reviewed in varying detail, from a variety of standpoints, and with diverse degrees of skepticism and acceptance. On the one extreme, some accepted it outright, unaware of, or simply ignoring the critiques of his genetic work. For example, in their well-known textbook, *An Introduction to Biochemistry*, Roger Williams and Ernest Beerstecher (1948: 404) stated with no qualifications that "Moewus and Kuhn's studies of the genetic control of sex and motility in *Chlamydomonas* is [sic] undoubtedly a classic of biochemical research."

Geneticists were more skeptical, and though many recognized the significance of Moewus's work, they demanded that crucial experiments be repeated. George Beadle exemplified this position. As mentioned in Chapter 3, according to Mrs. Kobb, Moewus had written to Beadle in 1940, and Beadle congratulated him "on the brilliant idea of using microorganisms for biochemical genetic analysis." Beadle did not mention Moewus in his celebrated paper of 1941. However, by the end of World War II, Moewus's work had received considerable attention in the Anglo-American literature. Beadle (1945: 43) summarized it in his first review of the literature pertaining to biochemical genetics when he stated:

"Without doubt the most remarkable series of studies in biochemical genetics is that of the German investigators, Moewus, Kuhn and coworkers on the flagellate *Chlamydomonas*."

Beadle (1945: 47) concluded his review with the following remark:

> It is unfortunate that after presenting so beautiful an interpretation and one which agrees with so many of the reported facts, one must introduce a note of skepticism. The facts reported and the interpretation are almost "too good to be true." As a matter of fact, Philip and Haldane (1939) have, on the basis of a statistical analysis of certain of Moewus' genetic results, made just such a criticism. . . . Moewus (1940[d]) has attempted to reply to this most serious criticism. Pätau (1941) has analyzed other data of Moewus with a similar conclusion (see also Ludwig, 1942). Sonneborn (1942) has subjected the Kuhn–Moewus work to a detailed and searching criticism and concludes that it is most important that the work be repeated by investigators working independently. The writer is in complete agreement with this sentiment.

At the other end of the spectrum were bitter antagonists – those investigators who actively resisted giving any recognition to Moewus's accomplishments. The attitude of the celebrated biophysicist, Max Delbrück, was exemplary in this regard. Since the 1930s, Delbrück had maintained a strong interest in problems of the nature of the gene and gene action (see Hayes, 1984; Kay, 1985). Yet, he refused to even consider the possible breakthroughs of Moewus. Delbrück wrote to Sonneborn (June 1, 1944):

> By mistake of our librarian I obtained a complete *photostatic* copy of Moewus' papers of 1940. Presumably I am the only person in this country with a private reprint of these papers. Even so I could not bring myself to read them carefully.
> Luria is in full swing doing experiments.

The reasons for this puzzling behavior are complex. Certainly political concerns have to be taken into consideration in order to understand the attitude of many workers toward Moewus and his work during the 1940s and 1950s. At the most general level, the anti-Nazi sentiment surrounding World War II provided an extremely hostile milieu for evaluating Moewus's contributions. As Joshua Lederberg recalls (in his letter to the author, June 26, 1985), there were rumors in the United States that "Moewus must have been quite thick with the Nazi regime" in order for him to have carried out his work during the war. The stakes were enormous: Threatened with Nazi and Fascist political domination during the war, one now had to consider the possibility that a center piece of the biological revolution was to be given to what many perceived to be Nazi science. John Raper, at the University of Chicago, summarized the possible biological significance of Moewus's work clearly when he wrote:

> The various investigations on sexuality and hormonal activity in the sexual processes . . . if the results are largely confirmed, would in all probability be recognized as one of the outstanding developments in the biological sciences of the present century. This work dealing with the sexuality and the precise mechanism of chemical control of sexual processes in *Chlamydomonas*, particularly *C. eugametos*, was intitiated in 1932 and 1933 by the German worker, Franz Moewus. (Raper, 1952: 474)

It must be remembered that many of the scientists who helped to foster the revolution in biology had been uprooted by nazism and fascism (see Fleming, 1969). These included the Nobel Laureate Austrian physicist Erwin Schrödinger, whose famous book, *What Is Life?* (1944), was so influential to J. D. Watson. The Italian biologist Salvador Luria fled fascism, migrated to the United States, and with Delbrück developed the famous phage group. Luria's interest was to use bacterial viruses as a means of isolating and examining the genetic material – "the secret of life." For his work on bacteria phage, Delbrück shared a Nobel Prize with Luria and Alfred Hershey in 1969. Delbrück himself had fled his native Germany in 1937 to work as a Rockefeller Fellow at the California Institute of Technology, before obtaining a position at Vanderbilt University, Nashville in 1940. His father, a professor of history, and his uncle, a professor of theology, at the University of Berlin had been active with important anti-Nazi academics.

At the same time, many of the pioneers in the development of biochemical genetics of microorganisms were Jews by birth if not by conviction: T. M. Sonneborn, Joshua Lederberg, Sol Spiegelman, Ruth Sager, and Ralph Lewin in the United States; André Lwoff and François Jacob in France. Perhaps, if Moewus had escaped Germany and had made it to the United States before the war, the attitude toward him might have been different. There is no question that because Moewus had carried out his work in Germany during the war, his perceived Nazi affiliations predisposed many geneticists to be very critical of his work. However, not just external political issues have to be considered. When attempting to understand the attitudes toward Moewus's work, one also has to take into consideration the interests of the participants based on their scientific training.

Delbrück, for example, was trained as a quantum physicist. He was one of the first theoretical physicists to move into the field of biology (see Hayes, 1984; Kay, 1985). His theorizing on the physical properties of the gene in the mid-1930s had led Schrödinger to write his influential book, *What Is Life?* Delbrück was interested in developing a "radical physical explanation" for the nature of the gene and gene action. In his view, theories of the nature of the gene and gene action "should be formulated without fear of contradicting molecular physics" (Delbrück, 1949/1966: 22). And he commanded a great deal of respect from young and not-so-young people interested in molecular biology. According to one of his disciples, Delbrück "deprecated biochemistry" and influenced his students to avoid it (see Benzer, 1966: 158). He saw a direct link between biochemistry and reductionism.

Delbrück was extremely critical of reductionistic explanations in biology, which he viewed as naive conceptions that left biology indistinguishable from physics and chemistry, devoid of its own laws (see Delbrück, 1949/1966: 16). Like many biologists of the 1940s and 1950s, he protested against what he claimed to be the biochemical view of the cell as a "sack of enzymes" (see Delbrück, 1949/1966: 22; Sapp, 1987). And like many biologists before him, he emphasized the importance of the organization of the cell (see Chapter 2). As Delbrück (1949/1966: 22) put it, "The enzymes must be situated in their proper strategic positions in order to perform their duties in a well regulated fashion." Delbrück had doubts about the merit of biochemical approaches to the nature of the gene and gene action. He doubted the value of explaining the relations genes–enzymes–product in terms of "complex mechanical models," which he compared to biochemical approaches. In short, Delbrück, as well as other physicists who migrated to the field of biology, sought to introduce new standards of explanation and rigor into what they saw as largely a descriptive science. Delbrück's methodological criticisms of the experiments of the *Neurospora* school pertaining to the simplistic, one-gene–one-enzyme theory have already been mentioned in Chapter 2. As Fleming (1969: 182) emphasized, Delbrück himself opted for a kind of genetics that was devoid of chemistry as the best way for understanding the secret of heredity.

Gossip

Despite Delbrück's reservations, the biochemical genetics of microorganisms became a legitimate area of inquiry during the 1940s and 1950s and proved to be an effective strategy for investigating gene action. Yet, as biochemical genetics emerged into a fast-flying field, the originality and pathbreaking nature of Moewus's work were mentioned less and less by geneticists. When geneticists did mention Moewus's work, most did so only to actively discredit it. Indeed, some biologists seemed to lean over backwards to make public criticisms, sometimes even trivial criticisms, about Moewus's work or his behavior. Gossip began to spread about Moewus's either refusing to send stocks or his sending dead ones to those who wanted to repeat and extend his work. The most well-known published criticism of this sort was circulated by the French protozoologist André Lwoff at the Institut Pasteur. Lwoff is well known today for his work on Protozoa, bacteria, and bacteriophages. This latter research was recognized with a Nobel Prize in 1965, which he shared with Jacques Monod and François Jacob.

Before examining Lwoff's comments about Moewus, it is important to point out that the French scientific community had a long tradition of conflict and competition with that of Germany. Since the defeat of France in the 1870 war with Bismarck's Prussia, French science had emerged in constant opposition to German science. This conflict continued throughout the twentieth century, fueled by the two world wars. Many of the scientific concepts about heredity and evolution in France were steeped in French patriotism and anti-Germanic sentiment (see Sapp, 1987a: 124–128). During the Second World War, when the German army invaded Paris, some French biologists such as Marcel Prenant and Jacques Monod went underground and joined the armed-resistance movement led by the Communist Party. Those biologists who were not members of the Communist Party, such as Boris Ephrussi, were active in England in Les forces françaises libres. Many others, however, simply tolerated the pressure of occupation. Lwoff was very active in the resistance group that operated from the Institut Pasteur (see Judson, 1979: 359–361). Several of his close friends were killed in concentration camps.

But again, one also has to consider Lwoff's special technical interests in Moewus's publications. Lwoff (1947) made some very hostile remarks about Moewus and his work in a paper on the problem of growth factors for protozoa. He was not interested in the problem of sexuality, but rather was concerned with the study of the "comparative physiology of the nutrition of protozoa." Lwoff was trained as a protozoologist under Edouard Chatton during the 1920s. During the 1930s, his main line of research concerned the elementary nutritional requirements of microorganisms. He first worked on ciliates, then moved to the still simpler bacteria, investigating factors essential for growth, which turned out, after detailed analysis, to be particular vitamins that had only recently been understood to be essential to the growth of animals. As mentioned in previous chapters, during the first decades of the century, the biochemical unity of living things was not at all clear. Lwoff's work of the 1930s on bacteria was later hailed as being significant for and supporting the startling conclusion that they possessed biochemical systems fundamentally comparable to cells of higher organisms (see Judson, 1979: 351).

Lwoff's paper of 1947 represented an attempt to bring together the work on protozoan nutrition which had been carried out during and shortly after the war. His objective was to discuss the similarities and differences in vitamin or growth factor requirements of various types of Protozoa. In Lwoff's view, studies of the physiological similarities and differences among Protozoa were late in coming. One of the reasons for this, he claimed (Lwoff, 1947: 101), was "the lack of protozoological knowledge by most bacteriologists, and a deficiency of bacteriological and biochemical knowledge by most protozoologists."

However, during the war, investigations of the nutritional requirements of Protozoa were carried out by several groups of researchers who were studying the relationships between plants and vitamins and growth factors required by bacteria. In 1945, a group at Harvard University had succeeded in growing malarial parasites in vitro. As Lwoff (1947: 101) pointed out, "The importance of Protozoa in pathology and in wartime problems, and the advances in rational chemotherapy have also stimulated scientists in this domain." Following World War II, the study of the nutritional requirements of Protozoa promised to be especially important for the new domain of biochemical genetics. As Lwoff (1947: 102) argued:

> An exact knowledge of growth factors is essential for the study of Protozoa in chemically defined media and therefore for the solution of numerous problems in the fields of pure and applied biology. To-day, the Protozoa are no longer the poor relations of microbial chemistry and one can foresee a rapid advance in our knowledge of comparative physiology.

Lwoff (1947: 103) stated that he would have liked to have adequately reviewed all the recent papers. However, in war-torn France, the collection of scientific journals was still incomplete, and his review could not hope to present a complete picture of the studies that had appeared between 1943 and 1947. He therefore limited himself to a discussion of thiamine requirements of certain groups of Protozoa. Conflicting results on the same species had already been reported, and Lwoff wanted to establish a culture collection in order that work done by others could be repeated and extended into a unified body of knowledge.

This was the context in which Lwoff (1947) alluded to Moewus and his work. He made no mention of the originality, scope, and possible importance of Moewus's papers for biochemical genetics or for understanding sex determination. Indeed, Moewus's synthetic and sweeping style of work on sexuality stood in direct contrast to Lwoff's meticulous experiments on growth factors. To Lwoff, Moewus's work only represented difficulties for the kind of detailed study of microorganisms he sought to establish. Moreover, it was illustrative of the problems to be encountered in work on microorganisms, without an established culture collection from which researchers could obtain stocks. Indeed, Lwoff dedicated only four short paragraphs to Moewus's work when he introduced it with the following remarks:

> As will be seen in this review, conflicting results have been obtained. And this calls for a remark which seems to us of great importance. It is understandable that, up to a certain point, different authors obtain different results in dealing with the same species. But it is regrettable that the results cannot be repeated because of inability to obtain the strain used. The following example shows the confusion that can exist because of the absence of a collection of type cultures. (Lwoff, 1947: 103)

Lwoff then proceeded to relate a story about Moewus, the Czechoslovakian protozoologist Ondratschek, and himself. He began by stating that Ondratschek in Prague had made a complete list of known species of *Chlilomonas, Polytomella*, and *Polytoma*, "including the celebrated *Polytoma Pascheri,* described by Moewus who studied the genetics of two remarkable varieties." Lwoff did not discuss the genetic results of Moewus. He mentioned only that Moewus had crossbred two varieties of *Polytoma* and studied the characteristics of their descendants. Without giving a reason, and without mentioning Moewus's results, Lwoff stated, "It is highly desirable that these results be analyzed statistically." Lwoff (1947: 103) simply referred to the statistical critique of Philip and Haldane. He stated, "Actually, such an analysis applied by Philip and Haldane to other experiments of Moewus has led to a conclusion which is extremely critical of Moewus's work." But he did not mention the other criticisms of Moewus's work which appeared before World War II. He also did not mention Moewus's first reply to Philip and Haldane's criticisms which had appeared in 1940.

Lwoff added only two new issues that might discredit Moewus and his work. He suggested that Moewus refused to send him stocks, and he claimed there was a discrepancy between the statements of Moewus and all others concerning the conditions (pH level) necessary to grow *Polytoma*. Lwoff stated that he attempted to obtain a strain of *Polytoma* from the Czechoslovakian biologist Ondratschek. However, Ondratschek had gone underground during the German occupation of Prague, and had subsequently disappeared. Moewus possessed the only strains of *Polytoma pascheri*, and Lwoff claimed that he had reneged on an agreement to exchange strains. Lwoff (1947: 103–104) wrote:

> It would be very worthwhile to be able to locate *Polytoma Pascheri*. Unfortunately, a few years ago, an agreement with Moewus did not lead to the results anticipated. In exchange for my strains, Moewus was not able to send me *P. Pascheri* which, at that moment, refused to grow. Moewus also obtained the hybrids of the two varieties of *P. Pascheri* and of two varieties of *P. uvella* in the medium utilized for growth of *P. Pascheri* at pH 4.5. In this medium, Moewus was able to obtain growth of his two varieties of *P. uvella*. Now there is one point on which all who are in the habit of cultivating *Polytoma* are in agreement: *P. uvella* does not grow when the pH is below 7.0. Consequently, it is to be feared that some difficulty might be encountered in trying to cultivate again these two varieties of *P. uvella* which are really so convenient for genetic work. In addition, one should note that while Moewus claims to have been able to cultivate *P. Pascheri* at pH 4.5, Ondratschek affirms that this species does not develop below pH 6.5. *Polytoma Pascheri* remains clothed in mystery!
>
> Now, it is also highly probable that considerable difficulties might be encountered by anyone wishing to cultivate the two varieties of *P. Pascheri* described by Moewus. For no one else, with the exception of Ondratschek, has ever had the occasion to see *P. Pascheri*. After the liberation of Prag [Prague], I tried in vain, to get in touch with "Reinhardt", the new name adopted by Ondratschek in 1943. I was told that Ondratschek-Reinhardt had disappeared from Prag, and his cultures were not to be found at the Botanical Institute.
>
> It would be desirable if authors who published an experimental study on a determined species, strain or clone, were required to deposit a culture in a collection of type cultures. It is to be feared that lacking this precaution, the scientific literature will find itself burdened with experiments which cannot be verified due to the "loss" of certain strains because of unforeseen and most unfortunate or fortunate, circumstances. (Lwoff, 1947: 103–104)

Lwoff's attack on Moewus's integrity is hard to understand, if political issues are not considered. Indeed, his account reads like a "good guy–bad guy" story. As discussed in Chapter 3, Moewus's work emerged from within a controversy over theories of sexuality between biologists at Prague and those at Berlin led by Hartmann. The theoretical dispute was now transformed into terms of war. But should we accept Lwoff's construction and insinuations uncritically? Based on Lwoff's account, are we forced to conclude that a crime had been committed by Moewus? Let us treat each issue in turn, for, without doing a great deal of mental gymnastics, one can easily offer an alternative account. First, regarding Lwoff's statement that Moewus "claimed" to grow *Polytoma pascheri* at pH 4.5. whereas his competitor, Ondratschek at Prague, "affirms" that this species does not develop below pH 6.5, Lwoff gave no reference to Moewus's statement concerning pH. It is not clear from Lwoff's account that Moewus actually staked this "claim" at all. Certainly, the appearance of pH 4.5 instead of 6.5 could simply be a typographical error in Moewus's publications. After all, Moewus himself may have carelessly overlooked it since his interest was in sexuality, not in growth factors per se, as was Lwoff's.

Lwoff clearly suggests that Moewus's failure to send stocks was a sign of Moewus's intention to deceive his colleagues. However, this interpretative leap is a large one and can be challenged in the best of circumstances. It is not uncommon for scientists to hold onto their cultures until they clear up their own problems. In fact, as will be discussed momentarily, when the American algologist G. M. Smith began to work on the genetics of *Chlamydomonas,* he was reluctant to send out stocks to others for this very reason. It is often seen as very reasonable behavior.

Scientists also sometimes maintain secrecy of this kind, which prevents others from doing supplementary research, in order to keep a competitive edge on their fellow researchers. Again, it is often seen as perfectly legitimate generally to avoid passing on information that could lead to anticipation by others (see Mulkay, 1976: 644). Of course, scientific secrecy of this kind is also prevalent in wartime, as it is today. However, Mrs. Kobb offers an alternative account of the failure to send stocks of *Polytoma pascheri.* She points out that the species *Polytoma pascheri* "was indeed a puzzle and an irritant because it seemed to support the Prague concept that sex is not an inherited character, but influenced by environmental conditions." Moewus had found this species during a field trip in 1935 to a mountain range dividing Germany and Czechoslovakia. Moewus took a sample of a green gelatinous mass, and from this he cultured *Polytoma pascheri* to use in his experiments. Mrs. Kobb writes (notes to the author, May 29, 1988) that "unfortunately the cultures could not be maintained over a long period. When others – it could have been the Prague group? – wanted a culture, there was no material left." She and her husband went back to the same place in 1936.

> Actually it was our "honeymoon"! We took not only a wooden box full of
> glassware and equipment along, but also a microscope. . . . We did not
> find the alga in any of the samples, nor turned it up upon further cultiva-
> tion of the crude samples in the laboratory. The same procedure was re-
> peated one year later [1937] – no success! It was definitely not bad will
> that F. M. did not send this species to other laboratories! And I am sure
> he explained this fact in his letters.

Whatever criticisms we might make about Lwoff's remarks, with the
anti-German sentiment following World War II, such published remarks
were enough reason for many geneticists to be supercritical of Moewus's claims. The social and political gossip about Moewus continued.
As Joshua Lederberg (interview, May 26, 1986) put it, whether Moewus
was actually a Nazi or not did not matter. He was perceived that way,
and he did carry out much of his research under that regime. One can
easily understand how people felt resentful of his having worked during
the war under Kuhn's patronage. Moreover, with the cloud of suspicion
already surrounding Moewus's work, one can easily understand that to
deny access to materials might amount to prima facie evidence of guilt.

On the other hand, many scientists were still reluctant to dismiss
Moewus's results as being entirely unreliable on these grounds. They
were quite uncertain as to their validity. The attitude of Joshua Lederberg is exemplary. In his first major published review of microbial genetics, Lederberg (1948: 151) used the kind of evidence provided by Lwoff
to delegitimize Moewus completely. In a very brief statement about genetic work on algae, Lederberg mentioned Moewus's work in the following way:

> Genetic work on algae has been overshadowed by the numerous contribu-
> tions of F. Moewus . . . on the chlamydomonads. Unfortunately, this
> "most remarkable series of studies in biochemical genetics . . ." (Beadle,
> 1945) has been challenged in almost every detail (Smith, 1946; Sonneborn,
> 1941; Lwoff, 1947). To these comprehensive analyses, the present writer
> can add no informed comment.

But had Moewus's work really been challenged in "almost every detail"? It is noteworthy that Lederberg did not refer to the original criticisms of Philip and Haldane. Moreover, like Lwoff, Lederberg did not
mention Moewus's response to Philip and Haldane's criticisms. Lederberg had relied on the reviews of others. In fact, Lederberg himself recognized the shortcomings of his published remarks about Moewus. The
following year, he wrote to Moewus (June 9, 1949), apologizing for the
"hasty treatment" of his work, stating that he had difficulty obtaining

Moewus's original publications. He asked Moewus to send reprints, and to keep him on his mailing list:

> Dear Dr. Moewus,
> I enclose reprints of a review article and of some of my experimental work which may be of interest to you. If I may apologize for the hasty treatment which was given to the genetics of algae, it would be partly based on the difficulty of obtaining here a comprehensive and definitive set of your publications.
>
> I would appreciate it greatly, therefore, if you could send me such of your works as are available still, and would place my name on your mailing list, a favor which I shall be glad to reciprocate.

Moewus (August 16, 1949) wrote back to Lederberg, thanking him for his offprints on bacterial genetics. He sent Lederberg copies of some of his papers and mentioned some of his recent work on *Chlamydomonas* genetics.

Relying on reports of others is a risky business. Lederberg's claim that publications of Lwoff, Smith, and Sonneborn had challenged Moewus's work in "almost every detail" deserves some examination. Lwoff's remarks about Moewus have already been discussed. The publications of Sonneborn have been examined previously. Sonneborn did not reject Moewus's work; instead, he continued to see great value in Moewus's work. Although Lederberg enlisted Sonneborn as an adversary against Moewus, Sonneborn was just the opposite. In fact, as will be discussed later, he became one of Moewus's chief apologists in the United States. He would not allow his criticisms, nor those of others, to be used to overlook what he regarded to be Moewus's great accomplishments. As we shall see, Lederberg's statement tends to simplify the claims of Smith as well. Smith actually attempted to repeat Moewus's experiments. However, he never rejected Moewus's work and never made statements that he had disconfirmed any of it. In fact, he made statements to the contrary.

The failure of replication

As mentioned in Chapter 5, because the production of experimental results involves elements of tacit knowledge or craftsman's skill, it often remains uncertain whether or not disconfirmations are due to the faulty technique of the replicator. In other words, it is often difficult to distinguish between a disconfirmation and a failed experiment (see Collins,

1974). This issue plagued the researchers who attempted to assess Moewus's work throughout the 1940s and early 1950s. But in biology there are additional problems, besides technical skill: Results based on work on one organism may not always apply to another. One had to ensure that the repetition was done on the same species and, indeed, the same strain of organism. Even then, as we shall see later, various other reasonable explanations for conflicting results could be entertained. A brief glance at Smith's attempts to repeat some of Moewus's experiments will help us to understand many of the difficulties geneticists had in reaching a decisive conclusion about the validity of Moewus's published results.

Gilbert Smith was recognized as a very honest elderly gentleman, a classical algal taxonomist and mycologist. He was well known for his two great books on algology. Several young researchers who turned to *Chlamydomonas* genetics took his courses at Hopkins Marine Station at Pacific Grove, Stanford University. Smith had a major teaching job; but unlike many scientists who burned the midnight oil, Smith had a rule that at five o'clock he went home. He did very conscientious research, coming to work early in the morning and leaving at 5 p.m. Smith did not want to put any graduate student on Moewus's work; it was too risky. The chances of failure were too great. A "failure" in this context would represent only the inability to confirm Moewus's results. However, Smith soon recognized that attempting to disconfirm Moewus's results by repeating his experiments was a dubious task. As we shall see, Smith quickly came to see that a lack of confirmation represented either an inclusive result or a new result, but it did not represent a "disconfirmation."

Smith took on the responsibility of repeating some of Moewus's work when he retired from teaching in 1946. He did some field work looking for *Chlamydomonas* species in Californian soil, carried out some laboratory work, and tried to do some checking up on some of Moewus's work. After isolating various soil alga, he began to focus his work primarily on *Chlamydomonas reinhardi*. This species subsequently became the most widely used by algal geneticists. During the late 1940s, when several young researchers including Ralph Lewin, Ruth Sager, and Paul Levine wanted to work on *Chlamydomonas*, several attempted to obtain strains from Gilbert. However, Lewin (interview, May 2, 1986) recalls that Smith was reluctant to give out strains until he established mating conditions and standard techniques. Around 1950, Smith released strains to Paul Levine and to Ruth Sager. In the meantime, Smith began to simulate Moewus's investigations on sexuality in *Chlamydomonas* in an attempt to evaluate his work.

Smith was very cautious in his interpretations of the way in which his work related to that of Moewus. He published his first report as his address as the retiring president of the Botanical Society of America. His paper (Smith, 1946), entitled "The Nature of Sexuality in *Chlamydomonas*," began with a careful review of the work of Moewus on sex determination. There was no mention of the criticisms that had been directed against Moewus's work. Smith mentioned only that it was controversial. He began his address simply, with the following statement:

> In 1933 Moewus published the first of an extensive series of articles on the nature of sexuality in, and the genetics of, freshwater algae. The algae he has studied include *Chlamydomonas*, *Polytoma*, *Stephenosphaera*, *Protosiphon*, *Monostroma*, and *Botrydium*. There has been widespread interest in the work of Moewus, especially that on *Chlamydomonas*, and the results he has reported have evoked considerable debate. This discussion will be limited to the series of Moewus' investigations centering around the nature of the substances that induced motility and fusion of gametes in *Chlamydomonas*. A brief preliminary report will also be given on sexuality in some species of *Chlamydomonas* collected in California. (Smith, 1946: 625)

There was no mention of the nature of the controversy; it was implicit. Smith summarized Moewus's work on sexuality. He was especially critical of the claim that carotenoids acted as sex substances in *Chlamydomonas*, at least in all *Chlamydomonas* species. After reviewing in detail the biochemical genetic descriptions of sexuality in *Chlamydomonas* provided by Moewus and his collaborators by 1940, Smith (1946: 628) simply wrote: "One can see from the foregoing account why the work of Moewus has aroused so much interest." But this statement raised the question of why others had not worked along similar lines and had not attempted to repeat and extend Moewus's work. After all, suggestions that Moewus's work should be repeated had echoed throughout the biological literature since 1939. In response to this, Smith pointed out the difficulties of trying to isolate strains of *Chlamydomonas* that regularly reproduce sexually so that genetic work could be possible. He mentioned the great, time-intensive effort Moewus had made in order to obtain his stocks. As Smith (1946: 628) wrote, Moewus (1940[b]: 420), stated that "he spent three years isolating material from more than five hundred samples of soil in obtaining an interfertile group of species."

But obtaining species of *Chlamydomonas* that regularly reproduce sexually represented only one of the difficulties of repeating Moewus's work. In order to attempt to confirm or disconfirm Moewus's experimental claims, one also needed to work on the same species of organisms worked on by Moewus. As Smith (1946: 628) pointed out, "Some argue that one must have members of a *eugametos* group before attempting to work in this field." Smith, therefore, attempted to isolate *Chlamydomonas eugametos* from samples of soil collected in California. He successfully collected three interfertile species of *Chlamydomonas.* Unfortunately, none of the strains that he isolated belonged to the *eugametos* group studied by Moewus. Nonetheless, Smith (1946: 628) believed it "worthwhile to study the sexuality of the Californian material and to compare the sexuality of these species with that recorded for the *eugametos* group."

The results of Smith's preliminary experiments with the three Californian species were striking. They differed from Moewus's reports about *C. eugametos* at every turn. For example, Moewus had studied the nature of sex substances in gametes most effectively by studying the effect of light on gamete motility. He reported that light was necessary to trigger the production of crocetin, which in turn made gametes motile. Moewus repeatedly stated that all members of the *eugametos* group do not become motile in darkness. When Moewus transferred motile cells from light to darkness, the motility-inducing substance did not remain effective for more than an hour after it was formed in the light (Moewus, 1938a: 759). In contrast, Smith reported that cultures of all his species did become motile in darkness even when they were cultured in darkness for several days. Gametes of his species could unite in darkness as well. Smith (1946: 629) concluded that it was therefore unlikely that the motility-inducing substance of the Californian species was crocetin.

"Thus far," Smith (1946: 629) wrote, "everything reported about the three heterothallic species collected in California is a negation of what Moewus reports for the *eugametos* group." However, Smith then mentioned that some of his observations did tend to agree with what Moewus reported. For example, he found that although sexual union could occur in darkness, light did have a major effect on sexuality. Perhaps there was some truth in Moewus's claim after all. Smith (1946: 629) did report evidence showing that gametic union in the Californian species was "more intense in light than in darkness." Smith returned to this issue in 1951, in a paper on "The Sex Substances of Algae." This time, he agreed with Moewus's claims about the necessity of light for inducing motility and the biosynthesis of sexual substances. He claimed that only a "small percentage" of cells in the Californian species become motile in darkness (Smith, 1951: 326). The difference between his results and those of Moewus, he asserted, was that although light was essential for the formation of sexual substances in the Californian species, in darkness these substances diffused from cells at a much slower rate (Smith, 1951: 327). Nonetheless, Smith became convinced that carotenoids were not involved in sexuality. Based on his studies of the effects of different kinds of light (red, orange, and green) on the formation of sexual substances, Smith (1951: 328) concluded that "for this species the sexual substances cannot be the combinations of *cis-* and *trans-*crocetin dimethyl ester that Moewus reports for the *eugametos* group."

Smith (1946) also attempted to repeat Moewus's experiments on the effect of a filtrate from one sex upon the gametes of the opposite sex. It will be recalled that when Moewus (1933b) did this under certain conditions, with *Chlamydomonas eugametos* and other types of algae, he held that filtrates from one sex had an effect on gametes of the opposite sex; a clumping reaction occurred. When Smith attempted this experiment, all of his results were negative. A hasty judgment might interpret this to be a clear refutation of Moewus's claims. However, this interpretation would ignore two crucial issues. One has been mentioned already: Smith was not working on the same species as Moewus. The different results could be a reflection of the different nature of the species employed in the experiments. Second, it could be possible that Smith's inability to confirm Moewus's reports was due to inadequate experimental procedures. Smith may not have followed Moewus's procedures to the letter. Smith himself was fully aware of both of these objections. After reporting his negative results on the effect of filtrates on sexual activity, he wrote: "No comment will be made because it is impossible to prove beyond all doubt that the failure to duplicate Moewus' results was not due to faulty technique" (Smith, 1946: 629). In view of these two difficulties, Smith was very reserved when interpreting his results. In his conclusion he wrote:

> The results obtained in this preliminary inspection of sexuality as found in three species in California are so incomplete that detailed comparisons with the results recorded by Moewus are not justified. However, sufficient progress has been made to show that the sharp correlation between illumination and development of sexuality reported for the *eugametos* group does not hold for the species as a whole. The question as to whether or not the sexual substances of species collected in California are fundamentally the same as those of the *eugametos* group remains one for further investigation. (Smith, 1946: 630)

Smith was persistent in his efforts to repeat and extend Moewus's work on sexuality. By 1950, he had begun to study the genetic side of sexuality in *Chlamydomonas* as originated by Moewus. He focused his work on just one species isolated in California – *C. reinhardi* – which "proved the most satisfactory for study of the inheritance of sexuality" (Smith and Regnery, 1950: 247). Again, Smith's results differed from those reported by Moewus. For example, on the basis of his studies of the inheritance of sexuality in *Chlamydomonas*, Moewus (1935a, 1936, 1938b) held that the genes for maleness and femaleness were not alleles. That is, they were not occupying the same relative position (locus) on homologous chromosomes. Instead, he claimed that they were at different loci on homologous chromosomes. In contrast, Smith and Regnery (1950: 248) concluded that "in *C. reinhardi* it is very probable that in this species the genes for sex are at the same locus and not at different loci as Moewus finds in the species he investigated."

The authors did not speak of refutation. However, they did suggest some faulty or inexact reporting of data in Moewus's genetic reports. For example, in the inheritance of characters, they pointed out that Moewus always reported that he obtained four clones from a zygote. However, they noted that germinating zygotes of *C. reinhardi* produce either four or eight cells. Moreover, not all the clones always develop. Quite frequently, some of the cells resulting from a zygote die. Moewus's papers, they argued (1950: 248), "do not show whether he always obtained four clones or whether he discarded material from zygotes where he obtained fewer or more than four clones." They did not develop the argument any further. Whether *Chlamydomonas eugametos* normally produced only four clones under Moewus's procedures was uncertain.

Smith was cautious. He never claimed an outright refutation of Moewus's work in any of his publications. Indeed, he remained quite uncertain about the validity of Moewus's results. The following year, Smith (1951) edited a well-known *Manual of Phycology*, subtitled *An Introduction to the Algae and Their Biology*. He wrote a chapter on "Sexuality in Algae" which included an extensive review of Moewus's work. In this review, Smith treated Moewus's work on sexuality in *C. eugametos* as if it were largely unproblematic and generally sound. The level of confidence Smith bestowed on Moewus's work can be illustrated by the expressions he used when describing Moewus's reports (italics mine):

> Moewus (1940[e]) *has found* that substances extracted from fertile sexual plants can make zoospores function as gametes. (Smith, 1951: 235)
>
> In heterothallic species female cultures in darkness are made sexually functional in darkness by filtrates from female cultures in light, but not by

filtrates from male cultures in light. The reverse *is true* for male cultures. (Smith, 1951: 236)

Further insight into the nature of substances excreted was obtained by studying the effect of light on filtrates from *Chlamydomonas eugametos*. (Smith, 1951: 236)

Moewus (1938[a], 1939[d]) *finds* that he can obtain the male and female copulation substances by mixing the precursor and the end product in definite proportions. (Smith, 1951: 237)

[B]y using various mixtures of *cis*- and *trans*-crocetin dimethyl ester, Moewus (1940[d]) *has shown* that the valence of gametes of the various species is due to the relative proportions of *cis*- and *trans*-crocetin dimethyl ester. (Smith, 1951: 238)

The very use of the discovery terms "finds," "has found," "has shown," "true," and "insight" in reference to Moewus's work only serves to strengthen its legitimacy and authenticity (see Latour and Woolgar, 1979: 77–81). Indeed, in his review of 1951, Smith did not mention the various other criticisms and often serious charges made about Moewus and the reliability of his reporting which had appeared by that time, by Philip and Haldane (1939), Pätau (1941), Ludwig (1942), Sonneborn (1941, 1942), Lwoff (1947), Thimann, Raper, and others. Instead, Smith attempted to normalize the controversy in terms of healthy skepticism. He stated that the skepticism concerning Moewus's work was due to its novelty and significance. He stated that it had yet to be confirmed by others, but that some of his own work and that of Moewus himself on other types of algae limited the scope of some of Moewus's claims. Thus Smith (1951: 240) concluded:

> Two points should be noted in connection with the foregoing account of sexual substances reported for the *Chlamydomonas eugametos* group. Except for certain biochemical phases, all of the published accounts are the work of a single investigator. Because the results obtained are so novel and of such great significance many biologists are unwilling to accept them unreservedly until they have been confirmed by independent investigators working along similar lines. Granting the correctness of what has been reported for the *C. eugametos* group, there still remains the question of how far it holds for other algae. In some other Chlorophyceae, including certain other species of *Chlamydomonas* (Smith, 1946), *Protosiphon botryoides* (Moewus, 1935[b]), various species of *Ulva* (Smith, 1947) and *Cladophora trichotoma* (Telonicher and Smith unpublished), gametes are liberated and function sexually even when thalli cultures have been standing in darkness for a day or more. Thus for these algae the sexual substances do not seem to be carotenoids formed in light as is the case with the *eugametos* series.

Thus, after five years of working on the same lines as Moewus, Gilbert Smith was unwilling to refute any of Moewus's claims. As we have seen, Smith's work began as an attempt to confirm or discomfirm Moewus. However, his original intention was modified for two reasons: It was still possible that Smith's failure to duplicate Moewus's results was due to Smith's own faulty techniques; it was also possible that his different results were due to the different species of organisms he employed. As a result, by 1951, Smith sharply transformed the aims and significance of his work – from a test of the validity of Moewus's reports to "granting its correctness" and questioning only the limits of its validity in other species.

Smith remained on good terms with Moewus. But among American geneticists, Sonneborn was undoubtedly more supportive of Moewus and his work than anyone else. As will be discussed later, in his view, Moewus had clearly anticipated much of the work for which Beadle and his group were beginning to receive wide acclaim in America. Sonneborn did his best to ensure that Moewus's work stayed in public view in American genetic circles. In his reviews of 1941 and 1942, Sonneborn had summarized how matters stood at the end of 1940. During the war, he of course had no idea of the progress that Moewus had made in his work.

A victim of circumstance?

Moewus continued to do research intermittently in Germany throughout the war. As noted earlier many biologists perceived Moewus to be a Nazi. However, Moewus was not a Nazi Party member. When I visited Mrs. Kobb in Heidelberg, she presented to me a document, the verdict of the *Spruchkammer* Moewus received in 1946. The document was the result of the nationwide "trial," carried out in accordance with the directives of the Allies following the Nuremberg trial. Each German man and woman had to fill out a long questionnaire which demanded information about the activities of any and all persons having a connection to the Nazi Party or involved in any political activity. The population was screened and classified into four groups: either not involved at all (class 0), cooperator (class I and class II), or war criminal or political criminal. Mrs. Kobb recalls that the great majority of government personnel, including university professors and high school teachers, were placed in class I or II; they received little or no punishment, no jail sen-

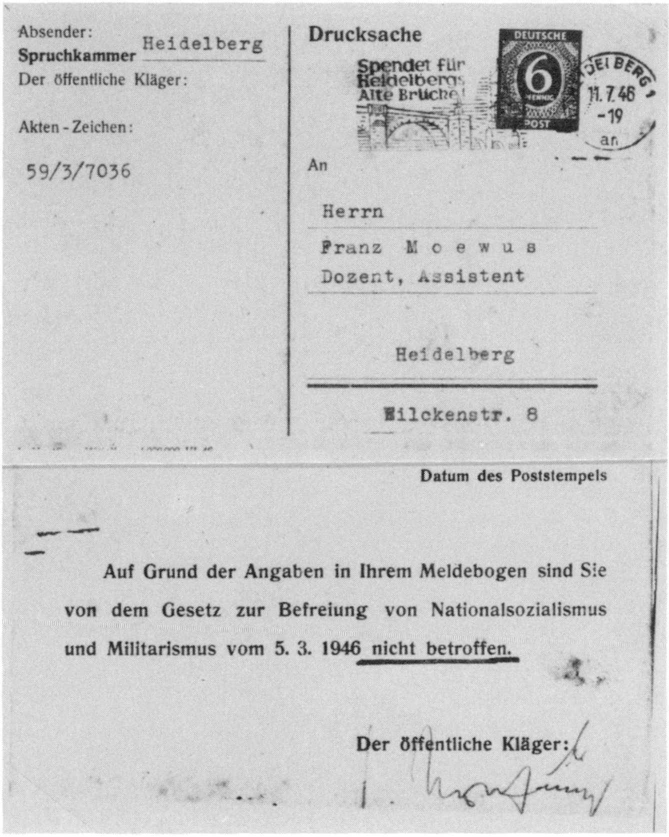

Absender: Heidelberg
Spruchkammer
Der öffentliche Kläger:

Akten-Zeichen:

59/3/7036

Drucksache

Spendet für Heidelbergs Alte Brücke!

DEUTSCHE 6 PFENNIG POST

HEIDELBERG 11.7.46 -19

An

Herrn

Franz M o e w u s
Dozent, Assistent

Heidelberg

Wilckenstr. 8

Datum des Poststempels

Auf Grund der Angaben in Ihrem Meldebogen sind Sie von dem Gesetz zur Befreiung von Nationalsozialismus und Militarismus vom 5. 3. 1946 nicht betroffen.

Der öffentliche Kläger:

Figure 5. Both sides of postcard notifying Moewus of his status regarding denazification. Translation: "On the basis of the information in your registration form, you are not affected by the law for the liberation from national socialism and militarism of 5.3.1946." (Courtesy Mrs. Liselotte Kobb.)

tences. But they were unable to work and were without pay until they got their verdict – and that could be two years or more. Very few people received a class 0 document. Moewus was one of the very few (see Figure 5). Moewus had not been a Nazi Party member; he had only hatred for the Nazi regime. Mrs. Kobb recalls that Moewus turned down several opportunities to become a professor at the newly installed German universities at Prague, Posen, and Strassburg, because it would have involved his becoming a member of the Nazi Party, and as a geneticist he would have had to teach human genetics, Nazi style (i.e., racism).

As suggested earlier, perhaps if Moewus had escaped Germany before the war, the attitude toward him would have been different. In fact, he did almost wind up on the American side during the war. In 1938, Moewus was invited to attend the Third International Congress of Microbiology at the Waldorf Hotel in New York City. Moewus was to present a paper on carotenoids as sexual stuff in *Chlamydomonas*. Mrs. Kobb recalls that there were four other Germans that were going, not as personally invited guests, but as representatives of Germany. They were all Nazi Party members. However, Mrs. Kobb recalls, prior to his proposed trip to New York, Moewus had no idea of the Nazi atrocities. When he went to Stuttgart to apply for a passport, he came back entirely shaken and shocked by what he had seen. There were Jewish people there waiting to get passports. Moewus saw Jews embrace in jubilation as some of them received their passports to enable them to leave the country. Moewus was ashamed, seeing for the first time how Jews were treated. Millions of Germans, of course, had no knowledge of these things.

Moewus received his passport and was on his way to New York; Liselotte took him to the train station in Heidelberg. She was in tears. As she said good-bye, she knew that war would break out tomorrow or the next day. Mrs. Kobb remembers that her friend, the "baker's lady," took her into her arms and said, "Be happy. He will be out of it. He will be saved. He won't be killed. Don't cry. It's good." That was on a Friday. Sunday food coupons were given out, and stores were open. Everyone rushed to buy what they could. Liselotte returned home. There stood Franz. The ship was called back.

Moewus worked in Kuhn's laboratory throughout the war. In 1938, after working with Kuhn for one year, Moewus attempted to apply for his *Habilitation* at the University of Heidelberg. This was a second thesis which one had to complete in order to be permitted to teach in the university. Moewus was not permitted to obtain it from the university. Mrs. Kobb believes that Franz Moewus could not apply for the *Habilitation* at the University of Heidelberg for political reasons: He was not a Nazi Party member. According to Mrs. Kobb, Auguste Seybold, professor of biology at the University of Heidelberg, would not allow him to do it. He declared that the University of Heidelberg did not need any new, special lectures in biology. Liselotte describes Seybold as a "hidden Nazi." He received his job at the University of Heidelberg in 1933. Heidelberg was the center of the Nazi student leadership; Seybold, Mrs. Kobb explains, was very much connected with them. However, as will be discussed later, Seybold was very critical and doubtful of the validity of Moewus's results. On the other hand, both Kuhn and Hartmann held only the greatest admiration for Moewus and his work. Moewus went elsewhere to obtain his *Habilitation*.

The Moewuses went to the University of Erlangen, a very small university in Bavaria. Franz Moewus had a geneticist friend who held a professorship in botany there, whom he knew from his days in Berlin. His dissertation for the *Habilitation* was on the response of algal gametes to chemical stimulus (chemotaxis). It was sanctioned by both Kuhn and Hartmann, who wrote letters praising Moewus's work. After completing his degree, Moewus was actually offered a job at Erlangen. However, Mrs. Kobb explained that he felt compelled to turn down the offer. The University of Erlangen was poor; Hitler was taking money out of the university system for his military buildup. The University of Erlangen simply could not afford to sustain a laboratory of algae research. However, they did have a great deal of garden space and asked Moewus to do genetic work on higher plants. Moewus decided to return to Kuhn's laboratory, even though Kuhn did not offer anything, not even a raise. Kuhn was very stingy regarding money matters. In fact, he paid Moewus out of his own research grants. Moewus had no official status at the Kaiser Wilhelm Institute in Heidelberg.

The Moewuses had left for Erlangen in October and returned to Heidelberg in May 1939. In August, war broke out; life changed drastically when drafting started. Government employees, teachers, everyone had volunteered for three-month military courses. Moewus did not. He was a pacifist. Kuhn also did not encourage anyone to take them. Moewus was lucky in this respect; at Erlangen, it would likely have been different. Moewus would not have had anyone as powerful as Kuhn to protect him from the draft. Kuhn was head of the *Forschungs* branch of the Council of Researchers for the entire country. He had managed to get his entire staff taken off the draft list; from the janitor to the assistant researchers, none were drafted. When drafting began, they started taking the younger men, and the older – those born in 1903. At the beginning of the war, they did not need Moewus's age group; they had sufficient others. Mrs. Kobb remembers how Moewus began to be disliked among his colleagues who had enlisted or had to go. However, staying out of the war was not as straightforward as that.

Two years later, Moewus and another member of the institute did receive notice that they were drafted. They were to meet the next morning with others at the train station in Heidelberg. Kuhn was out of town on business, but he left special instructions to contact him if anything should happen while he was away. Moewus sent out a telegram to contact Kuhn immediately, but Kuhn could not be reached. As his wife saw it, he had to go; not to do so would be extremely dangerous. The next morning, Franz and Liselotte said their good-byes. A few hours later, some Nazi Party officials in brown uniforms arrived at Liselotte's door asking about her husband. He had not shown up! Liselotte did not know where he was. They came again later in the afternoon. But she still had no idea where he was. Later that night, Moewus came home. He and his colleague had been hiding in the woods! By that time, Kuhn had received the message and sent word that Moewus was not to be touched. Moewus was saved from the war.

Heidelberg was a Red Cross headquarters and was not bombed during the war. It received only numerous shocks of bombings in nearby Mannheim and Frankfurt. As planes flew across the Heidelberg valley at night, one could watch them fall into beams of light from below; sometimes they fluttered into the horizon. It was a disturbing spectacle. When the fighting around Heidelberg became intense, Kuhn would often visit the Moewuses to seek company and consolation. Franz Moewus, however, simply buried himself in his work. He was able to work almost continuously, sitting at his desk reading journal papers and texts. In October 1942, Moewus was named dozent, enabling him to announce his lectures and teach at the University of Heidelberg as a faculty member. He received money based only on fees students paid to take his courses. He lectured in the Botany Department on the systematics of some higher plants, and on special and applied genetics in higher plants. At the same time, he carried out research at the Kaiser Wilhelm Institute.

During the war, researchers at the Kaiser Wilhelm Institute all had to work on so-called war projects. One wing of the institute was locked off; the Moewuses could never enter. In the remainder of the institute, small teams were formed and were given special projects. Moewus's project was the preservation of seed for farmers, to avoid waste and ensure that they would use just enough to get a proper spread of the grains, or whatever. He invented one bioassay after another in order to discover the optimal viable seed. There was no break in this kind of research throughout the war. Liselotte worked in the laboratory as well, when she could take time off from nursing her sick mother who lived with them. She was paid small sums of money in cash from Kuhn's grants. As Kuhn's secretary handed her the money, she advised her to slip it quickly into her pocket. Money, however, was hardly relevant during the war, of course, though it became extremely important to the Moewuses after the war. Franz Moewus also worked on inhibitors of germination in higher plants. These studies later became well-known as "the famous cress-root tests." During this war period, the influx of letters and publications from outside Germany stopped. There was dead silence across the ocean. Moewus worked in complete isolation.

After the war, Heidelberg was occupied by the Americans. City hall was an American city administration. As the American soldiers came to occupy the city, they claimed the nicest homes for their own. The Moewuses fled from their apartment, taking their furniture with them, hiding it so that it would not be destroyed by the American soldiers. Kuhn helped them to obtain a small apartment near his home. The Kaiser Wilhelm Institute was still open; some researchers could continue to work, to some degree, but many basic materials such as nutrients, agar, etc. were exhausted or in short supply. As the story goes (see Burk, 1973: 518), at the end of the war, Kuhn demonstrated to an American Army colonel's wife that she could turn green plants into red plants by adding triphenyltetrazolium chloride to the nutrient medium. As a result the institute was spared molestation.

At the same time, students began to return. However, Mrs. Kobb recalls that Moewus had serious difficulties finding students who would extend and broaden his own work. Many of the students had been drafted right from high school, and others had interrupted their studies for a number of years during the war. And, after the war, many had spent several years in prison camps until 1948–1949. Mrs. Kobb recalls that those who returned to university were interested primarily in financial security; therefore, they aimed for the *Staatsexamen* and could not afford the lengthy period of training required for the doctoral degree. Moreover, there was no grant money available for such studies. As Mrs. Kobb put it, "Unfortunately that was bad luck for Franz Moewus who had been waiting anxiously to find *Schuler* [students]." Indeed, from Moewus's point of view, the splendid isolation he had enjoyed working in Kuhn's laboratory during the war soon turned against him. He had no students to train in terms of his methods, to extend his work, and to correct his own mistakes. This lack of students only made his work more vulnerable. In effect, he found himself defending the work without foot soldiers.

But this problem was dwarfed by more immediate, life-threatening issues. During the war, there was normal distribution of food. But between 1945 and 1948, living conditions were almost intolerable. Everything disappeared; there was no transportation, no railroad; more importantly, food was in extremely short supply. Franz Moewus taught genetics with maize that he and Liselotte grew in the garden at the institute. It was experimental maize – pink and black – not designed for eating, and hardly edible. But the Moewuses were hungry; they shredded the corn and ate it. With little money, they had to support Liselotte's old mother, her brother who had returned from the war and was now homeless, three old relatives from Berlin, and the wife and son of her older brother who had been killed in the war. Franz Moewus lost a great deal of weight and suffered from a skin disease (psoriasis). He was given leave from his teaching duties. His upper torso was covered with scales. His wife ripped up sheets to wrap around his sores, but she had no pins and had to tie string around the sheeting to hold it in place. But when Franz walked, the sheeting unwound and trailed behind him. He was a pathetically comical spectacle.

The break from isolation

It was not until 1947 that Moewus and Sonneborn began to renew their correspondence and exchange their published papers and ideas. In August of that year, Moewus wrote to Sonneborn, describing the current status of his research. He mentioned his ideas on genic action and the nature of the gene. Moewus believed that he had cleared up two of the most important criticisms launched against his biochemical genetic work on *Chlamydomonas:* the statistical criticisms, and the striking claim that one molecule of sex substance could affect the functioning of gametes. Moewus also pointed out that they had only "circumstantial evidence" that the sexual substances in *Chlamydomonas* were carotenoid pigments. It will be recalled that Moewus and Kuhn were able to test the chemical nature of sex substances only by indirect means – by spectroscopy. Direct chemical identification was lacking. Moewus wanted more definitive chemical tests to be performed. He also briefly described the difficult working conditions, psychological stress, and the lack of basic materials such as agar to grow algae during and after the war:

> For 8 years now I have heard nothing more from you. We nearly saw each other in 1939, for I was already on a ship under way to the International Microbiologists Congress in New York. Unfortunately the powers that be wanted it otherwise, and the ship was ordered back. So I want to make

the attempt now, to again take up contact with you. In the first years of the war I was still able to work with *Chlamydomonas,* but then and after the war many things were not available, inner peace was lacking, many small things were lacking which are nevertheless required for work with algae, e.g. agar. So since 1943 I have been working in another area, namely, plant growth- and inhibitory-substances. I have developed a new auxin test which is more sensitive than the Avena test and which is simpler to execute. But this is not wholly satisfying for me, for I am time and time again drawn toward algae, with which my wife is also working independently. And especially with *Chlamydomonas* there is so much potential. Subsequent to the statistical uncertainties being cleared up, the unique 1 cell / 1 molecule reaction of sexual substances also, it will be worthwhile to continue working with genetics and physiology in order to attain the end goal, whether the gene is itself an enzyme or whatever. We have grounds to suspect that there are enzymes present in the cell nucleus which direct the formation of sexual substances. It would be decisive if the cell nucleus could be successfully separated from the cytoplasm, and the nuclear portion could be investigated. But unfortunately such experiments are not able to be carried out here at this time. Furthermore, procuring larger quantities of algae would have to be attempted in order to extract the sexual substances out of these themselves. If this were to succeed for at least a few, one would have found important points to go on. For up until now we have only circumstantial evidence, no genuine dcm onstration. While Prof. Hartmann was here a short time ago, he told me that you had obtained such nice results. I would be very grateful if you were to think of me on occasion. When I have received news from you I will also send you several offprints. (Moewus to Sonneborn, August 7, 1947)

Sonneborn was delighted to begin to exchange ideas and papers with Moewus. He saw many striking similarities between his own latest genetic results in *Paramecium* and those Moewus reported in *Chlamydomonas.* Sonneborn had identified a new cytoplasmic genetic factor, "kappa," in *Paramecium* which he believed to be very important for understanding the role of the cytoplasm in heredity (see Sapp, 1987a). It had several features in common with the results of Moewus that had provoked such controversy. For example, one molecule of the substance controlled by kappa could affect cells, and there was, statistically, close agreement between Sonneborn's genetic results and theoretical expectations. To Sonneborn, his own work on *Paramecium* lent support to the validity of some of the defamed reports of Moewus. Sonneborn quickly wrote back to Moewus, pointing out how much he appreciated his work. He emphasized that it deserved priority in the new domain of biochemical genetics and that some of his recent results in *Paramecium* related closely to those reported by Moewus in *Chlamydomonas:*

Dear Dr. Moewus,
I was very happy indeed to hear from you and to know that you are still carrying out your beautiful and important researches. I am sending you at
once copies of all available reprints of my work since last I sent you a set. I
will not therefore, tell you about what we have been doing since I hope the
reprints will reach you safely and you will soon be able to see the full accounts. We are now investigating mainly antigenic characters in *Paramecium* and are finding some very extraordinary and interesting things. We
have been able to "mutate" certain types experimentally by exposure to
corresponding antibody. Every exposed animal mutates and thereafter re-
produces true to the new type, quite unlike Jollos' old results and ours.
These new mutations persist indefinitely through successive autogamies.
The breeding analysis, however, shows that the antigenic differences are
controlled by cytoplasmic factors, just as the killer character is. I have not
yet published any of these results, for our study is still in progress, but it
will soon be far enough along for me to get the first paper out. There are a
number of other very interesting things in progress in our laboratory by my
students and associates. I wish we could get together to talk things over
some time, for I feel that your work and mine have much in common. I may
get to the Genetics Congress in Stockholm next summer, and it would be a
great pleasure if I could meet you either there or elsewhere.
    I hope you will not resent the criticisms I have published about your
work, for I believe that I admire your work as much as, if not more than,
any person in this country. I spend much time discussing it with my students and never fail to point out how your work preceded and set the
pattern for much of the work of Beadle and his group in this country. I
am especially interested in your one cell/one molecule reaction, for we
have run into very similar results with respect to the killing action [of the
kappa-controlled substance on] *Paramecium*. One molecule of this can kill
a sensitive *Paramecium*. In our case, the statistical results agree very closely
with theoretical expectations. There will soon appear in *Physiological Zoology* a paper on this from our laboratory by Miss Austin, and I, too, am
preparing a paper on the same subject. Needless to say, your views on
the presence of enzymes in the nucleus are of enormous interest and I
shall be anxious to see your publications on this when they appear.
    With very best regards and best wishes for success in your further stud-
ies. (Sonneborn to Moewus, September 9, 1947)

Sonneborn had no idea of the difficult conditions in which Moewus
lived and worked. On October 13, 1947, Moewus wrote to Sonneborn,
pointing out that he and his wife were able to work again on *Chlamydomonas* and other algae. He only cryptically mentioned his empty
stomach.

Dear Professor,
My sincere thanks for your friendly letter and the transmission of your
offprints. I have read them with great interest and congratulate you on

the nice results. In the same mail I am sending you several of my papers. . . . So now I can present your papers in a colloquium with students and let them be discussed. As I wrote to you in my letter of 7.8 [August 7], it is my most heartfelt wish to dedicate myself totally to algae again. That is to be fulfilled now and I will, together with my wife, again work more with *Chlamydomonas* and other algae. I have just spent 4 weeks on the North Sea island Spiekeroog, on which there is a research station of the University of Hamburg, and there was able to collect several marine algaes. . . . As always I continue to be interested in the sexual substances of other algaes.

The reading of scientific articles is not straightforward, here with us, for with hungry stomachs it is difficult for one to concentrate. That holds true for scientific work and for the conducting of lectures. . . . Just as your papers arrived, a care packet also arrived, that Prof. Christensen in St. Paul sent to me. It was the first that I have received. I already know Prof. Christensen from my time in Berlin, where he was working with Prof. Kniep. The most pleasure for me came from the coffee. A cup of this magnificent beverage works wonders and one can work right through the night. But regretfully, the tin is already nearing its end. Hopefully you will not hold it against me when I describe everything in such detail and I please ask of you, if you on occasion. . . .

At the moment I am writing several smaller papers collectively that carry the main heading: "On the Genetics and Physiology of Nuclear and Cell Division." . . .

In the hope that you will occasionally let us hear from and maybe even see you. . . .

Sonneborn sent a contribution to a fund collected by American geneticists for aid of German scientists. He requested that something be sent to Moewus if it met the approval of the group. There was an investigation of each individual recipient before the group decided to send a package. Sonneborn encouraged Moewus to continue his work on *Chlamydomonas* and kept Moewus up to date on his own current research on cytoplasmic inheritance. In the winter of 1947, he had traveled to Stanford, where G. M. Smith, who had taken up the study of sexuality in *Chlamydomonas*, gave him a demonstration of the mating reaction first reported by Moewus. Sonneborn went over Smith's results in some detail and, as he wrote to Moewus (January 7, 1948), it seemed to him that Smith was confirming the phenomena that Moewus reported.

Your reprints also arrived and I am very glad to have them. It has been over a year since I last went over with students your beautiful studies, but I shall cover that ground again in the fall semester and I shall try to bring your studies up to date.

Recently I had the occasion to travel to California where I met, at Stanford University, Dr. Smith who, as you probably know, has taken up the study of sexuality in *Chlamydomonas*. He demonstrated for me the mating reaction which I was very glad to see. I went over his results in some detail and was pleased to find that the more broadly his investigations branch out, the more he is confirming the phenomena that you have reported. If you are not already in correspondence with him, I am sure you would enjoy hearing what he is doing and would recommend that you write to him. . . .

You will be interested to know that the work on cytoplasmic factors in *Paramecium* is developing rapidly. My colleague, Dr. van Wagtendonk, has established that the killer poison is probably a deoxyribonuclear protein. I have discovered a method of transferring kappa to sensitive animals by means of exposing them to the broken up bodies of killer animals. . . . My associate, Mrs. Le Seur, and I have carried out a very extensive investigation of antigenic traits and have shown that these also are dependent upon cytoplasmic factors which are independent of kappa. We have been able to change permanently the hereditary antigenic type of the organisms in 100% of animals exposed to specific antibody. . . .

Moewus was glad to learn that some aspects of his controversial work on sexuality were being confirmed. He was also grateful that he was going to receive a CARE package. The next month he wrote to Sonneborn, congratulating him on his research and pointing out that after direct chemical identification of sex substances was achieved, he would bring the *Chlamydomonas* work to an end. In this letter, Moewus began to allude more directly to the poor living conditions in Germany. Indeed, from the description he gives, it is difficult to understand how he could have done any work at all:

Sincere thanks for your friendly letter of 7.1 [January 7]. It was a pleasure to hear from Dr. Smith . . . and that you were able for once to see the group formation for yourself. I immediately put myself in communication with Dr. Smith, and I hope that we will be able to exchange our experiences. First of all, I want to tell you that we have just succeeded in isolating the [first] algae hormone in crystallized form from dried algae. . . . After this success we are now going to work on isolating further algae hormones. And I hope, that with this the work on *Chlamydomonas* will come to something of an end.

On your investigations and results in *Paramecium*, I sincerely congratulate you. The mail which you considerately sent last year was a great pleasure for me. One can see how the intensive engagement with a particular object leads to fundamental knowledge. . . .

I myself am at the completion of my growth- and inhibitory-substance experiments, after I finish off writing an article about a new growth-substance test. At the same time I am working on the genetics of Chlorogonium. . . . I am in contact with Dr. Darlington, London; he has offered to publish the results in *Heredity* in the English Language. For the general state of journal publication here is still very low, so that it is almost impossible to publish articles. . . .

Such small articles as the one enclosed, do appear here but even that takes almost 9 months! Larger articles are not to be accommodated in any journal because paper is not available.

I am sincerely thankful to you that you did not feel that my wishes for occasional improvement of the living conditions were disagreeable. Rest assured that I was reluctant to write to you in that way and I thank you for having initiated something in this regard.

Although I do not want to make too much of a demand on your time, I nevertheless would like to write briefly about the situation. We scientists belong to the normal users, who are thus dependent on the officially pre-scribed food rations. So we receive, e.g., in the whole of February not one gram of fat – that is, neither butter, oil, nor margarine. We get 100g of meat weekly – that is, 14g daily. Imagine for yourself the portion of 14g that one is to consume daily. So the result is that there is a very, very great deficiency in fat and protein. And on sheer strength alone it is not possible for the human organism to build everything. One can only wonder at how one can put up with it all. But I suffer very much from it, because since 1945 I have a skin disease, psoriasis, the cause of which is not bacteria, fungus, nor virus. Because of this disease I have not been able to sleep an hour for months. It must be the manifestation of a deficiency, but regret-fully one knows nothing about this very widespread disease. But despite all this one can overcome all these shortcomings with willpower, which is made even harder through barely being able to heat one's room at home because there is no coal, etc. I don't want to bore you. But you can imagine how letters like the one you have written spur on new work output time and again, and for that I am very grateful. (Moewus to Sonneborn, February 10, 1948)

By September of the next year, Moewus's work on sex substances in *Chlamydomonas* was drawing to a close. Sonneborn and Moewus contin-ued corresponding and sharing manuscripts. On September 16, 1949, Moewus wrote to Sonneborn, alluding to the discussions that would arise in Germany over his *Paramecium* work. He also mentioned that most of the important sexual substances in *Chlamydomonas* had been suc-cessfully identified by direct chemical means and that he was now turn-ing his attention to some difficult problems relating to sexual phenom-ena in higher plants.

I thank you sincerely for the last post of papers. All articles interested me greatly. A student in the Zoological Institute of Prof. Ludwig will re-port on all your *Paramecium* papers on two days in November. I am already looking forward to the discussion that will arise. After all, it is an area that you have developed with laborious effort and have brought to such great success.

> The *Chlamydomonas* work has now come to something of a close, subse-
> quent to the most important substances . . . having been isolated. . . .
> Instead I occupied myself in Spring with the auto-sterility of flowering
> plants. That is quite a tricky problem. One case I have at least been able
> to explain already. . . . (Moewus to Sonneborn, September 16, 1949)

Sonneborn (November 10, 1949) immediately wrote back to Moewus,
attempting to dissuade him from discontinuing his work on *Chlamydomonas*. Moewus and Sonneborn had different research strategies. Moewus, it seems, was very interested in problems related to sexual phenomena in plants. To pursue his interest, he was willing to switch
readily from one organism to another – to different genera of unicellular
organisms, and to different genera of higher plants. Sonneborn, on the
other hand, liked to exploit his knowledge of one organism to the full-
est, using it as a technology for investigating whatever phenomena he
could. *Paramecium* was especially suitable for researching cytoplasmic
inheritance and nucleocytoplasmic interactions (see Sapp, 1987a). In
Sonneborn's view, Moewus had done great work in domesticating *Chlamydomonas* for use in biochemical genetics. He should continue to work
on similar lines, using this newly created technology for investigating
various other problems. This seemed to Sonneborn the logical way to
proceed: Stick with one organism and know it well, and try to achieve
the most exhaustive genetic analysis possible of one organism rather
than turning to different techniques to investigate other phenomena in
other organisms. But Moewus himself seemed to recognize the merits
of working systematically with one organism. As he had written to Sonneborn (February 10, 1948), when celebrating Sonneborn's work on *Paramecium:* "One can see how the intensive engagement with a particular
object leads to fundamental knowledge." Yet, for whatever reason,
Moewus seemed determined to abandon *Chlamydomonas*. All Sonneborn
could do was to emphasize again to Moewus how much he appreciated
his work on *Chlamydomonas* and the priority it deserved in the field of
microbial biochemical genetics. He also attempted to persuade Moewus
to follow his own work on extranuclear inheritance, and use *Chlamydomonas* to analyze photosynthesis associated with chloroplasts in plants.
He wrote to Moewus:

> I was very pleased to have your letter of September 16 telling me about
> the student in the Zoological Institute who discussed our *Paramecium*
> work. I would be very glad indeed if you would pass on to me any of the
> suggestions or criticisms which came out of that discussion. We are always
> glad to get criticisms.

I am very sorry to hear that you are bringing the *Chlamydomonas* work to a close. I had hoped from your earlier papers that you would continue with *Chlamydomonas* along lines that you suggested you might carry out. I refer particularly to what I thought were your great plans to analyze the complete morphology and physiology of the organism from the same point of view and by the same methods that you had used in analyzing the control of sexual processes. This, it seems to me, would have been a monumental undertaking and of the very greatest importance. I thought perhaps you would include an attempt to analyze photosynthesis. It seems to me that it would be a great loss to science not to have you continue work of that kind.

I have probably written to you before how very greatly I admire your work. I have lost no opportunity to convey the importance of your investigations to the students in my various classes. As you probably know, some workers in our country have received on this side of the Atlantic great acclaim for work that they have done on other microorganisms, as if they had initiated the whole idea of analyzing the genic control of enzymatic activities. I have taken pains to point out how thoroughly you anticipated this approach and how beautifully you carried it out in even earlier years.

Although I am sorry to see you shift to another type of problem, yet I know that whatever you undertake to accomplish will yield results of great interest, and I wish you the very best success in all your work. (Sonneborn to Moewus, November 10, 1949)

In summary, Sonneborn found Moewus's ideas to be stimulating and the biochemical and genetic investigations of *Chlamydomonas* to be pathbreaking. There is no doubt that this view was shared by many of his students as well.

7. Fostering a new generation of geneticists?

> I read all of Moewus and the stuff with Kuhn etc., and believed it. . . .
> The fact that he had two-strand crossing over didn't bother me. I really
> wasn't really very sophisticated then. I guess a fraud at this level I just
> believed wasn't possible. (Watson, interview, May 13, 1986)

Moewus and Watson: prelude to the DNA story

During the 1940s and early 1950s, Sonneborn often discussed Moewus's work with his students. Many of those who attended his classes would become well-known geneticists, such as Elof Carlson, D. L. Nanney, J. R. Preer, and the Nobel Laureate James D. Watson. Watson went on to investigate with Francis Crick the structure of DNA, "the secret of life" – a phenomenal discovery widely acclaimed as one of the major triumphs of twentieth-century science. For their contributions toward elucidating the structure of DNA, Watson, Crick, and Maurice Wilkins were awarded a Nobel Prize in 1962. Since that time, the story of the discovery of DNA has become a well-told tale.

Watson himself wrote the first account of the discovery of the structure of DNA. His little book, *The Double Helix* (1969), has become a classic among accounts of modern science. It is a personal account, and one of the first books about the nature of modern science which reveals to outsiders the human and social side of laboratory life. The story of *The Double Helix* tells of Watson's youthful arrogance, his sexism (see Sayre, 1975), the struggle for recognition, and the resulting secrecy in molecular biology during the postwar years. Watson's account never pretended to be complete. He recollected events and personalities only of the years between 1951 and 1953. His education at Indiana University, his relations with Sonneborn, and the controversy surrounding Moewus were not included. Yet, both Sonneborn and Moewus were highly influential to Watson's interest in the nature of the gene and gene action.

During the late 1940s, Indiana University emerged as a major center of genetic research. H. J. Muller, who was awarded a Nobel Prize in 1947 for his work on the artificial production of mutations by x-rays, continued to work on *Drosophila* genetics. Salvador Luria worked on bacteriophage and transformations in bacteria. Sonneborn worked on *Paramecium* genetics. However, *Paramecium* did not prove to be a highly useful organism for biochemical genetics. As mentioned earlier, the biochemical genetics of microorganisms required an organism that could be grown on a well-defined chemical medium. *Paramecium* eats bacteria; it could not be grown on an artificial synthetic culture medium. *Paramecium* did prove to be an extremely useful tool for analyzing nucleocytoplasmic relations and problems of cytoplasmic inheritance, however. These problems remained the focus of Sonneborn's work. Sonneborn himself maintained an interest in the larger problems of biology – the role of the environment in genetic expression, the spacial, structural organization of the cell, and the nature of genetic development. These issues were usually ignored by those researchers who focused solely on the nature of the gene and how it controlled biochemical reactions.

Of the three leading biologists at Indiana mentioned above, Sonneborn was recognized as the outstanding teacher with a large and loyal following. The following views of Watson (1966: 239) were representative of those of young genetics students at Indiana:

> During my first days at Indiana, it seemed natural that I should work with Muller but I soon saw that *Drosophila's* better days were over and that many of the best younger geneticists, among them Sonneborn and Luria, worked with micro-organisms. The choice among the various research groups was not obvious at first, since the graduate student gossip reflected unqualified praise, if not worship, of Sonneborn. In contrast, many students were afraid of Luria, who had the reputation of being arrogant toward people who were wrong. Almost from Luria's first lecture, however, I found myself much more interested in his phages than in the paramecia of Sonneborn. Also, as the fall term wore on I saw no evidence of the rumored inconsiderateness toward dimwits.

Watson eventually settled into the phage group led by Delbrück and Luria and completed his doctoral research under the direction of Luria. Watson had great admiration for European science and for German theoreticians (Watson, interview, May 13, 1986). He has often claimed that the book *What is Life?*, written by the Austrian Roman Catholic physicist Erwin Schrödinger (1944), had a decisive influence on his own interest in the nature of the gene. As mentioned in Chapter 4, Schrödinger's famous book theorized on the gene in the light of principles of quantum mechanics which had been most prominently developed by the German physicist Max Delbrück. Watson (1966: 240) wrote about Delbrück, "The prominent role of his ideas in 'What is Life?' made him a legendary figure in my mind." By the spring of 1948, Watson had become polarized toward both "finding out the secret of the gene" and joining the phage group of Luria and Delbrück (Judson, 1979: 47).

As mentioned in Chapter 6, Delbrück and Luria opted for a combination of "pure genetics" and physics to explain the nature of the gene. Delbrück himself doubted the value of biochemical approaches, believing as he did that a new quantum revolution in biology was necessary before the complex nature of the gene could be understood. Luria, on the other hand, seemed to realize that "it is impossible to describe the behavior of something until you know what it is" (Watson 1968: 23). Yet Luria himself was not willing to learn chemistry; instead he suggested that Watson be sent to work with a chemist after he completed his doctoral dissertation on phage genetics in 1950. Viruses were understood by many to be a form of "naked genes"; about one half of the bacterial virus was DNA and the other half protein. By the time Watson completed his doctoral thesis, it was clear to him and many others that DNA – not protein – was the "essential genetic material."

Watson was sent to Copenhagen to work with the chemist Herman Kalckar, who was interested in nucleotide chemistry and was supposed to be receptive to the need for "high-powered theoretical reasoning." However, the plan failed. Watson recalled, "Herman did not stimulate me in the slightest." Kalckar was interested in the metabolism of nucleic acids, which seemed to be of little immediate concern to understanding how genes duplicate themselves (see Watson, 1968: 23–24). According to Watson (1968: 28), it wasn't until later, when x-ray diffraction images showed that DNA could be crystallized and therefore had a regular structure, that he became interested in chemistry. When the structure of DNA was known, he reasoned, one would be in a better position to understand how genes work.

This may have been the first time Watson became interested in chemistry as it pertained to the structure of the gene. However, it did not mark the first recognition by Watson that "the combined techniques of chemistry and genetics could yield real biological dividends." In fact, Watson was first introduced to the fruits of combining biochemistry and genetics through the work of Moewus. And he found it stimulating indeed. As he recalls, "There was Beadle and Tatum, but the thing about Moewus was that it was much more exciting. It was a bigger story" (interview, May 13, 1986). Watson first heard mention of Moewus's work in Muller's course. However, Muller had only a minor interest in biochemical genetics. As one of his students, Elof Carlson, points out, "Muller had only a luke-warm interest in Beadle's work because he thought it ignored the more complex relations of genes to characters and he didn't appreciate the simplicity of a one-gene, one-enzyme model" (letter to the author, September 2, 1986). Muller was generally indifferent to Moewus's work also because none of it was pertinent to his classical genetic interests in the gene and gene mutation.

Watson became interested in Moewus's work when he was a student in Sonneborn's class. Sonneborn, on the other hand, was an enthusiast of new approaches to genetic analysis, especially those that involved developmental processes, sexuality, or nucleocytoplasmic interactions. He applauded Moewus's ideas and the possibilities his work held out for biochemical approaches to genetics. Sonneborn's microbial genetics course attracted and excited large groups of students. Sonneborn held a small informal seminar with students that met weekly at his home in Bloomington to discuss recent genetic literature which was bursting from the new domain of the physiological genetics of microorganisms. Watson attended regularly for two years. In fact, Watson carried out his first major study of the literature on gene action in Sonneborn's course on the "Genetics of Microorganisms," and it centered exclusively on the work of Moewus, Kuhn, and their collaborators.

Sonneborn was known to work regularly 70 hours a week, divided between his research and his teaching. The energy he focused on teaching is amply illustrated in the syllabus to his postgraduate course on the genetics of microorganisms in 1949. The course surveyed the literature on the genetics of *Neurospora*, yeast, bacteria, *Chlamydomonas*, and, of course, *Paramecium*. Sonneborn strove to give students a thoroughly critical understanding of the genetics literature. It gave accounts of various alternative theories of genic action, pointing out their possible significance and the difficulties of constructing a general model of genic action. Sonneborn challenged students to devise ways to test them critically. Sonneborn's teaching strategy also involved a heavy historical dimension, focusing on originality and innovations. He introduced students to the development of microbial genetics and critically discussed new concepts and techniques and what was novel about them. The work of Beadle and Tatum, and of Moewus, received high priority. With regard to the *Neurospora* and *Chlamydomonas* studies, Sonneborn wrote in his syllabus:

> The closest approach to the *Neurospora* studies was made by Moewus and his associates on the unicellular green alga, *Chlamydomonas*. A detailed and thorough study was made of genes controlling many biochemical reactions involved in the development of mobility and sexuality. The methods

employed were in detail parallel to those employed in the *Neurospora* work. Moewus set forth plans for the extension of this sort of analysis to every aspect of the organism – morphological and physiological – and gives an account of 42 gene loci, all mapped in the chromosomes, which could be used in the analysis. . . . All appeared in 1940, the year before the first paper on *Neurospora* biochemical genetics.

The *Chlamydomonas* work and the *Neurospora* work, comparable in scope and design, differ greatly in magnitude from all other work in biochemical genetics. (Sonneborn, "Syllabus on Genetics of Microorganisms," 1948: 41)

In the fall of 1948, Watson wrote a paper for Sonneborn's course, entitled "The Genetics of *Chlamydomonas* with Special Regard to Sexuality." Watson, who was teaching a course in scientific German at Indiana University, was able to do a thorough study of the reports of Moewus and his collaborators written in the 1940s. In his paper, Watson devoted a great deal of attention to the physiological results pertaining to sex determination and, based on Moewus's data, offered a novel interpretation for the genetic control of sexuality in *Chlamydomonas*. Unlike Lederberg, who said that he could not easily obtain Moewus's original papers, Watson had little difficulty. Watson described his attitude toward the *Chlamydomonas* papers clearly in his introduction:

This outline was originally intended to be a digest of the three known reviews on Moewus' work on *Chlamydomonas*. However, after completely writing a very inadequate outline, my curiosity was raised to see whether any possible later work by Moewus and Kuhn might clarify a quite messy and incomplete picture. Fortunately, I was able to locate without too much trouble a number of recent papers on *Chlamydomonas*. It appeared to me that this later work made quite untenable the hypothesis of Moewus and Kuhn which appeared in 1940. However, I have yet to find any paper by them in which a new hypothesis is offered. It therefore seemed profitable to me to see whether I could form a new hypothesis that would be consistent with the empirical situation. To do so, I felt that I would have to reread many of the original papers. Unfortunately lack of time prohibited me from reading any of the papers before 1940. I was able to quickly read the 1940 papers which were apparently a good summary of his work till then. My general impression of them was that they were very sloppy and inaccurate. There are many contradictions in these papers, of which space will permit me to mention only a few. The general impression however is still that the large majority of the observations were reported accurately. The main difficulty in analysing Moewus is to know what information to throw out as inconclusive or even false.

I must apologize in two main respects, one that due to my attempt to find a new meaning for the genetical work, I have badly ignored the problem of sexuality as approached from any method other than biochemical. The first part of the outline is merely a bad rehash of your reviews. My

second point for apology is more important. Any real attempt to hypothesize on the Moewus data should involve a much better knowledge than I
possess at present. It is quite probable that my hypotheses are quite untenable due to facts which I have ignored or overlooked. (Watson, unpublished manuscript, 1948: 1)

Watson began his account by emphasizing criticisms on both genetic
and biochemical aspects of the *Chlamydomonas* papers. Regarding the
former, Watson noticed a major discrepancy between the results of
crossing-over reported by Moewus and those reported by others who
worked on the genetics of microorganisms. "Crossing-over" is a means
by which genes get shuffled in gametes. It contrasts with "linkage,"
which is defined as the tendency for genes in the same chromosome or
linkage group to enter the gamete in the parental combinations. It is the
means by which genes enter the gametes in combinations other than
those of the parents (recombination). Exchange of genes or crossing-
over between members of chromosome pairs occurs in most plants and
animals. Since the first decade of the century, geneticists reasoned that
the percentage of crossing-over between two loci is roughly propor-
tional to the distance between these points on the chromosome. Therefore, crossing-over data can be used to determine the relative positions
of genes on chromosomes.

Watson pointed out that, based on crossover frequencies, Moewus
was able to construct chromosome maps for each of *Chlamydomonas's* 10
chromosomes. However, when crossing-over occurred, Moewus
claimed that all four descendants of a zygote revealed crossing-over.
Such results, Watson reasoned, were compatible only with two-strand
crossing-over. Moewus's work on other flagellates also supported only
two-strand crossing-over. This situation, Watson (1948: 3) noted, was
highly unusual since it was the only case in the genetics literature that
revealed "two-strand crossing-over instead of the orthodox four-strand
crossing-over."

Moewus himself was inconsistent in his reports in this regard. Thus
Watson continued:

> In his 1940[a] paper Moewus states without giving any evidence that four-
> strand crossing over does occur in *Chlamydomonas* and may in fact be the
> usual type. However, he fails to even suggest how this is compatible with
> his earlier published reports of only two-stranded crossing-over. While he
> stated (1940[a]) that he would publish further details, it appears that up
> to the present time he has not. (Watson, 1948: 4)

(Sonneborn noted on the margin of Watson's paper that this apparent
discrepancy could be accounted for by a difference in temperature on

germination.) It might also be mentioned that, as Jennings (1941: 750) noted, Moewus (1938b) himself discussed the difficulties arising from his data as to crossing-over. He claimed that this was not the normal state of affairs; it was pathological. Under certain conditions, he reported, crossing-over occurs according to the normal four-strand scheme. He promised a future account of investigations showing this.

Watson also noted the statistical criticisms of Philip and Haldane, who claimed that Moewus's data on crossing-over and segregation were much too close to the theoretical result than would be expected on the basis of chance alone. Unlike several others who reviewed Moewus's work, Watson noted that Moewus (1940[a]) had replied to Philip and Haldane's criticisms, arguing that the crossing-over data applied only to certain gene loci found on the sex chromosome. Finally, Watson (1948: 4) alluded to the statistical critiques of Pätau (1941), Ludwig (1942), and Wright (letter to Sonneborn, January 29, 1944). Watson (1948: 1) stated that he did not have time to analyze critically Moewus's (1943) paper responding to these charges.

Watson outlined the biochemical aspects of the work of Moewus and Kuhn on sex determination and gave a detailed description of their model for the genic control of hormone production. He then detailed Sonneborn's (1942) criticisms of the Moewus–Kuhn hypothesis with regard to inconsistencies and experimental incompleteness (e.g., none of the hormones had yet been isolated in pure form and crystallized). However, in his treatment of the work of Moewus and Kuhn, Watson went further than others had to date. He argued that subsequent work done by Kuhn, Moewus, and Löw (1942a,b; 1944) contradicted several of Moewus's earlier observations and interpretations. Without going into detail, it seemed clear to Watson that various parts of the original hypothesis of Kuhn and Moewus concerning the relationship between "termones" (sex-determining hormones) and "gamones" (hormones for copulation) were completely false. It is enough to state that based on their new results, Watson offered a new hyopothesis for genic control of sexuality in *Chlamydomonas*.

Despite all the difficulties with the *Chlamydomonas* work, Watson had the greatest admiration for the work of Moewus and his collaborators, as the concluding statements in his essay demonstrate:

> In conclusion it should be stated that this hypothesis lacks the basic simplicity of the original Kuhn–Moewus idea. However I do believe that it accounts for more of the empirical facts and has fewer apparent contradic-

tions than does the original scheme. In spite of these changes the Moewus–Kuhn work on sexuality remains a first-rate undertaking of [a] most gigantic magnitude. However with regard to the genetic analysis (Moewus) one often has the partial feeling that some of the statements reported as facts were merely wishful thinking. It is hard to imagine how all of [the] work reported was ever done. Certainly some experiments were either never done or reported falsely. Again one is forced to repeat the usual statement that work of this great importance with regard to gene action should be repeated independently. (Watson, unpublished manuscript, 1948: 23)

Watson did not state exactly what he believed was merely "wishful thinking," what experiments were "never done or reported falsely."

In his investigations of *The Path to the Double Helix* (1974), Robert Olby (who kindly sent me a copy of Watson's *Chlamydomonas* paper) isolated a passage from Watson's critique of the work of Moewus and Kuhn as being crucial to an appreciation of Watson's early scientific training. Although Olby only cryptically mentioned the work of Moewus and Kuhn, and Watson's critique, he claimed it to be "the first clear expression by Watson of his recognition of the fundamental importance attaching to the chemistry of gene action" (Olby, 1974: 302).

As Watson later recognized (interview, May 13, 1986), Moewus and Kuhn's work contributed greatly to his intellectual interest in the problem of the gene and genic action. Indeed, one might add that it was also significant to the support given to Watson during his early research career. A few years before Watson reported his work with Crick on the structure of DNA, when he was applying for research fellowships, Sonneborn wrote references for him. In 1949 and 1950, when reviewing his applications for fellowships from the National Research Council, Sonneborn repeatedly evaluated Watson in part, in reference to the work of Moewus and Kuhn. Sonneborn wrote about Watson:

> In my graduate course, Watson wrote a term paper in which he analyzed a large and complex set of German papers on biochemical genetics. This is a masterly job, the core of it being worthy of publication. He included a theoretical interpretation that differed from that of the authors and indeed seemed more reasonable. In the seminar at my home, we discuss current data from our research group; Watson's questions, comments, criticisms and suggestions are always worthy of respect and attention. No one contributes more to the group than he does. . . .
>
> In my opinion, Watson is not only highly superior, but uniquely so. I would rate him the equal or superior of any graduate student I have known. I predict for him outstanding success in a research career. He is the kind of student for whom the grade A is too low; I have had to give him A^{++}. I cannot recommend him too highly. (Sonneborn to C. J. Lapp. February 25, 1949)

In his reference of 1950, Sonneborn wrote:

> Watson is extremely alert, intelligent, and acute. He sees quickly how his own special techniques and approaches can be applied to the problems of others. At this stage of his development he has fully mastered the type of experimental analysis prevalent in virus genetics and now needs to get better balanced by training in quite different approaches.
>
> As a student in my course, he wrote as a term paper a critical analysis of an intricate but important series of papers. This was by far the best term paper ever written for me and was worthy of publication: it was a real contribution to the field.
>
> I expect Watson rapidly to become an outstanding investigator. (Sonneborn to C. J. Lapp, January 1950)

Moewus's work was exciting and theoretically stimulating to Watson; there is no question that it fostered his interest in the "secret of life." However, Watson himself never did experimental work on algae. But to those young geneticists who did come to focus their work on the genetics of algae, Moewus's work was less than inspiring. To them, it presented only obstacles to progress in algal genetic research. The experiences of Ralph Lewin and Ruth Sager are particularly revealing of the reluctance of young scientists to repeat and extend Moewus's work on *Chlamydomonas*.

Moewus and Lewin

"I did not want to work in muddy water." (Ralph Lewin)

Lewin was born and brought up in England. He completed his bachelor's degree in biology at Cambridge University, where he learned a great deal of mycology and became interested in algae. He later completed his Ph.D. in the United States at Yale University during the late 1940s. At that time, much of microbial genetics was done with the bread mould, *Neurospora;* Lewin wanted to do similar work with an alga. He soon recognized that among algae *Chlamydomonas* was probably the most suitable. They grew fast, could be easily maintained in pure culture, and included heterothallic (sexual) species. Lewin's doctoral dissertation (1950) was entitled "The Life Cycle and Genetics of *Chlamydomonas moewusii* Gerloff."

Since that time, Lewin's work has been largely on *Chlamydomonas*. Today he is a leading algologist at the Scripps Institution of Oceanography and is well informed on the Moewus affair. In the late 1940s and early

1950s, however, Lewin was more than reluctant to engage in the Moewus controversy, and he recalls that because of Moewus, he was also reluctant to work on *Chlamydomonas*. In the late 1940s, many graduate students preparing to work in microbial genetics read up on *Neurospora* and *E. coli* strain A12. The exemplars in biochemical genetics were the papers by Beadle and Tatum and by Lederberg, and, as Lewin recalls, "When you looked around there were papers by McClintock and Stadler and then you turned up these rather odd papers from Germany." Lewin's German was good, but he also had the useful reviews of Sonneborn, Thimann, Jennings, and others. Like many graduate students who were interested in microbial genetics, Lewin had to look into Moewus's reports.

According to Lewin, it was clear to him even before he worked on *Chlamydomonas* that Moewus's reports were dubious and had to be avoided. Yet *Chlamydomonas* was promising as a tool for investigating physiological and genetic mechanisms in higher plants. Like other microorganisms, it possessed a comparatively short life cycle, and biochemical genetic analysis was possible. As Lewin explains:

> I was very reluctant to work on *Chlamydomonas*. I wanted to work on something different. I did not want to work in muddy water. That's why I wanted to work on the red alga – start with something completely different.
>
> For a long time I fought against this business. . . . The business of doing what Moewus was doing and having to read all that and having to evaluate him, having to repeat and having to refute or support. I didn't want to get into it, so I tried to find some other organism. There was no obvious other organism. . . . I was stuck with *Chlamydomonas.*
>
> When I was getting into the doctoral thesis I had the feeling that this was the best thing there was. You've got to work with a unicell that pairs; it has to be something in which you have heterothallism so you can control the time when they mate. . . . (Lewin, interview, May 2, 1986)

Lewin's strategy for working on *Chlamydomonas* was simply to ignore Moewus's work as much as possible. He explains:

> The first thing to do is to make sure you know how to handle the organism, go through some of the experiments already reported by Moewus and when you have them under your belt, then go on from them to turning up other things. . . . And that's what I didn't want to do.
>
> Under normal circumstances you write to Moewus [to] get the cultures; you check this, check that; you find out – all right, there's media already worked out for you, you've got the strains already worked out for you,

mating techniques already worked out for you. Wonderful. Then you go on from there. That's the logical way you proceed with a white rat or maize or anything. I didn't. I had to start with nothing.

I used *Chlamydomonas moewusii*. No one had worked on it [except Moewus]. It was safer to say that no one had worked on it, so then I could work out my own nitrogen concentration, light regimes, temperature regimes, mating systems and so on. I was not bound by, at each step saying, "We tried Moewus's media, we had to do this; we tried the light, we had to do this." . . . (Lewin, interview, May 2, 1986)

Indeed, as will be discussed later, the culture methods developed by Lewin and other microbiologists in the United States were quite different from those used by Moewus. Mrs. Kobb (notes to the author, May 29, 1988) remembers that when American biologists later saw the Moewuses' techniques in Woods Hole (see Chapter 11), they were struck by the differences in culturing techniques:

> To the great amazement of the American scientists visiting our lab in Woods Hole, our cultures were not grown on an artificial, well-defined culture medium. Since the Berlin times, *Chl. eugametos* was cultured on a semi-defined medium. It was prepared with a sterile garden soil decoct, distilled water, and four mineral salts, no vitamins or nitrogens added. For solidification of the media, when needed, this medium was prepared with the addition of raw, natural sugar (*Stangenagar*). Natural *Stangenagar* became rare during the war and was unknown in USA microbiological labs. We had great trouble tracing that material to an import firm in Manhattan. It was raw agar in the form of threads.

In his doctoral dissertation, Lewin carefully sketched out the details for controlling the life cycle of *Chlamydomonas eugametos* and showed that the organism was suitable for genetic study and so on. His thesis also included an "historical review." As discussed earlier, several biologists, including Jennings and Sonneborn, had credited Moewus for providing some of the first demonstrations of Mendelian principles in microorganisms. He was, of course, highly celebrated also for his outstanding publications in biochemical genetics. Lewin, however, had a different attitude toward Moewus. He did not attribute any discoveries to Moewus. Instead, he credited the algologist Adolf Pascher with making a major genetic breakthrough:

> Pascher (1916) succeeded in crossing two species of *Chlamydomonas*, and even in obtaining segregation of certain characters. This achievement is in fact a landmark in the history of genetics, since this was the first occasion in which the segregation of genetic characters in the immediate products of meiosis, as postulated by Mendel, was confirmed in any plant (Pascher, 1918). (Lewin, unpublished, 1950: 102)

Although Lewin credited Pascher with showing Mendelian principles in unicellular algae, it was not clear if Pascher's experiments were fabricated. The validity of Pascher's reported results was questioned by Hartmann (1934), and many algologists including Lewin himself later doubted Pascher's ability to cross the two different species of *Chlamydomonas* (Lewin, interview, May 2, 1986). Pasher did not follow up on his experimental studies. During the following 25 years, he went on to become a leading taxonomist of freshwater algae. Pascher was a German, living in Czechoslovakia. During the German occupation of Czechoslovakia, he collaborated with the Nazi regime as chief director of science in Czechoslovakia – as Kuhn had done in Germany. Pascher committed suicide in 1945. The origin of the genetics of microorganisms was indeed elusive. As Lewin (1954: 107) later wrote in despair: "Since doubt has been cast on Pascher's results by Hartmann (1934), and on those of Moewus by a number of workers . . ., the information available in this field can hardly be described as illuminating."

When writing his doctoral dissertation of 1950, however, Lewin overlooked the difficulties with Pascher's claims, and, as mentioned, overlooked what many others perceived to be Moewus's major contributions. In fact, Lewin mentioned only one of Moewus's papers (Moewus, 1940a), and he did so only to discredit it. Indeed, his disbelief in Moewus's work is amply illustrated in the language he used. Unlike Smith and Watson, Lewin did not use discovery language such as "it was found," "demonstrated," "shown," "revealed," etc. Instead, he used expressions such as "Moewus published . . .," "goes on to say," etc. Like Watson, Lewin dwelled on Moewus's claim for the existence of two-strand crossover. Lewin's own work supported the usual hypothesis of four-strand crossover. Lewin wrote:

> Moewus (1940[a]) published linkage maps for the ten chromosomes of *Chl. eugametos*, based on an analysis of over 100,000 zygotes, in which 42 morphological and physiological characters are described and genetically located. . . . [He goes on to say that under certain conditions one may also obtain 4-strand crossovers, to be dealt with in a subsequent paper.]
>
> Under the conditions of the present investigations of *Chl. moewusii*, zygotes yielding four different clones were usually obtained . . . and thus no support is lent to the 2-strand-crossover hypothesis. (Lewin, unpublished, 1950: 103).

Lewin followed this criticism with other statistical and genetic criticisms pertaining to data reported by Moewus in the paper in question and some other genetic anomalies:

Moewus's paper (1940[a]) is criticized by Pätau (1941) on statistical grounds. Ludwig (1942) extended Pätau's calculations, and arrived at the value of $p = 1.5 \times 10^{-22}$ for Moewus's results. Harte (1948) pointed out some apparently unique genetic features, viz:

1. A 2-strand crossover frequency never exceeding 48% would indicate, according to current accepted theory, a high percentage of univalents, and thus a low viability would be expected. Moewus, however, indicated that almost all the zygospores germinated regularly.
2. A complete absence of triple crossovers.
3. An apparently *negative* crossover-interference; when two crossovers occurred in one chromosome pair, they always arose close together. (Lewin, unpublished, 1950: 103–104)

Lewin concluded his review with some rather sharp comments about Moewus's ability, or rather inability, to observe certain morphological features of *Chlamydomonas* described in his papers: "The present author is particularly sceptical of the reported ability of Moewus to observe five differences in the form and position of flagella, stigma, etc., while the 200 spores remain within the optically distorting spore wall" (Lewin, unpublished, 1950: 104).

When appreciating Lewin's comments and the discrepancies in reported observations about the morphology of *Chlamydomonas*, it is important to know whether or not Lewin and Moewus were examining the same kind of organisms. In order for a biologist to attempt to disconfirm or confirm a colleague's work and extend it so as to construct a genetic profile of an organism, it is obvious that they need to know if they are indeed working on the same organism. A common and useful classification is essential. However, classifying *Chlamydomonas* species was by no means a straightforward affair. Indeed, it led to considerable confusion in the 1940s and 1950s. Moewus carried out most of his biochemical genetic investigations of sexuality on a species that because of its faculty for copulation, he called "*Chlamydomonas eugametos*" (Moewus, 1931). Lewin, on the other hand, worked on a species of *Chlamydomonas* that differed at least in name from that used by Moewus. Lewin worked on *Chlamydomonas moewusii*, which he believed to be the same species as *Chlamydomonas eugametos* of Moewus. But was it the same as *eugametos*? Moewus believed it was not.

The origin of the name of the species *Chlamydomonas moewusii* therefore requires some examination. The name emerged out of a controversy between the algologists at Prague and those at Berlin. Moewus's description of *Chlamydomonas eugametos* did not agree with the culture material that was later described as *C. eugametos* by algologists at Prague (e.g., see Czurda, 1935). The German algologist Johannes Gerloff (1940) described and coined the name *Chlamydomonas moewusii* in an attempt to clarify the situation. Gerloff attempted to correct the confusion by constructing a major classification of *Chlamydomonas* taxonomy. He did this in his doctoral thesis entitled "The Variability and Systematics of the Genus *Chlamydomonas*," which he completed at the Friedrich-Wilhelm University in Berlin in 1940.

The discrepancies in the literature pertained to morphological descriptions of *Chlamydomonas eugametos*. For example, whereas Moewus claimed that *Chlamydomonas eugametos* had "a thick membrane without papillae," Czurda (1933) at Prague, who received his strain from Moewus, reported that it had "an evident papilla and a thin membrane" (Gerloff 1940: 319) But the conflicting descriptions were not as clear-cut as this. In his studies of variability, Moewus (1933a) had mentioned the possible formation of papillae in his *C. eugametos*. However, apparently such forms were rare. Moewus himself attempted to help Gerloff to understand the conflicting descriptions of the species. He suggested that three different "races" of *Chlamydomonas eugametos* might be involved. Gerloff (1940: 319–320) reported:

> One finds in these papers a number of conflicting and confused statements, the clarification of which is greatly desired. According to a personal communication from Moewus, some of these difficulties can be explained by the fact that different "races" are concerned. Thus the race from which the description was made is constant: it possesses a thick membrane and lacks papillae. Likewise the "race" supplied to Czurda, which differs from the type in its thin membrane and anteriorpapilla. The investigations of variability which Moewus published in 1933 were carried out with a third *variable* "race." (English copy obtained from Ralph Lewin)

Moewus suggested that a definitive classification of different kinds of *Chlamydomonas* could be made only on genetic grounds: by investigating their ability to be crossbred, through a study of genetic variations and mutations. However, Gerloff found this suggestion to be impractical. He argued that if one accepted Moewus's concepts, a division of the genus *Chlamydomonas* would be impossible, since one could never be sure that one was dealing with a constant "race" of a species. If one accepted Moewus's approach to classification, he reasoned, in order to arrive at a definite differentiation of the various species of *Chlamydomonas*, it would be necessary to investigate variability, hybridization, and mutation within all described species. "Only then," Gerloff argued, "could one establish genetic relationships and proceed to the establishing of a good species."

There were hundreds of different species of *Chlamydomonas,* some of which differed markedly in morphological characters (Gerloff reported 321 different species). Considering that the majority of described species had only rarely or had never been found more than once, such a genetic undertaking would require "an infinite time." Thus he argued against classifying *Chlamydomonas* species on genetic grounds. Instead he attempted to do it on morphological grounds, which he claimed were both more "natural" and more expedient. As Gerloff (1940: 320–321) argued:

> In this way a systematic division and understanding of the species of *Chlamydomonas* is rendered simpler, and sociological and ecological investigations are likewise facilitated. Such a systematic classification would express more closely the natural relationships, in my opinion, than the establishment of different "races." The individual species of these group-species are often to be distinguished from one another only with great difficulty. Which characteristics one wishes to select for the specification of the separate forms depends essentially on their variability. The less variation they exhibit, the better are they suited for the delimitation of species. Thus the presence of a papilla, among other features, should be considered a character of specific importance.

Gerloff himself was able to isolate from different soil samples cells that agreed "more or less" with the papilla form described by Czurda. And, based on the above considerations, he named the papilla form, which Czurda had obtained from Moewus, as *Chlamydomonas moewusii.* He distinguished this species from *Chlamydomonas eugametos,* which lacked papillae. Moewus, he claimed, carried out his experimental work on *Chlamydomonas eugametos,* not on *Chlamydomonas moewusii.*

Despite Gerloff's attempt to make a distinction between *Chlamydomonas eugametos* and *Chlamydomonas moewusii,* during the 1950s Lewin conflated the two names. He identified *C. moewusii* as *C. eugametos* and referred to Gerloff (1940) for having made the identification. Lewin (1957: 874) explained:

> Gerloff described *C. moewusii* in an attempt to establish a legitimate name for the controversial "*C. eugametos*" of Moewus. However, for convenience the latter name is retained here when reference is made to Moewus' original observations.

Lewin himself did not obtain his species of *Chlamydomonas* from Moewus. If it was, in fact, different from that used by Moewus, as is suggested here, then this may account for his inability to confirm some of Moewus's observations, at least with respect to the morphological descriptions of *Chlamydomonas.*

Figure 6. Franz and Liselotte Moewus, Stockholm 1950. (Photo courtesy Ralph Lewin.)

In all of his writings, Lewin mentioned Moewus only in reference to the criticisms he and others offered. In one of his first publications on *Chlamydomonas*, Lewin (1952: 233) concisely summarized the state of affairs in *Chlamydomonas* genetics:

> Though it is of some importance to investigate physiological and genetic processes in green plants by the use of a micro-organism with a comparatively short life-cycle, there have been very few genetic studies carried out with algae. The bulk of the literature in this limited field is the work of Franz Moewus, and his results, almost without exception, have remained unconfirmed.

It was difficult for Lewin to avoid treating some of Moewus's work during the early 1950s. In fact, Lewin met Franz and Liselotte Moewus at an international botanical congress in Stockholm in 1950 (see Figure 6). Liselotte remembers the Stockholm meeting well:

> Franz was well-esteemed at that time. I knew they were already considering the Nobel Prize at that time. There was a small circle he was invited

to. . . . They were sort of interviewing him. And he got annoyed. He didn't know what for. . . . They put him on the grill. He clammed up. These were the situations that he would clam up – if someone asked sharp questions, and he knew they were destructive. There was one man [who asked,] "Why didn't anybody say anything about the statistics in Germany? Why didn't anybody [say something] about the genetic work?" I said, "Why should anybody say anything? There wasn't anybody but him." But Franz would not say, "Hartmann did not understand that." I could not go too much against my husband. He couldn't say, "Hartmann doesn't know a damned thing about chemistry." How could he? They attacked. There was one man who initiated this, and he said, "For God's sake, Mr. Moewus, after all, it's about the Nobel Prize!" That was a remark among the voices that was lost. I was there; I heard it; and I didn't tell him. (Mrs. Kobb, interview, May 25, 1987)

Lewin recalls that he actually confronted Liselotte Moewus with some of the difficulties found in their work. One of the most striking features of Moewus's work, for those who worked with *Chlamydomonas*, was the sheer extent of the analyses. For example, in one of Moewus's papers of 1940, he claimed to have analyzed more than 100,000 zygotes – a number that seemed difficult to imagine. Lewin recalled:

When I asked about all these analyses, when I said "Well, that's an awful lot of analyses to have done isn't it?" And Liselotte said, "Yes, and you know, Franz had no help apart from me, he had to do it all himself." (Lewin, interview, May 2, 1986)

It was clear to Lewin that Liselotte Moewus was an intelligent person, and if the work was indeed fabricated, she also had a hand in it. Indeed, as will be discussed later, many participants in the controversy had doubts as to *which Moewus* was playing the leading role in the investigations.

Although Lewin was very critical of Moewus's work in his published accounts, his private letters to Moewus present a different picture. Several months after the Stockholm meeting, Lewin wrote to him, asking for stocks and pointing out that some of his work was being confirmed by others. He maintained cordial relations with him, as the following letter he wrote to Moewus (April 17, 1951) illustrates:

Dear Dr. Moewus,
You were kind enough to send me copies of several reprints some weeks ago, which I have not yet acknowledged. Thank you very much for having remembered me. I an sorry I did not write earlier.
 I have recently had sent to me *Chlamydomonas* material (from Mr. I. A. Cooper, of Staten Island, N.Y.) which contained abundant zygospores

and mating stages, and which morphologically and in cell size agrees very well with the *Chl. dysomos* which you described in 1931. Since it grows very well, I have been considering a few experiments with this species, as a change from *Chl. moewusii,* and should be most grateful if you would let me have any further information which you may have on this species. Am I right in assuming that it was described from field collections, and was not obtained in pure culture?

While talking with people at the International Congress in Stockholm, I offered to send various cultures and bits of information to people, and I'm afraid that some of these have still not been sent for various reasons. I suppose you must be much more occupied than I. However, if you do have time, I should be most grateful if you would let me have a slide showing cytological (nuclear) stages in *Chl. eugametos,* which you were kind enough to offer during a memorable *Chlamydomonas* dinner party high above the town of Stockholm, and a culture of *Stephanosphaera,* if you still have this in culture. If you have been able to locate the address of Dr. Johannes Gerloff, I would also appreciate this. I do hope he is well, though I have seen nothing published by him since his monograph in 1940.

I have heard something to the effect that Richard Starr had the privilege of working in your laboratory recently, and greatly envy him his valuable experience. I had occasion to write to him some weeks ago, and was surprised when his reply came from Cambridge. I had not known hitherto that he had left Vanderbilt University.

Joyce is still working on the culture of freshwater diatoms, and asks to be remembered to you both. Please give our best wishes to Mrs. Moewus.

Yet, already in 1951, Lewin wanted to write a major critical review of Moewus's work on *Chlamydomonas.* He asked a young Swedish geneticist, Nils Nyborn, at the University of Lund to collaborate with him on a paper for *Quarterly Review of Biology.* Nyborn had spent four weeks as a visitor in Heidelberg to learn the techniques of using *Chlamydomonas* as developed by Moewus. He had read almost all the papers of Moewus to date. Yet, Nyborn did not know what to think about Moewus and his work. Moewus was generally known to have been a quiet, soft-spoken and extremely pleasant person. Moreover, the very novelty of his work – its synthetic nature, drawing on algology, biochemistry, and genetics – made it difficult to assess. Even though Nyborn had spent a considerable time with Moewus in the laboratory and had made a close study of his papers, he still felt incompetent to make judgments about Moewus's work.

He was unwilling to write a collaborative paper with Lewin. He wanted to wait until more facts were in before he made public statements about it. Nyborn did suspect that some of Moewus's work was "biased and polished." At the same time, he did not want to engage in what he considered to be the sometimes "most unscientific stoning" Moewus received from Lwoff and others. On the other hand, Nyborn could not afford to spend time repeating Moewus's work, even though Moewus had given Nyborn some of his cultures. Nyborn was a young geneticist who had just received his first grant to continue to work on *Chlamydomonas*. He now needed to publish original work. He found himself in an awkward position: He was unwilling to make public criticisms of Moewus's work because of a lack of definitive information, and he felt his career would suffer if he spent valuable time attempting to repeat some of the work in order to obtain the required facts to make a fair judgment. Yet if he did not speak out against Moewus, he was worried that he would be perceived by some as a "suspect figure," a "co-criminal." It was a "damned if you do, damned if you don't" situation. All he could do was to suggest that someone else might be able and willing to repeat some crucial aspects of Moewus's work and that he could make sure he or she received the appropriate cultures. Nyborn (April 15, 1951) explained to Lewin:

> As to your offer to write for the *Q. R. Biol.* on genetics of algae together with you, I must say that I don't feel too eager. Frankly, I do not consider myself competent at all to treat these things, generally, and the Moewus thing particularly.
>
> I have penetrated Moewus' works, almost all of them I think, and I was together with him for some time, but I really feel very uncertain what to say or think about him. While I am not prepared to join the almost general and sometimes most unscientific "stoning" that he has been subject of by e.g. Lwoff, Pringsheim, Smith, Lederberg and others, I am also not willing to say that I am convinced that all his published results should be right. On the contrary, I feel quite convinced that all his results cannot be true . . . and I think, therefore, that it would be difficult, even unwise, to review his works in detail, if future would come to show several of these results to be – to put it mildly – polished and biased.
>
> So I am afraid I can not accept your invitation, I am sorry. But thank you, anyhow. The last half year I have been completely absorbed with our works on mutations in barley. The only thing I have been doing with my algae is to ask for a grant for these works and I just got it a few days ago. I also got Smith's strains of *Chl. reinhardi*. They look very nice, grow much quicker than the *eugametos*, I think, and copulate easily. Smith has, however, put me in a somewhat difficult situation by saying that I have promised to send him Moewus' cultures in exchange. I really never did promise anything like that, then I could also have promised you! I am going to write to Moewus shortly saying that I am going to "spread" his strains. He did promise strains to several persons himself, and I am going to say:

"I heard you promise yourself, so I am sure you do not object against my doing it." What could he say?

Moewus wrote me some weeks ago, that he had just had a visit by an English student . . . who among other things did *Restgametenversuche* together with Moewus. He obviously had more fortune in seeing Moewus work than had I. Perhaps you should try to get into contact with this chap to hear his opinions on the matter.

Moewus also seems to be working with paralysed and flagella-less mutants at present?!

Well that's it. Concerning my work on algae, I think, I am going to delimit them to mutagenesis purely, chemical radioprotection, radioresistant mutations and similar things, to begin with. Not so much more genetics and the normal physiology. Unfortunately, and I do regret it, I cannot afford repeating Moewus' works. It would certainly be most interesting! And I have a weak feeling that some people will come to look at me as a suspect figure, a cocriminal, as long as I refrain from doing this. But I hope that the strains will come to the U.S. somehow, so that the necessary tests can be done, and so that Moewus can meet his doom or reparation.

. . . P. S. It would be rather easy to repeat Moewus' most elegant works on *Forsythia*, accounted for during the congress. If they would prove to be true, one would perhaps have to believe in his rutin-mutations in *Chlamydomonas*. If not. . . .

I will have no time, no possibility to repeat these works, but perhaps you could get somebody interested!!!

Lewin wrote a review of Moewus's work himself. However, it did not appear until three years later (Lewin, 1954), when the situation concerning Moewus and his work had changed dramatically. In fact, as we shall see later, Lewin's review led him into a public controversy with Hartmann himself.

Moewus and Sager

"I just wished the Moewus problem would go away." (Ruth Sager, interview, May 15, 1986)

Ruth Sager shared Lewin's general attitude toward Moewus; Moewus was a problem; his work muddied the waters. It was something that had to be avoided. Like Lewin, Sager began to work on *Chlamydomonas* in the late 1940s. However, unlike Lewin who was interested in sexuality, Sager was interested in investigating chloroplast inheritance. She wanted to use *Chlamydomonas* as a useful tool to dissect the genetic apparatus controlling photosynthetic functions. Sager's doctoral dissertation was on corn genetics. She was a student of the well-known cytogeneticist Marcus Rhoades at Columbia University. Rhoades's cytogenetic work in corn was considered second only to that of Barbara McClintock.

After completing her Ph.D. at Columbia University in 1948, Sager decided that one of the most interesting problems in corn genetics was chloroplast inheritance. Various genetic studies on higher plants, carried out since the first decade of the century, suggested that the chloroplast might possess a genetic system comparable to that of nuclear genes. Prior to World War II, most of the genetic work on extra-nuclear inheritance had been carried out by German investigators working on various genera of higher plants. But the evidence for a general role of the cytoplasm in heredity was often criticized by American geneticists, who upheld the predominant, if not exclusive role of the nucleus in heredity. In fact, the only prominent investigations of cytoplasmic inheritance carried out in the United States prior to World War II were done by Sager's mentor, Marcus Rhoades (see Sapp, 1987a).

However, with the rise of microbial genetics during the 1940s and 1950s, the genetic role of the extranuclear constituents of the cell – such as chloroplast, mitochondria, centrosomes, and kinetosomes – began to attract a great deal of attention. The work of Sonneborn on *Paramecium* and Ephrussi and others on yeast, suggested that the role of the cytoplasm in heredity might be of greater importance than had been previously recognized by geneticists. Chloroplast inheritance in higher plants provided the most clear-cut example of the role of cytoplasmic genetic elements. However, genetic analysis comparable to that done on chromosomes had not been performed. Sonneborn had suggested, as early as 1949, that Moewus do this work with *Chlamydomonas* (see Chapter 6). Moewus did not take up Sonneborn's suggestion. Instead, the work was left to Sager. Sager became quickly convinced, by what was going on in microbial genetics, that the way to develop more refined genetic analysis of chloroplast was to use a microorganism rather than a higher plant. The question then was, Which microorganism? In the summer of 1949, she went to Stanford University to take a course in microbiology offered by the celebrated microbiologist C. B. van Niel. Van Niel was well-known for his work on bacterial photosynthesis, and every summer he offered an introductory course on bacterial photosynthesis and energy metabolism in yeast. Ralph Lewin had taken van Niel's course the previous summer. For Sager, it was a great experience. It was van Niel who recommended using *Chlamydomonas*. Van Niel had long recognized the importance of *Chlamydomonas* for biochemical genetics. As mentioned in Chapter 4, van Niel had reviewed some of

Moewus's work on biochemical genetics in *Chlamydomonas* in the most celebratory way in 1940. Sager recalls that "van Niel was absolutely unambiguous." *Chlamydomonas* was the only organism to work on (Sager, interview, May 15, 1986).

At that time, it will be remembered, the algologist Gilbert Smith was working on *Chlamydomonas* at Stanford University. As discussed in Chapter 6, Smith was recognized as a great algologist, who had turned to *Chlamydomonas* genetics because of Moewus's claims. Smith isolated his own strains of *Chlamydomonas* and worked out the life cycle in the laboratory and, in Sager's view, got the first "clean" segregation data, of mating-type segregations as a single-gene difference, in the laboratory. This, to Sager's knowledge, was the first clear-cut piece of genetics that had been done on *Chlamydomonas*, and it was the only genetics that had been done on *Chlamydomonas* – except by Moewus. As discussed previously, Smith himself never stated that he had disconfirmed Moewus's results. Nonetheless, in Sager's view, Smith's work was a clean contradiction of what Moewus had been saying about "the existence of various intersexes." Sager found Smith's work to be very appealing because it was "very straightforward." Smith was kind enough to give Sager a pair of his mating strains, and all of Sager's later work on *Chlamydomonas* was done on that pair she received from him (Sager, interview, May 15, 1986).

Sager recalls that she had learned of Moewus's work only through Smith's papers, and "had a vague awareness that there was some German work that was very peculiar." But she was not attracted to Moewus's work; she was not particularly interested in mating-type control or sexuality in algae; she was interested in using *Chlamydomonas* as a tool to study chloroplast inheritance. Between 1949 and 1955, Sager worked at the Rockefeller Institute for Medical Research in New York. She spent her first two years as a Merck Fellow of the National Research Council, investigating chloroplast biogenesis. She spent her final four years at the Rockefeller Institute as an assistant to Sam Granick, a senior researcher who worked on biochemical pathways in chlorophyll production. Granick wanted to use genetic mutants for study, and Sager's appointment allowed her to work on genetic mutants affecting chloroplasts. *Chlamydomonas reinhardi* proved to be an excellent tool for this kind of investigation. It possessed one chloroplast and several mitochondria per cell. It exhibited a readily controlled sex cycle and was able to carry out all stages of its life cycle on a chemically defined medium, so that biochemically distinct mutant phenotypes could be detected.

During the 1950s and 1960s, Sager developed techniques for readily acquiring and analyzing various genetic mutations which she attributed to chloroplast genes. Her pioneering work on chloroplast genetics culminated in her well-known book, *Cytoplasmic Genes and Organelles* (1972). Although her work on chloroplast inheritance was not directly related to that of Moewus, as will be discussed later, Sager herself would cross paths with Moewus during the summer of 1954 at the Marine Biological Laboratory in Woods Hole, Cape Cod. That encounter would, in effect, mark the beginning of the end of the controversy surrounding Moewus (see Chapter 11).

8. Raising the stakes: A Nobel Prize for Moewus?

> For having twice pioneered in two of the most important developments in modern genetics, Moewus clearly merits high recognition, regardless of any criticisms that might be made against his data.
> (T. M. Sonneborn, 1951: 496–497)

By 1950, the significance and validity of Moewus's work on the biochemical genetics of sexuality in algae lay in a precarious state. There had been no collective or organized attempt to evaluate the authenticity of his published results. The many biologists who did publish criticisms of Moewus's work did so in an individualistic manner. The disparate attitudes toward Moewus's work reflected a great diversity of theoretical, technical, and social interests. The detailed criticisms of Moewus's work were scattered in various journals and monographs. They included statistical criticisms of his numerical genetic data pertaining to segregation and crossing-over, criticisms of morphological descriptions, theoretical objections to his biochemical and genetic interpretations, etc. Indeed, often interpretative differences were so entangled with factual differences, it is difficult, if not impossible, to sort them out. Moewus's work on the biochemical genetics of *Chlamydomonas eugametos* remained controversial. Table 1 lists the principal criticisms directed against Moewus and his work as well as objections to these criticisms as set forth *by scientists* in their discussions thus far.

The question of the reliability of scientific results is a complex issue. As discussed throughout this book, we are not speaking of *objective* facts, for the facts scientists report are influenced by the questions they ask and the theories they use. As discussed in Chapter 5, when discussing Moewus and Mendel, what data a scientist "chooses" to ignore depends upon his or her experimental intentions: the theory he or she hopes to illustrate, and the theory he or she is not interested in illustrating. But the theory-ladenness of observations presents only one difficulty for assessing the "reliability" of scientists' reports. There is also the problem of the "technique-ladenness" of observations – the scientific objects, equipment, procedures, and skills that are involved in the production of scientific results. As mentioned earlier, when describing

175

Table 1

Criticisms of Moewus	Objections to criticisms
Data are too good to be true statistically.	Mendel's results are also too good to be true.
Failed to send stocks.	—
Results cannot be repeated.	Lack of technical skill of competitors; there was no attempt to replicate based on same species of organism.
Omitted data from his papers.	—
Too many data examined.	Represents pioneering work on the genetics and biochemistry of micro-organisms.
Gave faulty biochemical and genetic descriptions.	Endorsed by Hartmann and Kuhn.

Smith's attempts to confirm or disconfirm Moewus's claims, his different results may not be a sign that Moewus observed poorly or reported falsely. They may only be a reflection of different techniques used and/or organisms employed.

Many biologists overlooked these difficulties when they evaluated Moewus's work, and in so doing they came to conclude that Moewus's work was wholly "unreliable." Many made the next conceptual leap that any discrepancy was due simply to his intention to deceive his colleagues. All this, however, is to say nothing about why so many biologists were originally so critical of Moewus's work. As we have seen, the original skepticism toward Moewus's reports was due to several issues. The very novelty and significance of Moewus's work, the statistical critiques combined with the political atmosphere around World War II – all predisposed biologists to be supercritical of it.

The unorthodox advocate

Since 1939, when Philip and Haldane first made charges against Moewus, there was an avalanching effect. Finding and reporting flaws in Moewus's work became "standard practice" in genetics. Indeed, with few exceptions, those biologists who reviewed Moewus's work were more interested in tearing down the edifice he helped to construct than in attempting to understand it, correct it, and alter it accordingly. Sonneborn was among the exceptions. As we shall see in the present chapter, in 1951, against a great deal of resistance, Sonneborn made a heroic public attempt to ensure that Moewus's work received proper credit and proper evaluation by American geneticists.

If one were only to take into consideration larger political issues when attempting to understand the attitudes of biologists toward Moewus and his work, then Sonneborn's support for Moewus might appear to be a great paradox. Sonneborn himself was Jewish. But unlike Delbrück and others, he was far removed from the events in Europe. However, antisemitism was not a German or Nazi invention, of course. Sonneborn himself had experienced prejudice first-hand in the United States during World War II because he was of Jewish extraction. In 1939, he rejected a job offer as associate professor at Johns Hopkins University, partly because of the antisemitic environment at that university. The president of Johns Hopkins University, Isaiah Bowman, a famous geographer, made the situation explicit for Sonneborn. Bowman told Sonneborn that he would never be made head of the department because as a Jew he would be subjected to "irresistible pressures" to take Jews into his department which, according to Bowman, would make the non-Jews leave. As Bowman saw it, such an appointment would "ruin the department" (Sonneborn, 1978, unpublished autobiography: 65–71). It was not just that Sonneborn could not be expected to be promoted in rank, however; finding a position anywhere in the United States during the late 1930s seemed to present problems for him. F. B. Hansen, one of the officials of the Rockefeller Foundation (see Chapter 9), viewed the situation in 1939 as follows: "Jennings finds small sums from year to year to support Sonneborn who is rated highly as a geneticist but almost impossible to place in a permanent post because of strong Jewish traits" (Hansen, diary, July 1, 1939).

One cannot call upon larger political, religious, and "racial" issues to account for Sonneborn's support for Moewus. Instead, one has to consider his intellectual interests and the kinds of research he carried out. As emphasized earlier, Sonneborn was an enthusiast for new approaches to genetic analysis, especially those that involved developmental process, sexuality, or nucleocytoplasmic relations. Because he was a student of H. S. Jennings in the 1920s, Sonneborn had the greatest respect for German protozoology led by Hartmann. In many ways, the work in Hartmann's laboratory paralleled that carried out by Sonneborn and Jennings at Johns Hopkins University. The work done in both laboratories was pioneering in the development of microbial genetics. Sonneborn, like Moewus, worked on the genetics of microorganisms long before they became popular organisms in genetics. Both Sonneborn and Moewus worked on the development of sex types in microorganisms. Sonneborn also worked on the difficult problem of *Dauermodifikationen*, a problem which, Sonneborn always emphasized, was first investigated in Hartmann's laboratory. As mentioned in Chapter 3, this phenomenon of *Dauermodifikationen* involved environmentally induced, hereditary changes in microorganisms. During the 1940s and 1950s, its study promised to be very important for solving problems of development; it also challenged some classical Neo-Darwinian views of evolution. The problem of cytoplasmic inheritance was central to Sonneborn's work, and it was also central to genetic work carried out in Germany between the two world wars.

Sonneborn's sympathies and respect for the great pioneering work of German protozoologists were extended to Moewus. Although he doubted that Moewus's results were "quantitatively" sound, he believed that the essential "qualitative" aspects of Moewus's reports on the genetics and biochemistry of sexuality were reported correctly and were of the greatest significance. Instead of destroying Moewus's work, he attempted to build on it and refine it. Watson's review of Moewus's work served only to strengthen Sonneborn's perception of the importance of Moewus's work and his resolve to ensure that it was not rejected hastily.

Moreover, if Moewus's work was generally sound, then it deserved priority recognition over that of Beadle and Tatum in developing biochemical genetic technology. Sonneborn never believed that Moewus was the source of Beadle's ideas, nor even of his own work on mating types in *Paramecium*. However, he had great admiration for Moewus's ideas and techniques and appreciated the potential of his work for biochemical genetics. Sonneborn also perceived the vigorous activity that began to emerge around the study of *Chlamydomonas* in the United States to be largely due to the pioneering work of Moewus. Algologists, such as Smith and others, who began genetic work on *Chlamydomonas*, he argued, most likely knew nothing about how to carry out genetic investigations in algae prior to Moewus's work.

Sonneborn recognized clearly that his opinion about the novelty, significance, and reliability of Moewus's work was not shared by many of his American colleagues. However, he was not a person to flinch under the pressure of public opinion. In fact, Moewus's work was not the first controversial work from Hartmann's laboratory that Sonneborn defended against American critics. He had also defended the work of Victor Jollos pertaining to the problem of *Dauermodifikationen*. Jollos was a leading biologist at Hartmann's division of the Kaiser Wilhelm Institute for Biology in Berlin. He was forced to leave Germany with the rise of nazism and was invited to the United States as one of the many refugee scholars of Jewish extraction. Jollos arrived at the University of Wisconsin in 1934 and soon found himself in a milieu that was hostile to both him and his work. By 1940, he was in a desperate situation, having no laboratory facilities, and no means of subsistence.

As I have discussed elsewhere (Sapp, 1987a: 62–65), the issues underlying Jollos's tragic plight in the United States are complex, and some of them highlight the cultural and intellectual differences between American and pre-Nazi German biological communities. At the University of Wisconsin, Jollos had managed very quickly to antagonize almost everyone with whom he had come in contact. First, Jollos's theorizing on heredity and evolution was criticized in the United States. By the 1940s, most American geneticists had come to assume that mutations occurred randomly; natural selection controlled the direction of evolution. Jollos believed his experiments indicated that certain environmental conditions also could have an immediate *directing* influence on the mutations they produced.

Almost immediately, he found himself embroiled in a controversy with American geneticists. Some geneticists attempted to repeat some of Jollos's experiments but were unable to confirm some of his novel results concerning the effects of the environment on heredity. Instead, they defended the exclusive role of natural selection in directing the course of evolution. Second, several students who took Jollos's courses at Wisconsin criticized him for the way in which "he rode rough-shod over any opposing points of view." Finally, and not least importantly, there was also considerable antisemitic feeling at the University of Wisconsin. Jollos could not expect to have his position continued. On February 6, 1940, he wrote to Sonneborn in despair:

> Dear Dr. Sonneborn,
>   May I ask you if you know, perhaps of a job for me? I'm in a desperate situation, having no laboratory facilities and no means for further existence. There is no prospect for me in Wisconsin, and I have no direct contacts elsewhere.
>   I would gladly accept *any* post that would secure a moderate existence and the possibility to do some scientific work. I'm not afraid of a heavy burden of teaching, and I imagine that I could be of some use, as I have had teaching experience in various fields of zoology (general zoology, parasitology, cytology, invertebrates) besides protozoology and genetics. . . .

Sonneborn deeply appreciated Jollos's work; he tried to do all he could to help him find a job. As he wrote to Jollos (February 12, 1940),

> You must believe me when I tell you that I am prompted to take these efforts for you because I am a great admirer of your work. For years I must confess I was skeptical about much of your work, but in recent years my own work has led me more and more to appreciate the greatness of yours. It is this that makes me dread the possible tragedy of having a man with your great gifts and powers cut off from the opportunity of using them for the good of the science we both serve.

Sonneborn knew well of the issues that confronted Jollos in the United States. Over the next five months he wrote to various universities and funding agencies, describing Jollos's desperate situation and searching for help for him. But all was to no avail. Jollos died the next year, leaving his family in poverty.

Sonneborn continued to extend Jollos's work on *Dauermodifikationen* throughout the 1940s and 1950s, and his own tedious and extensive work, supported by a strong institutional base at Indiana University, had great success in bringing various unorthodox genetic phenomena to the attention of American geneticists. He was an outspoken and forthright individual who, by the 1940s and 1950s, possessed a great deal of authority as a brilliant geneticist. As one who worked on unorthodox problems, Sonneborn was also highly reflective about methodological issues in genetics. He was also no stranger to controversy, and he actively participated in public debates. For example, during the late 1940s and 1950s, Sonneborn's ideas of cytoplasmic inheritance were persistently attacked by Beadle, Sturtevant, Muller, and others who defended the predominant, if not exclusive role of nuclear genes in heredity (see Sapp, 1987a).

During the 1950s, Sonneborn frequently and skillfully defended the importance of cytoplasmic inheritance from those who attempted to trivialize its role in inheritance. Defenders of the "nuclear monoply" in heredity frequently referred to the relatively few genetic examples of cytoplasmic inheritance, in order to support their claims of its minor significance in nature. Sonneborn (1951: 199) countered this argument by suggesting that the few examples of cytoplasmic inheritance may only reflect the minimal extent to which geneticists have investigated cytoplasmic inheritance. He then pointed out the theoretical and technical biases of orthodox geneticists' observations and reported results. He claimed that there was no method available for the study of cytoplasmic inheritance comparable to the Mendelian method in simplicity, power, and reliability. This, he argued, resulted in considerable selection in the examples of heredity studied and reported in the literature. The "simpler cases that yield readily to a familiar methodology are preferentially attacked and reported. Complex or less standard cases tend to be put aside or interpreted formally in terms of accepted theory" (Sonneborn, 1951: 199)

Cytoplasmic inheritance and problems of the relations among genes, the cytoplasm, and the environment were indeed complex problems. In this study, it was quite common to find phenomena that were not replicated consistently. Sometimes it seemed that only nuclear genes were involved; at other times, there seemed to be extranuclear influences. The early work on yeast genetics by Sol Spiegelman and Carl Lindegren at Washington University is a case in point. When yeast genetics first emerged into prominence, in competition with the work on *Neurospora* and bacteria, Lindegren found numerous strange non-Mendelian phenomena that were inconsistent, and he reported various results that could not be reproduced. The situation was best described by Spiegelman, when attempting to encourage Boris Ephrussi to work on yeast genetics in Paris:

> I should like to express my fervent hope that your laboratory is prosecuting yeast genetics with some vigor. I say this because since I have returned I have had the opportunity to go over the Lindegren data in the raw, of the past year and a half quite carefully. The whole business is in a pitifully confusing state. What makes it particularly disturbing is that constancy and reproducibility of results is apparently difficult to attain. Crosses which at one time give beautifully clear-cut Mendelian segregations with respect to a given enzyme will go completely haywire when redone at a subsequent time. Other heterozygotes which have consistently yielded non-Mendelian segregation ratios will suddenly and unpredictably clear up and yield only 1 : 1 ratios. It is quite evident that some as yet unknown factor or factors are working to disturb normal segregation of enzymatic characters. (Spiegelman to Ephrussi, November 16, 1946)

It was these kinds of results that, Sonneborn suspected, many geneticists would overlook in their investigations and reports. They would work and report only on those phenomena for which there were readily available techniques and easy solutions. Sonneborn was interested in the forefront of knowledge making, where standard techniques and coherent theory were lacking.

Sonneborn's active participation in public debate was perhaps no more evident than during the celebrated Lysenko affair. During the late 1940s, geneticists in the Soviet Union, led by T. D. Lysenko and his followers, denounced Morganist genetics. They called it abstract, Fascist, racist, and incompatible with Soviet science and dialectical materialism. To Lysenkoists, all parts of the cell interacted together with the environment in the control of hereditary characteristics. From this they developed a somewhat Lamarckian theory of the inheritance of acquired characteristics. In 1948, Lysenkoist views were endorsed by Stalin himself. Those who supported the Mendelian-chromosome theory in the Soviet Union were deprived of their laboratories and sometimes of their lives. Many geneticists in the West denounced Lysenko as a charlatan. The controversy around so-called Communist genetics raged throughout newspapers, popular magazines, and scientific journals throughout the late 1940s and 1950s. Geneticists had become accustomed to reading politicoscientific debates, and many actively engaged in them. Stalin's Russia was complemented by McCarthy's America. Communist "witch hunts" were extended into scientific circles, and those who were sympathetic to some of Lysenko's views were brought to public attention (see Sapp, 1987a: 163–191).

Sonneborn himself became actively involved in the Cold War polemics. Cytoplasmic inheritance was perceived by many to be supportive of Lysenkoist doctrines. Sonneborn's work, in particular, was singled out as being in favor of Lysenkoist views. During the late 1940s, he held the powerful position of president of the Genetics Society of America. He quickly constructed a special committee to counteract Lysenkoist propaganda. Yet, although Sonneborn had no sympathies for official Marxist views, he did believe that some of his work supported the inheritance of acquired characteristics. However, while Lysenkoists in the Soviet Union recklessly disregarded the whole of Neo-Mendelian genetics, Sonneborn did not. Instead, he discerned various mechanisms of inheritance that operated parallel with Mendelian inheritance.

Sonneborn's published response to the Lysenko affair in several respects paralleled his response to the Moewus affair. Many Western geneticists resorted to name-calling to dismiss Lysenkoist views altogether. It was claimed that the inheritance of acquired characteristics was "superstition put forward as politics." Sonneborn, however, adopted an alternative approach. As in the Moewus controversy, Sonneborn wanted the political and scientific dimensions of the debate to be sorted out. He wanted to ensure that the babies were not thrown out with the bath water. In the case of Lysenkoism, the babies were the inheritance of acquired characteristics and extranuclear inheritance.

In his paper entitled "Heredity, Environment, and Politics," published in *Science* (1950: 529), Sonneborn argued that "the political support given to a biological theory and its agreement with a political philosophy may be irrelevant with respect to its scientific validity." First, he attempted to account for the attitude of those geneticists who viewed the inheritance of acquired characteristics as "an outmoded superstition." Whether right or wrong, Sonneborn reasoned that their attitude was at least understandable in view of the record:

> Of the many previous attempts to demonstrate experimentally the inheritance of acquired characteristics, all have failed. In most cases the attempts yielded negative results. When positive results were claimed, the work later proved to be fraudulent, indecisive, or incompletely performed; repetition with unobjectionable methods always failed to establish the claims. No wonder most geneticists consider the matter closed. (Sonneborn, 1950: 529)

He then carefully examined the theoretical premises and experimental foundations of Lysenkoist views. He asked Lysenkoists to provide detailed descriptions of their procedures so that the validity of their results could be tested by others. He point out that his own work on *Paramecium* did provide evidence that supported the inheritance of acquired characteristics. However, he carefully distinguished the methodological and theoretical foundations of his work from that of Lysenkoists.

Advocating unorthodox ideas was a perilous business, and Sonneborn was fully aware of the risks. Since his days as a student under Jennings at Johns Hopkins University, he had been schooled in the dangers of associating himself too closely with belief in the inheritance of acquired characteristics. He knew of the failed attempts to demonstrate the inheritance of acquired characteristics. He also knew of the tragic fates of some of those who had associated themselves with this view. When Sonneborn was a student, the most well-known experiments attempting to prove the inheritance of acquired characteristics were those of Tower at the University of Chicago and those of the Austrian Paul Kammerer. Both had come to be regarded as perpetrators of "fraud." As Sonneborn remarked in his autobiography:

> At my final examination for the Ph.D. degree, Jennings seized the opportunity to alert me at least to one of the dangers of such an association. He asked me who in the 20th century had claimed to obtain positive results on this, and to describe and criticize their work, and to tell what became of them. The latter was the payoff: all had come to a bad end – Tower had gone crazy, Kammerer had committed suicide, etc. etc. At the end of my account, Jennings said, "Let that be a lesson to you" or something to that effect. That was a sobering thought; but it had the effect only of putting

> me on guard, not diverting my attention from the possibility. (Sonneborn, unpublished, 1978: 19)

Indeed, throughout his research career, Sonneborn remained no stranger to unorthodoxy and its corollary – controversy. During the early 1950s, as the controversy over Lysenkoism continued, Sonneborn was already tooling up for another battle, this time to ensure that Moewus's work received proper recognition from American geneticists, to ensure that it was not dismissed on nationalist or political grounds. A major symposium on the nature of the gene and gene action, entitled Genes and Mutations, was organized at Cold Spring Harbor in the summer of 1951. The symposium was organized by Milislav Demerec, who deliberately selected a theme similar to that discussed at the Cold Spring Harbor Symposium of 1941 (see Chapter 4).

In the Foreword to the publication of the symposium papers, Demerec pointed out the "striking progress" that had been made in the previous 10 years. Essentially, there were two major developments. First, the concept of the gene was changing. "Ten years ago," Demerec (1951: v) wrote, "they were visualized as fixed units with precise boundaries, strung along chromosomes like beads on a thread, very stable, and almost immune to external influences. Now, however, they are regarded as much more loosely defined parts of an aggregate, the chromosome, which in itself is a unit which reacts readily to certain changes in the environment." Many more agents besides x-rays and ultraviolet light had been found to induce gene mutations. The second major change Demerec emphasized concerned the organisms used in gene studies. "In 1941, about thirty percent of the Symposium papers reported research carried out with *Drosophila*, and only six percent of the papers dealt with microorganisms; whereas this year only nine percent of the papers relate to *Drosophila*, and about seventy percent to microorganisms" (Demerec, 1951: v). The Cold Spring Harbor meeting of 1951 heralded a major revolution in gene studies. Many of those who pioneered the development of molecular biology were present.

Sonneborn had been asked to give the final general lecture. In order to rectify what he considered to be Moewus's neglect in America, Sonneborn decided to devote about half his time to expounding the genetic contributions of Moewus. On April 25, 1951, he described the circumstances in a letter to Moewus:

> I am now nearing the completion of a series of 30 lectures on Sexuality and Genetics in Algae in which of course an account of the researches conducted by you, Kuhn and your collaborators has played the most

prominent part. My students all share my admiration and enthusiasm for your magnificent accomplishments. Your results give new insights into many fundamental problems and raise a host of fascinating questions. I wish, as I have so often in the past, that we could meet and discuss these matters.

It is my plan eventually to use these lectures as part of a book on the genetics of microorganisms. But my entire manuscript will probably not be ready for another two years. Meanwhile, I feel obliged to call the attention of my American and British colleagues to your exceedingly important discoveries. As you must know, the old criticism of Philip and Haldane and Sturtevant's dictum have resulted in a most unfortunate neglect of your work by English-speaking geneticists. Perhaps my two reviews of 1941 did not help matters; but I tried to present an objective picture including both pros and cons. My current study of your work, however, has convinced me more than ever of the preeminent importance of your work, and I feel that it is necessary – not for you, but for all of us – that your work receive the study it deserves in my country.

I am planning to discuss your work under conditions that cannot fail to attract attention. On June 7–15 there is to be a Symposium on some current problems in genetics at Cold Spring Harbor and I have been asked to give the final general lecture. This seems to me to be a fine opportunity to present, among other things, my views on the significance of your findings. I am also giving a series of lectures in Chile in July and August and plan to devote some of those lectures to the same subject. (T. M. Sonneborn to Franz Moewus, April 25, 1951)

Moewus quickly wrote back and endorsed Sonneborn's proposal to defend his work. He also addressed the question of Philip and Haldane's critique and the lack of impact of his rebuttal. He pointed to the shortage of paper in Germany and the lack of international correspondence during the war, and alluded to language problems in understanding his papers. At the same time, he mentioned that he would be leaving Germany, and algae, to study growth hormones in Australia:

I know that Haldane's criticism has done much damage. Regretfully the rectification appeared at a time when no exchange of offprints was possible (1943), and when we here because of the paper shortage were also hardly able to obtain offprints. I find that lamentable, but it cannot be changed.

I am pleased to hear that you are making "propaganda" for my papers. I would very much like to hear, after the symposium in Cold Spring Harbor, what criticisms are made so that I can take up [a] position to [respond to] them.

Last winter I did a series of mutation experiments and am now in the final stage of the genetics tests. I then want to write up the paper and want to attempt it in the English Language this time so that the understanding will be easier. Since this year I will very probably move into the English-speaking area (Australia), this would be good practice for me. Although I will not be able to continue work on my algae there, I will be able to turn to my second area of work, that of growth – and inhibitory – substances, hopefully under better conditions, with greater intensity than is possible here with Prof. R. Kuhn. Prof. Kuhn is just now in Philadelphia for 6–8 weeks, where he will be giving lectures and where he may decide to move.

It is regrettable that we are not able to meet for discussion with each other. Out of just such discussions each participant learns more than what is possible to glean from reading papers. Let us hope that we will still meet, for I admire your papers extraordinarily. They demonstrate to everyone how substantial progress can be achieved systematically. (Moewus to Sonneborn, May 20, 1951)

Despite Moewus's enthusiasm for the "propaganda" Sonneborn would make for his papers, Sonneborn was not optimistic about the response he was going to receive. To him, the attitude of those biologists who ignored or discredited Moewus's work – such as A. H. Sturtevant and Norman Horowitz along with Beadle, at the California Institute of Technology – was due to irrational issues. Sonneborn knew that the rumors about Moewus's failures to send stocks were being used as assaults upon the validity of Moewus's work. He also recognized that the breakup of Kuhn and Moewus might also be judged as a sign of Moewus's lack of credibility. Sonneborn wrote back to Moewus, warning him to take precautions. The reasons Moewus had for wanting to leave Germany will be discussed later. It is enough to know here that to Sonneborn, Moewus's plan to abandon algal research threatened to be a major catastrophe to biology. He calculated that Moewus must have had hundreds of different cultures, and he wanted these stocks to be preserved and the work on them continued. Again he tried to dissuade Moewus from bringing his *Chlamydomonas* work to a close:

Your good letter of May 20 just arrived. Thank you very much for the reprints. I shall put them to good use. Of course I expected that you would not have available all I requested and am surprised you have as many as you say you sent me.

I shall write to you after the Cold Spring Harbor Symposium and tell you what happened. I am *sure* that I will find, not suggestions or reasonable criticisms, but closed minds and prejudice. Do you know a German by the name of Lang who is now visiting the California Institute of Technology? I hear he recently reported some of your work at a seminar there and was greeted with intense skepticism. If you know him, you might write him and get more details. As you know, Sturtevant and Horowitz are bitterly opposed to you and will not consider seriously anything you do. Beadle, who is a much fairer person, has been greatly influenced by their judgments.

If I could find the time to do so, I should like to publish a critical review of your work since 1940. In going over this for my lectures during the present semester, it seemed to me that your results were of great significance in relation to a number of problems not discussed in your papers and I should like an opportunity to point that out. Also, a number of questions and criticism arose in connection with your results. I have not the time now to go into all this, but if I ever organize my analysis perhaps I can do this while in Chile this summer – I will send it to you.

I am greatly distressed to hear that the partnership between you and Kuhn is about to dissolve and that your work on the algae may come to an end. This seems to me like a major catastrophe to biology. What will become of your invaluable collection of stocks and mutants? If these are lost and not turned over to the several people who are intensely interested in repeating and extending your work, it will look like scientific murder. I do hope you will make sure that Smith at Stanford, Lewin in England, and the young man (whose name I cannot remember) at Vanderbilt College in Nashville, Tennessee, get the stocks. If necessary, I should be glad to try to keep your stocks alive here until they are put into the hands of people who can use them. Indeed my students and I might wish to undertake some work on them ourselves. One of the often repeated criticisms about you, circulated by Lwoff, has been your failure to provide him with the stocks he wished. It would seem to me most important for the good of your name and work that you make sure your important materials are preserved and made available to those who want them.

The separation and possible migrations of you and Kuhn will also doubtless give rise to much speculation and gossip. I hope both of you take all possible opportunities to make the matter clear.

If you go to Australia, will you take the eastern or western route? If you go west, I greatly hope you will plan to visit me. I should consider it a great privilege and opportunity to have you here for a week or more. You did not say when you would be leaving. I shall be in South America July 5 to August 15, but will be here for practically all the rest of the year. I think it would be worth your while to visit a number of laboratories in this country on your way if you can do so.

I hope you will understand that the unrestrained comments I have made above are made because I feel it is of the utmost importance that you and your work be correctly understood and appreciated. (T. M. Sonneborn to F. Moewus, May 30, 1951)

In his paper delivered at Cold Spring Harbor, entitled "Some Current Problems of Genetics in the Light of Investigations on *Chlamydomonas* and *Paramecium*" (1951), Sonneborn attempted to evaluate Moewus's position in the history of the development of several domains of major interest – the genetics of microorganisms, biochemical genetics, sexuality, and a number of special but theoretically important genetic problems. Sonneborn (see Figure 7) began his paper with a discussion of the significance of Moewus's work for understanding cytoplasmic heredity,

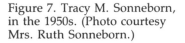

Figure 7. Tracy M. Sonneborn, in the 1950s. (Photo courtesy Mrs. Ruth Sonneborn.)

nucleocytoplasmic relations, and protein synthesis, and allied it to his work on *Paramecium* and to that of Ephrussi on yeast. He then tried to show how Moewus's work deserved priority recognition in the new domain of microbial genetics and how suggestive it was for future investigations on other microorganisms.

Sonneborn saw no major conflict between the important discoveries he saw in Moewus and Kuhn's work and the statistical, biochemical, and genetic criticisms launched against it. In his opinion, the biochemical points at issue were, for the most part, "questions of interpretation and were irrelevant to an estimation of the reliability of the observations." Moreover, by 1950, Moewus and Kuhn had found and corrected some of their mistakes themselves. "Their picture of the biochemical genetics of sexuality in *Chlamydomonas* in 1950," he argued, "is very different from the picture they presented in 1940. The changes, resulting from their discoveries, serve in my opinion to strengthen rather than weaken confidence in the work" (Sonneborn, 1951: 497).

Moreover, Sonneborn considered the work to be too elaborate to be fraudulent. To him, who knew Moewus's work perhaps better than anyone outside of Germany, it seemed inconceivable that anyone could stage such an ambitious fraud. The chemicals controlled by genes, which were said to be in control of sexuality, had been reportedly isolated from the organism. The properties of the enzyme systems involved had been described in elaborate detail. Furthermore, this work was certified by a Nobel Prize winner in biochemistry, Kuhn, who together with his students did most of the biochemical work. To Sonneborn, the biochemical analysis seemed to verify the genetic work to some extent. Sonneborn found it difficult to imagine "how a biochemist could extract enzyme systems and active substances that fitted into a sensible biochemical *and* genetic pattern from cultures of precisely the expected genotype and not from others, unless the genetic analysis was essentially as correct as the biochemical analysis." The complex, detailed, and integrated biochemical and genetic work done by Moewus and Kuhn led Sonneborn to argue that it was either a "gigantic hoax" or it was "in most respects sound" (Sonneborn, 1951: 497).

After carefully and systematically reviewing the work of Moewus and Kuhn, and the criticisms launched against it, Sonneborn claimed that one of the 1940 papers written by Moewus represented the first one in which the biochemical activities controlled by the genes were discovered. He further argued that there was not sufficient ground for rejecting or ignoring Moewus's work and called for an evaluation by independent repetition by others, not by what he saw as a "campaign of calumny and gossip":

> While the work still includes certain difficulties (which I plan to point out in detail elsewhere), I cannot see sufficient ground for rejecting or ignoring it. The great need, as all critics agree, is for independent repetition by others of some key experiments using the same strains. Thus far no such repetitions have appeared. Now, however, a few of Moewus' strains have been sent to other laboratories and attempts to carry out independent repetitions of certain phases of the work should be possible. It is exceedingly important that this be done. Meanwhile, however, the remarkable suggestive and instructive results set forth by the Heidelberg team should hardly be treated as if they did not exist. (Sonneborn, 1951: 498)

In short, Sonneborn was of the opinion that Moewus had made great contributions, regardless of whether his critics were correct about the errors in his results. In his view, the foundations of modern genetics of microorganisms were laid when he and Moewus and a few others first learned to domesticate microorganisms for genetic purposes:

> An objective evaluation of the work of Moewus must first of all recognize his outstanding contributions which are beyond dispute. Moewus was one of a small, scattered and at that time unappreciated, group of investigators, largely isolated from the main current of genetic study, who laid

the foundations of the modern genetics of microorganisms during the decade beginning about 1927. (Sonneborn, 1951: 496)

Moewus's subsequent genetic studies, he argued, made *Chlamydomonas* one of the "genetically best known materials," and Moewus led the way again "by initiating the first extensive systematic investigation of biochemical genetics in any microorganism. That he fully appreciated the importance of this type of study is evident from many statements in the four remarkable papers." Thus, Sonneborn argued: "For having twice pioneered in two of the most important developments in modern genetics, Moewus clearly merits high recognition, regardless of any criticism that might be made against his data" (Sonneborn, 1951: 496–497).

Sonneborn's attempt to give credit to Moewus's work as major discoveries and to generate interest in the controversy surrounding his published results met with a mixed reception. On the one hand, he was severely criticized by a large group of American workers for taking Moewus seriously. There were many bitter antagonists among American geneticists. They included the influentials at the California Institute of Technology, A. H. Sturtevant and Norman Horowitz (Figure 8). Horowitz (letter to the author, August 4, 1986) recalls that he found Sonneborn's attitude toward Moewus difficult to understand:

> I was present at Cold Spring Harbor in 1951, when he gave a brilliant defense of Moewus in the face of overwhelming odds. I thought at the time that Sonneborn perhaps felt sympathy for Moewus because Sonneborn, too, worked with an organism that presented many unusual features that set it apart from the species commonly used in genetic research. The difference, of course, was that there never was any question of Sonneborn's integrity.

But, could the attitude toward Moewus by Beadle's group have been due to competition for recognition and prestige? Horowitz (letter to the author, August 4, 1986) recalls:

> There was certainly no competitive feeling in the Beadle group toward Moewus. Partly, this was because none of us took him seriously. In addition however, I think most of us felt that we were so far ahead in the search for an understanding of gene action that there really was no competition.

On the other hand, according to Sonneborn, the European geneticists tended to agree with him as to the stimulating and important nature of Moewus's contributions. This was also true of some of the younger American geneticists such as Joshua Lederberg (Figure 8), who had

Figure 8. Left to right: N. H. Horowitz, K. C. Atwood, and J. Lederberg; Cold Spring Harbor, 1951. (From *Genes and Mutations, Cold Spring Harbor Symposia on Quantitative Biology 11:* vii, 1951.)

overlooked the significance of Moewus's work prior to the Cold Spring Harbor Symposium (Lederberg interview, May 26, 1986).

Despite the opposition he received, Sonneborn continued to hold firm in his belief that most of what Moewus had reported was sound, though he was convinced that Moewus did polish off his data to some extent. However, as emphasized in Chapter 4, this was not uncommon in genetics research. There is a tendency in genetics for individuals to neglect results that do not fit an analytic hypothesis and report only on those that are consistent with it. Sonneborn knew of these difficulties; he had a very sophisticated understanding of the biases influencing genetic experiments and geneticists' reports. As discussed earlier, many of his reflections of genetic procedures stemmed from his interest in unorthodox ideas of extranuclear inheritance and the inheritance of acquired characteristics.

These kinds of biases in experimental procedures and reports are more noticeable at revolutionary times in science when strikingly new technical procedures and concepts are being negotiated. But they are ubiquitous during "normal" periods of scientific practice as well, even though they are rarely discussed. To a large extent, it is a matter of personal judgment, a skill that is acquired with experience and social interaction in the domain, to know precisely how much and what sort of data to include when making convincing arguments based on observations that are to be transcribed into the tables, charts, and diagrams of short articles. Moewus could not be excluded for biased practices of this universal nature – regardless of whether they were conscious or unconscious.

To Sonneborn, it seemed that Moewus was somewhat inexperienced in these matters; he had made some quantitative and judgmental errors as well as misinterpretations when writing up his work – as might be expected of someone who was pioneering a new domain and isolated from the mainstream of genetics. As mentioned earlier, Sonneborn also believed that Moewus had been "dishonest" in some of his reporting of data. Nonetheless, he did not believe Moewus's "dishonesty" or biases in reporting even remotely approached that of others who refused to acknowledge the significance of Moewus's work. Sonneborn saw no evidence of major deception in Moewus's work and took a great personal risk in defending it.

Upon returning from the Cold Spring Harbor meeting, Sonneborn (June 18, 1951) wrote to Moewus, describing the reactions to his paper. He pointed out the risk he had taken in defending Moewus and warned him of some major criticisms that would soon appear in publication. At the same time, he coached Moewus in terms of what he perceived to be his "mistakes":

> I returned today from the Cold Spring Harbor meeting and found a packet of 19 reprints from you. I am very glad to have these and wish to thank both you and Professor Kuhn for sending them. I hope you will continue to send me the newer papers as they become available. . . .
>
> You asked for a report on the reactions to your work at the Cold Spring Harbor meeting. I devoted about half my paper to certain aspects of your work and was severely criticized by a large group of American workers for taking your work seriously. On the other hand, the European geneticists present, especially Gustafson and Levan of Sweden and Hadorn of Switzerland, were much more sympathetic; and some of the younger American workers also agreed with me as to the stimulating and important nature of your contributions.
>
> The whole situation is extremely unfortunate. There were many stories about how you had refused to send cultures or had agreed to do so and then failed to see the matter through. Now that Smith of Stanford and Nyborn of Gustafson's laboratory have some of your cultures, this situation is of course improving. But you have many bitter antagonists: Lewin of Yale, Provasoli of New York, Horowitz of Pasadena, and a number of others. The general attitude of these people – and of others who are now working on *Chlamydomonas* – is that they are going to ignore your work completely. This seems to me an indefensible point of view, and I protested against it repeatedly in conversation and in my formal paper. I

pointed out especially the great credit due to you for initiating the bio-chemical genetics of microorganisms and for visualizing in one of your 1940 papers the immense possibilities in the genetic analysis of microorganisms. I also selected from your various papers four or five points which bore on major points under discussion at the conference and attempted to show how they deserved priority recognition and how suggestive they were for future investigations on other organisms. I am sure that I shall suffer for coming out so strongly in your behalf, but it seemed to me that someone in this country was obliged to do so and, since I know your papers so thoroughly, I felt obliged to take the risk myself.

With all my admiration for the importance of your work, I must never-theless say in frankness that your *numerical* results still bother me and that I do not have confidence in them. One of the most troublesome observa-tions you ever reported was the conversion by visible light of 1% *cis*- to *trans*-dimethyl crocetin per minute. This was brought up again by Horo-witz and Lederberg at the Cold Spring Harbor meeting and will be at-tacked by Raper, I am told, in a review on Sexual Hormones in Microor-ganisms. Have you ever repeated and checked this? It is being used as one of the most deadly arguments against the credibility of your work.

I have also found in your data quite a few other numerical difficulties, even in the papers since 1940. There are clearly errors within the tables you published. One also finds difficulties in evaluating the data on in-duced mutations. It is quite important to know precisely how much sur-vival and how much death occurred among the isolates used in the various mutation studies. I refer both to the 1940 papers and to the later work on induced mutations in the rutin experiment.

After going over your papers this year I strongly suspected that you have misinterpreted the action of the *mot* gene and probably also the *F* and *M* genes. I have an alternative interpretation of the action of *F* and *M* which I will send you some day when I write it out fully. I referred to it in my Cold Spring Harbor paper.

You see I feel very free to send you these criticisms because you must know now, after my public statement of support of your work, that in spite of the difficulties I do appreciate what you have done. To reassure you further, I may say that I have engaged the help of a very prominent English biologist to try to get the Rockefeller Foundation and the Max Planck Institute to do what they can to make sure that you get adequate support for continuing your work on the algae. Whether this will come to anything or not I do not know, but I should hate to see you give this up and go to Australia.

9. Big biology: The politics of the Rockefeller Foundation

> This is mainly to remind you to see what you can do with Weaver and the Max Plank Institut about Franz Moewus of Heidelberg. As I told you, he has in my opinion been the outstanding leader in the development of biochemical genetics and in spite of criticisms on the quantitative aspects of his work, his discoveries seem to me to be among the most important in modern genetics. (T. M. Sonneborn to Julian Huxley, June 5, 1951)

In the summer of 1951, the collaboration between Moewus and Kuhn had finally dissolved. Moewus had only a minimal teaching position at the University of Heidelberg and had no official status at the Kaiser Wilhelm Institute (which changed its name after the war to the Max Planck Institute). The living conditions were far from favorable in war-torn Germany, and funding for science in Germany was in short supply. Many scientists attempted to find jobs elsewhere, preferably in the United States, where science was highly supported and various new specialties were more advanced. Kuhn was visiting the University of Pennsylvania and was exploring job possibilities there. Moewus was planning to embark on a journey to Australia. Moewus saw little immediate future in Germany in heading his own institute and training students. His work remained in doubt, and the controversy surrounding it was well known at the University of Heidelberg. In fact, because of the doubts about his work, Moewus was refused promotion to the title of professor.

Professor Seybold, head of the Botanical Institute, kept the criticisms about Moewus's results fresh in the minds of the faculty at the University of Heidelberg. Seybold repeatedly wrote to Moewus, advising him to send his stocks to well-known scientists. Seybold had serious doubts about some physiological results reported by Moewus. As he wrote Moewus (May 20, 1949), "I will not pretend to pass judgment on all of your scientific work. But from the beginning it was rather uncertain that one molecule of crocetin will activate one cell. These doubts I have told you again and again and discussed with you in 1938 or 1939." He wrote to Moewus again the following month, emphasizing the need to have his cultures sent to colleagues. "As long as this is not done," Seybold wrote Moewus, "all the doubts about the correctness of your results will not stop" (Seybold to Moewus, June 21, 1949). On February 3, 1951, Dean Ludwig wrote Moewus, pointing out that in the faculty meeting of the previous month, Moewus's promotion was again reconsidered. Ludwig stated that Moewus did not respond to Seybold. Ludwig wrote to Moewus, "Only you can answer all the doubts. As long as this is not done, the faculty will not ask for your professorship."

Moewus (February 9, 1951) responded immediately to Ludwig's letter. He stated that he had, in fact, responded to the questions posed by Seybold, and he sent a copy of the letter in question. Moreover, Moewus emphasized that the situation had changed, because his work on sexuality and genetics was taken up by many laboratories in the United States. He pointed out that he had given a report at the International Microbiology Congress in Stockholm in 1951. He also added that he had sent cultures to several people, including G. M. Smith at Stanford, Harold Bold at Vanderbilt University, Pringsheim at Cambridge, and two others in Germany who told him that "*Chlamydomonas* was excellent for class work." Moewus then requested permission for special leave from his teaching duties to go to Sydney, Australia for one year beginning September 20, 1951. He was granted leave until the summer semester of 1953.

Although the faculty at Heidelberg were willing to let Moewus leave for Australia, to Sonneborn, Moewus's plan to abandon his work on algae and go to Australia threatened biology with a major tragedy. It seemed to Sonneborn that Moewus must have had hundreds of mutant stocks, and he wanted those stocks to be saved and for Moewus to continue to work on them. Sonneborn spent the period between July 3 and August 17 in South America as visiting professor in the Department of Parasitology, University of Chile. But before leaving, he attempted to secure assistance from Warren Weaver, director of the Natural Sciences Division of the Rockefeller Foundation to try to help save the situation.

Approaching the officials of the Rockefeller Foundation for help was a good idea. During the 1930s and 1940s, the Rockefeller Foundation had greatly fostered the development of microbial genetics and the integration of biochemical and physical approaches to genetics. Founded in 1913, and reorganized in 1928, the Rockefeller Foundation, more specifically the Natural Sciences Division, developed research grants for individuals and projects, and mastered the art of conducting a large program of relatively modest grants. As Kohler (1978: 481) has pointed out, the organization and style of the program of the Rockefeller Foundation served as a model for establishing the mode of operation of federal science agencies such as the National Science Foundation in the United States after World War II. Beginning in 1932, Warren Weaver was the director of the Natural Sciences Division. His principal aim was the development of "molecular biology." In fact, it was Weaver himself who coined the expression "molecular biology" in 1938 (see Kohler, 1976). Weaver himself was a former physicist, and by the early 1930s, genetics seemed ripe for applying the techniques of physics and chemistry.

The Rockefeller Foundation's support for biochemical genetics would prove to be one of Weaver's most successful initiatives in the incubation of the field known today as molecular biology. The Rockefeller Foundation played an instrumental role in building up the Biology Department at Indiana University during the 1940s. By the early 1950s, this department developed from obscurity into one of the primary genetic centers in the world, with such leading figures as Sonneborn, H. J. Muller, Ralph Cleland, Marcus Rhoades, and Salvador Luria. The Rockefeller Foundation also gave substantial financial support to Stanford University, where Beadle worked. Against this background of broad general support for departments, Weaver also gave special grants to outstanding individual teams when the opportunity arose. In fact, Weaver claimed great responsibility for developing the team led by George Beadle and Edward Tatum. In 1941, when Beadle and Tatum reported their first study on the biochemical genetics of *Neuropora*, Warren Weaver wrote:

> A brilliant young geneticist, namely Beadle, has turned up one of the most important and exciting leads which has developed for a long time in the ultra-modern field lying between genetics and biochemistry. The discovery has resulted, in major part, from circumstances which we helped to create; and also in major part from the fact that we put with Beadle several years ago a very able young biochemist whom we had specifically trained for this sort of work.
>
> The question of major help in connection with Beadle's discovery will not arise very soon, since we have already made a grant-in-aid which will enable him to explore the situation during the course of the next year and will enable him to learn just how big and promising the discovery really is. (Weaver to R.B.F., December 24, 1941)

The Rockefeller support for Beadle and his group was significant in allowing them to pursue fundamental problems of gene action during the war. The foundation had great appeal to biologists of the 1940s, insomuch as it had a degree of flexibility in making new awards quickly and permitted work on fundamental problems to a degree that was equaled by few funding agencies. As Kohler (1976: 296) argued, "a decline in the invocation of utility" accompanied the Weaver program of molecular biology. The case of Beadle and his work is illustrative of the attraction of the Rockefeller Foundation grants program from the point of view of the scientist. It will be recalled that the *Neurospora* technology promised to be important for determining how the gene works, but it could also be used for quantitative assays of known vitamins and amino acids and had a potential for discovery of new vitamins and new amino acids. To continue work on these promising leads in an adequate fashion seemed to require more funds than Stanford University was willing to provide. This raised the issue of funding from the drug industry, the establishment of patents, and so on. Almost immediately, Beadle was solicited by Merck and Company, who offered to fund the entire project in return for patent rights (see F. B. Hansen, diary, 1941).

Beadle had no interest in patents or personal financial profit. He was interested in fundamental theoretical problems of the nature of the gene and gene action. His first preference was a grant from the Rockefeller Foundation, which would free him of all obligations other than to work hard and publish freely his results. The Rockefeller Foundation gave the grant without delay. This money, and further support from the Rockefeller Foundation, enabled Beadle to invite Bonner, Horowitz, and Mitchell to his group at Stanford and move on to the California Institute of Technology (see Beadle, 1966: 29–30). There had been great difficulties at Stanford University under the troubled leadership of the president of the university. Beadle and his group felt forced to leave, and the Rockefeller Foundation provided necessary support.

When Beadle and his group moved to the California Institute of Technology in 1945, the interests of Beadle and Weaver began to find a solid institutional basis in which to develop. Beadle's biochemical genetics school became part of a triumvirate. It cooperated with the research program led by Linus Pauling, which focused on the structure of large macromolecules such as proteins and nucleic acids, and with the "informationist" school led by Max Delbrück, which concerned itself with bacteria and its viruses. These groups collectively were dedicated to attacking what they considered to be the great fundamental problems of biology and medicine: the structure and nature of proteins, and of nucleic acids; the structure of the gene; the mechanisms of inheritance, cell division, and growth; the structural basis for the physiological activity of chemical substances; and the structure and properties of antibodies, enzymes, viruses, and bacteria (see Beadle and Pauling, unpublished outline of their program, 1946).

The influence of the Rockefeller Foundation was not limited to the United States, however. It also extended to European science, especially that carried out in France and Germany. Before and after World War II, Weaver's division played a leading role in supporting the development of biochemical genetics in France, at the Rothschild Institute for Physico-chemical Biology led by Boris Ephrussi (see Sapp, 1987a). For the Rockefeller Foundation to help support Moewus's work on biochemical genetics, or to provide conditions for its full evaluation, in principle, seemed only logical. In fact, before the war, the Rockefeller Foundation had given considerable fellowship support to Kuhn's institute of biochemistry at the Kaiser Wilhelm Institute for Medical Research in Heidelberg. In 1949, one of Weaver's colleagues, Gerard Pomerat, who was in charge of the foundation's European activities, had visited Kuhn's institute to assess laboratory needs and the possibilities of resuming fellowship activity.

The Rockefeller Foundation officials kept detailed, almost standardized diaries of their conversations and experiences with leading scientists; these diaries frequently also included personality profiles of past recipients and possible recipients. Many commentators on fraud in modern science have been critical of the breakdown of the masters–apprentice relationship with the development of huge research laboratories. They have argued that research directors tend to be less involved in, and less scrutinizing of the research programs they supervise (see, e.g., Broad and Wade, 1982). It is appropriate, then, to take a closer look at the institutional setting where Moewus conducted most of his work. Pomerat's report (diary, May 20, 1949) provides a good impression of how he viewed the work being done there. It also provides a glimpse at the huge laboratory factory where Moewus worked with Kuhn. In his diary, Pomerat wrote that

> Kuhn's main field of research interest centers about problems of sexuality and fertilization in plants, animals and unicellular organisms, including cross-fertilization, self-sterility, etc., all from a genetical and biochemical standpoint. [They] have isolated crystalline substances from both male and female lines which seem essential for fertilization – probably affect action of genes on certain enzyme systems.

Pomerat did not mention anything about the controversy surrounding the validity of these reports or the authenticity of the work. He alluded to other work in "chemical genetics" concerning the study of alkaloid glucosides. Some glucosides had been obtained from potatoes which, when injected into certain beetle larvae, inhibited their development; and others were held to be effective against tomato wilt. He then mentioned Moewus, who helped with this work besides doing his own research on *Chlamydomonas:*

> K.'s geneticist, *Dr. Moewus* (met only briefly because giving lecture), helps with this but also works on problems of crossing-over and sex determinism in the one-celled organism *Chlamydomonas.* [He] developed this when working with Professor Hartmann. [He] speaks some English; looks like good chap. [I] might speak about fellowship next visit. K. says his chem. genetics team collaborates with Professor Rudorf at Voldagen near Hannover. . . . K. has another team collaborating closely with *Hartmann* at Tubingen, and is starting another joint program with *Ludwig* here.

In addition to this work, Pomerat noted that Kuhn had another unit conducting studies "on bacteriostatic substances, on the biochemistry of TBC [tubercle bacillus], and the problem of redox indicators." The latter could quickly determine the germination capacity of plant seeds and could also be used to indicate liver damage caused by various substances. There was still a third unit studying mushroom poisons and isolating some of them in crystalline form. In addition, there were two independent researchers: One was a chemist; the other, a physicist, was studying the effects of ultrasonics on cancer. She had found that certain enzymes that were bound to certain substances in cells could be set free with sonication. This work promised to provide a useful tool in enzymology. Pomerat's description of the working conditions and diversity of research gives a good indication of the extensive and diverse research operation that Kuhn headed, his responsibilities, and the conditions of the institute after the war:

> K.'s only official post is that of Director of this K.W.I. here, but because he was once a professor he holds an honorary professorship at the University of Heidelberg. Doesn't give regular lectures, but runs a University seminar on problems of natural science in medicine. Has four dozenten under his direction, who give lectures also in several University departments, plus eight assistants and some lesser technical aids. In addition supervises work of nine doctorands. . . . We visit most of the labs which are rather well equipped. In fact, K. says his most serious need is for money with which to buy pure chemicals; [they] would have to be obtained abroad in most cases because German manufacturers have not yet felt market good enough to warrant resumption of production. What a strange situation in the former Mecca of [the] chemical industry!
> . . . K.'s own primary needs are for chemicals, but he says next [he] would place optical equipment, centrifuge, monochrometers, etc. Some could not be purchased in Germany; e.g., Beckmann, apparatus for paper chromatography, etc. [They] could do a lot of repair and refurbishing of

some equipment they now have if they had money to repair it. Lastly, [they] need money to fill up blanks in journal and book collections.

The professors at the Kaiser Wilhelm Institutes complained about their appalling financial situation. Some claimed that if they did not go into debt continuously, they would have to close their institutes. But Pomerat had reservations about placing Kuhn on the first list of men to be helped by the Natural Sciences (NS) Division of the Rockefeller Foundation. Essentially there were two difficulties. First, if Kuhn or any other candidates were offered a fellowship to go overseas, they could not give a guarantee that they would return. The second problem concerned Kuhn's relations with the Nazi regime. Pomerat retained "some doubts as to whether K. is politically clear and clean enough to warrant placing him among the first group of men to be helped by NS. Outside opinion is divided." Pomerat concluded his report with the following comments: "Kuhn looks somewhat weary and perhaps a little 'beaten' – at least his eyes are always sad, his speech is soft and slow – rather tired or old. Nevertheless, he is clear in his discussion, although neither aggressive nor very forceful" (Pomerat, diary, May 20, 1949).

In 1953, Pomerat made another visit to Germany to inspect the laboratories of the new Max Planck Institute for Biology. After the war, the Kaiser Wilhelm Institute for Biology was renamed the Max Planck Institute and had moved from Berlin to the medieval village of Tübingen in southern Germany. The laboratories of the new institutes rivaled anything in the United States in terms of facilities and equipment. Pomerat visited the laboratories of Georg Melchers, which were concerned with investigations of plant physiology, genetics, and plant virus investigations. Pomerat (December 1, 1953) remarked on the splendid new buildings and well-equipped laboratories. He also visited the laboratory of Max Hartmann and admired the new lavish air-conditioned rooms, elaborate culturing conditions, and friendly atmosphere in the laboratory:

> The two or three largest rooms are completely air-conditioned so that plant and animal cultures can be continually maintained (and worked upon) at 19°C. There are marvelous home-made little cabinets to house hundreds of rows of small culture jars of marine algae and marine worms under controlled lighting conditions, and everywhere one has the feeling that these living organisms must be "quite happy" here. (Pomerat, diary, December 2, 1953)

Pomerat took a moment to describe Hartmann, his relationship to his assistants, and the breadth of his research:

Age 77 and at least as late as last year still a mountain climber, Professor Hartmann remains young and vigorous in character. He is interested in and keen about his younger people and their response is equal to it in warmth and in spontaneity. He does not speak or understand English, but he understands French and his German is readily geared down to G.R.P.'s [Pomerat's] ear so there is a reasonable chance to get acquainted and to savour the flavor of the place before the young people come in to complicate the reception with English! Professor H. continues to be active in research over a wide area of subjects and as evidence of his recent work gives G.R.P. reprints of studies on fertilization and sexuality in marine algae, on the life cycle and mating phenomena in a microscopic marine alga, [on] polyploid nuclei in Protozoa, and on the races of lizards which occur in the Balearic Islands, etc. – all good, long, well-illustrated papers! (Pomerat, diary, December 2, 1953)

There was no mention of Moewus and the controversy around his work. If Moewus had once been the "blue-eyed boy" of Hartmann's laboratory, he now had been replaced by Karl Grell, who was Hartmann's chief assistant. Grell was recognized as one of the few good cytologists of Protozoa and Protista in Germany. At the recommendation of Sonneborn, the Rockefeller Foundation planned to offer Grell a fellowship to visit the laboratories of Sonneborn and others working on Protozoa in the United States. As will be discussed in Chapter 11, Grell's visit to the United States would lead him into direct confrontation with Moewus.

Pomerat was extremely impressed with what he observed at the institutes where Moewus had worked. As will be discussed later, others criticized Moewus for jumping from one problem to the next; they claimed that he had hid behind the diverse nature of his work. Some would later censure Kuhn and Hartmann for failing to be critical of Moewus's work and maintaining poor supervision over it. Pomerat, on the other hand, only celebrated the extensive nature of the work he found in these laboratories. The only institutional critique he made concerned the relationship between the research institutes and the universities, between research and teaching. Indeed, the Rockefeller Foundation officials made similar criticisms of the research conducted outside the universities in France as well (see Sapp, 1987a: 181). With regard to the Max Planck Institutes, Pomerat (September 22, 1953) made the following remarks:

From the point of view of pure research in Germany the Institutes are probably unique. On the other hand, the criticism is frequently made, and I think justifiably made, that the Max Planck Institutes are robbing the universities of their best men, are taking out of the universities first-class research men who might also have played a significant role in the inspira-

tion and guidance of graduate students, and are capturing a dispropor-
tionately large fraction of funds that might be available for the support of
research within the university system.

The Rockefeller Foundation officials had no idea about the heated controversy surrounding the significance and reliability of the biochemical
genetic results that Kuhn published with Moewus until Sonneborn
brought the matter to their attention in 1951. They also had no idea that
Moewus was considered by some to be the leader of this new domain
and that he was planning to abandon *Chlamydomonas* and move to Australia. Sonneborn encouraged his influential friend, Julian Huxley, then
director of UNESCO, to bring the matter up with Weaver to see if he
could help. Huxley was visting Sonneborn and his wife, Ruth, in
Bloomington and was meeting with Weaver in New York to discuss
other business. Sonneborn (June 5, 1951) wrote to Huxley in New York:

> Dear Julian,
>
> We all miss you greatly. Ruth is lost without her woodland companion!
> This is mainly to remind you to see what you can do with Weaver and
> the Max Planck Institut about Franz Moewus of Heidelberg. As I told you
> he has in my opinion been the outstanding leader in the development of
> biochemical genetics and in spite of criticisms on the quantitative aspects
> of his work, his discoveries seem to me to be among the most important
> in modern genetics. It would seem to me a great tragedy for science to
> have this important work come to an end by Moewus' removal to Austra-
> lia. His numerous mutant stocks of *Chlamydomonas* are in themselves of
> countless value and *must* not be allowed to disappear from use. Anything
> you can do to save the situation will be a real service to science.

Huxley's well-known book of 1942, entitled *Evolution: The Modern Synthesis,* is a masterly synthetic work. Huxley had broad interests and tal-
ents to match. However, his memory was notoriously poor. Although
he mentioned Moewus to Warren Weaver, he had forgotten what the
difficulty was about. Weaver was well aware of Sonneborn's work and
recognized him to be one of the major authorities in the genetics of microorganisms. In fact, the Rockefeller Foundation officials often consulted Sonneborn about the career plans of researchers in his field.
Weaver took an immediate interest and wrote to Sonneborn promptly
(June 15, 1951):

> Julian Huxley spoke to me the other day about your interest in Franz
> Moewus of Heidelberg; but Julian really did not remember what the diffi-
> culty is.
> I can, of course, have no idea at this juncture as to whether or not we
> could be helpful, but I would at least like to know what the dilemma is,
> how it has come about, and what you think could and should be done.

Sonneborn (June 19, 1951) wrote back immediately, describing how he viewed the situation and what might be done:

> About Franz Moewus of Heidelberg, the principal points are as follows.
> Moewus was a student of both Kniep and of Max Hartmann in Berlin in the late 1920's. At that time he began investigating sexuality in Algae. His results, published in a long series of papers, constitute the chief experimental basis for Hartmann's famous theory of Relative Sexuality. The status of this subject up to 1943 is summarized in Hartmann's book *Die Sexualität*.
> . . . About 1938, Moewus moved to Heidelberg where he began to turn intensively to the genetics of sexuality in Algae. He quickly laid the foundations of biochemical genetics of micro-organisms, by developing the basic concepts and methodology as well as supplying the facts concerning biochemical genetics of sexuality in one of the unicellular green algae, *Chlamydomonas*. Moewus' first paper along these lines appeared three years before Beadle and Tatum's first paper on biochemical genetics in *Neurospora*, and, a year before their first paper appeared, Moewus published a series of most remarkable papers dealing with the general theory of biochemical genetics with its application to *Chlamydomonas*. Yet he has seldom received credit due to him for his path-breaking investigations.
> The reasons for this neglect may be complex, but the chief reason is that Philip and Haldane published a decisive criticism of his numerical results. They maintained that the data were far "too good," with the implication that the data were not completely reliable. Several years later [1943] Moewus showed that most of the data examined by Philip and Haldane were not what they had assumed them to be; . . . This I am convinced was true, but still I believed Moewus did "polish off" his data to some extent. It is clear from his published tables that he makes arithmetical mistakes in calculating X^2 and even in simple addition. As you may know, Mendel's results were found to be "too good" also, by the same type of analysis!
> I have published criticisms of some of Moewus' results on other grounds. Still others have pointed out the difficulty in his reported linear transformation of *cis*-dimethyl crocetin to the trans isomer by a photochemical reaction. The linear transformation is held by some to be impossible on physical grounds. I am also somewhat doubtful of his quantitative reports on mutation rates under the action of mutagenic agents.
> I have now told you the worst. On the other side one must consider the following. Ever since 1938, he was working in close collaboration with Richard Kuhn, Nobel Prize winner in biochemistry. Kuhn and his students have done most of the biochemical work, while Moewus has done the biological and genetic part. The chemicals reported to be active in the control of sexuality have been isolated from the organism. The properties of the enzyme systems involved have been reported in detail. I am unable to see how this could be done if there were any essential qualitative faking in Moewus' part of the work.
> But the campaign of calumny and gossip, once begun, has continued to roll along. Smith at Stanford, Lewin at Yale, Provasoli at the Haskin's

Laboratory and Legar [*sic*; Sager?] at the Rockefeller Institute (working with Granick) are all now working on *Chlamydomonas*. They would not have done this without the background provided by Moewus, yet they are either opposed to Moewus or deliberately ignore his work, as if it did not exist. There are stories about Moewus' refusal to send his cultures to others or his sending dead cultures. And so on.

I spent a large part of the last summer going over Moewus' and Kuhn's work and, while fully aware of the difficulties, came to the conclusion that his results are among the most important in modern genetics in many ways. He doesn't always fully appreciate their genetic importance, but the discoveries and results are there and others can supply whatever is lacking in the way of appreciation of significance. I felt so keenly the importance of this work that, in the paper I gave last week at the Cold Spring Harbor Symposium, I devoted about half of my time to pointing out some of its genetic importance. As a result, I learned from Professor Gustafson of Sweden that in 1948 he sent one of his assistants (Nyborn, as I remember) to work with Moewus, to learn his methods, and to bring back cultures for study of certain problems. Gustafson said Moewus was very cooperative and that in the work of Nyborn all results thus far obtained confirmed Moewus' claims – Hadorn of Switzerland also said he received cultures from Moewus and that they behaved as claimed. Glass of Hopkins informed me that Moewus had also recently sent some of his cultures to Smith at Stanford. This made a very different picture from the one commonly circulated in this country.

I have been in correspondence with Moewus and learned from him recently that he intends to migrate to Australia, permanently, as I understand it. He said he will there be unable to continue his work with the Algae, but will instead work on growth hormones. There was no tone of regret in the way he communicated this information. But my response was quite different. It seems to me a sort of scientific murder to let Moewus' invaluable material be lost. He has many strains of *Chlamydomonas*, both wild and laboratory mutations. How many of these are in the hands of other people now I do not know. You can imagine how many there must be when I tell you he has mapped over 70 gene loci and has a number of multiple alleles at many of these loci. He must have hundreds of different cultures. Confidentially, I don't think Smith of Stanford would know what to do with these even if he had them. For very few organisms is the genetics so extensively known as for *Chlamydomonas*, thanks to the efforts of Moewus.

It thus seems to me that one of two things is needed. Either every effort should be made to make conditions in Germany attractive enough to keep Moewus there and to induce him to continue his work; or *all* of his many stocks should be deposited somewhere where they will be properly maintained and available to investigators as needed. After talking to Huxley, it seemed that the best solution would be to try to prevail upon the Max Planck Institute in Berlin to give Moewus an adequate and attractive setup. While I am not working on *Chlamydomonas* and, so far as present plans go, do not intend to do so, yet the preservation of the *Chlamydomonas* culture seems so important to me that I would undertake to keep the cultures going, if no other alternative exists. Cleland is adding a young algologist, Starr, to his staff and may be interested in getting the cultures, or you may know of some better place. I hope you do.

These are the main facts and problems. If there is any further information I can give you, please write me for it. I leave July 3 for South America where I will be collecting paramecia and lecturing in Santiago until August 15.

Weaver wrote back, stating that he was going to turn the problem over to Pomerat, but he was not optimistic. As he explained to Sonneborn (June 25, 1951):

I am going to turn this matter over to my colleague, Dr. Pomerat, who has primary responsibility for our activities in Europe. I am really not at all optimistic about our being able to do anything. We have as yet done so little in Germany since the war, that it would be rather spectacularly curious for us to pick out, for one of our very first postwar grants, a case which is obviously so controversial that any action we might take would certainly be questioned by many. But I can assure you that we will give the matter serious consideration.

I am going to be in Bloomington Saturday, June 30, and Sunday morning, July 1, attending a meeting of the Directors of the A.A.A.S. I think it unlikely that there will be time to see you, but if there is anything pressing which you wish to say, I would at least want you to know that I was going to be there.

The Rockefeller Foundation officials were very prudent in their support of scientists. In understanding their reluctance to support Moewus or to ensure that his experimental work be repeated and properly evaluated, one has to take into consideration several issues. First, the primary funding strategy of the Rockefeller Foundation was to support established researchers or young researchers who had shown great promise – to make "the peaks higher." Moewus's work was controversial. By providing funds for Moewus to leave Germany, it might appear that they were, in effect, endorsing his work, by favoring it above that of others in Germany. One also has to take into consideration the larger political context in which the foundation operated during the early 1950s. The postwar years were politically troubled years inside the United States and inside biology. For the Rockefeller Foundation officials, in addition to the difficulties of supporting German scientists who might be suspected of war crimes, there was the threat of communism, and also the rise of Lysenkoism. These difficulties became extremely important in their foreign science funding policies. The threat of supporting Communist science was particularly complex for the Rockefeller Foundation officials in France, where many leading biologists were Communists (see Sapp, 1987a: 181–191). There were difficulties with supporting some geneticists in the United States as well.

In fact, at the very moment Sonneborn was writing to seek support for Moewus, and to ensure that he was not dismissed to the periphery of science by going to Australia, George Beadle wrote to Weaver in an attempt to save Michael White from a similar fate. White was well recognized as a brilliant cytogeneticist. He was a former student of J. B. S. Haldane, who in 1947 had left University College, London and had moved to the United States to take up a position as professor of zoology at the University of Texas. However, White ran into trouble during the McCarthy period. From 1932 until the beginning of 1935, White was a member of a student organization affiliated with the Communist Party in Great Britain. The University of Texas was a state-run university, and, during the period of McCarthyist hysteria, no state funds could be provided to anyone who had been a member of any Communist organization. The situation was well explained by George Beadle, when writing to Warren Weaver for help:

> Mr. M. J. D. White, Professor of Cytology at the University of Texas, is in an impossible situation. . . . It seems he was a Communist Party member in England back in the 30's. He became convinced that this was all wrong in 1936 and quit the party. After this he apparently actively opposed Communism. The Texas legislature has tacked on to the appropriations bill a rider providing that anyone refusing to sign an oath that he or she is not now and *never has been* a Communist cannot be paid from the funds covered in the bill. You can see where this puts White. It looks like sure ruin professionally either way for him . . . the only solution is for White to leave either temporarily until the law, which I should think would certainly not stand up, could be changed, or permanently. (Beadle to Weaver, May 15, 1951)

Beadle pointed out to Weaver that he and some of his colleagues at the California Institute of Technology were willing to take White on for a year. However, the Trustees of the California Institute of Technology would have to be told the story, and Beadle believed they would run into trouble with a "certain extreme group of trustees." On the other hand, Beadle suggested, "if an outside agency were to propose that we help in what is obviously a very unfair situation to White there is a good chance that our Trustees would be willing to O.K. an appointment here." "The question is simple," he wrote Weaver, "Do you see any way to help?"

Weaver (May 18, 1951) promptly wrote back to Beadle, informing him that he was not interested in sticking his neck out to support White. As he explained:

> I do not know Mr. White. I have no information whatsoever concerning his general reliability. I have no basis whatsoever for convincing myself of what I assume is probably the case – namely, that he has completely abandoned his communistic ideas. But it would certainly be only under very extraordinary circumstances, where I had personal convictions based on direct personal experience and evidence, that I would feel justified in exerting the slightest amount of pressure or suggestion. To do so in cases of this sort is a questionable and dangerous procedure at best; and to do so when I have no real basis for doing so seems to me, I am sorry to say, quite impossible.

White subsequently left the United States and moved to Australia, where he eventually became "Foundation Professor of Genetics" at Melbourne University.

Sonneborn received a reply from Pomerat several weeks after Weaver responded. The reply was final; the Rockefeller Foundation officials were willing neither to help Moewus in Germany, nor to help in depositing his stocks elsewhere. Pomerat (July 30, 1951) wrote to Sonneborn:

> Since my return from Europe Dr. Weaver and I have had an opportunity to talk about your letter of June 19, 1951, in which you explore with us the possibility that the Foundation might be able to do something which would encourage Dr. Franz Moewus to remain in Germany or, failing that, that it might make it possible to transfer all of Moewus' *Chlamydomonas* cultures to some important laboratory outside Germany.
>
> As Dr. Weaver indicated on June 25, our division is re-establishing its program of support for science in Germany along very modest and very carefully considered levels. Under these circumstances we feel that it would be most unwise for the Foundation to contribute toward influencing the career plans of any German scientist. As far as doing something about getting the *Chlamydomonas* cultures shipped to some laboratory outside of Germany is concerned, we also feel that this is something which the Rockefeller Foundation is not at present prepared to undertake.
>
> I know how very much you have taken Moewus' problems to heart and I very much regret having to send you this disappointing reply. On the other hand, this has been a year of extraordinarily heavy demands upon our relatively restricted resources and so it is a task I have had to undertake a good many times in the last few weeks.

Moewus was on his way to Australia. He seemed only glad to have the opportunity of leaving Germany and Heidelberg, where he had become boxed in after the war. Whereas Sonneborn perceived Moewus's plans to go to Australia as a major tragedy for biology, Moewus had a much more optimistic attitude. He was one of the first scientists to leave Germany and looked forward to Australia with great expectations. The Moewuses packed all their belongings, had their furniture sent to Australia, and intended to remain there indefinitely. They set out for Australia via England, where they were hosted by one of England's leading cytologists, C. D. Darlington. Since the late 1930s, Darlington maintained a great interest in problems of gene action. In 1938, he had been invited to attend a small workshop on gene action, held in Klampenborg just before the German invasion of Austria. Organized under the auspices of the Rockefeller Foundation, it was the first of its kind designed to bring physicists, chemists, cytologists, and geneticists together to attack problems of the physical and chemical basis of heredity.

Darlington himself, however, did no experimental work on gene action. And as Lewis (1983: 127) remarked, "Darlington based his scientific work firstly on hypothesis and secondly on observations and facts. . . ." Darlington was outspoken in his claims of placing a priority of theory over observation in scientific work. As he put it, "Hypothesis has often proved more reliable than the facts of direct observation," and "All of us like to be supported by earlier observations but few like to be anticipated by earlier ideas" (Lewis, 1983: 127). Darlington also warned scientists to know their audience, especially their enemies. "For next to friends, enemies, if well chosen, are the best stimulus to research" (Lewis, 1983: 118). German biologists were never among Darlington's enemies. He held only admiration for German botanists, and Moewus was among them. Darlington visited Kuhn's laboratory in Heidelberg in the late 1940s, and, Mrs. Kobb recalls, he admired Moewus's culturing techniques of maintaining zygotes instead of keeping agar cultures of various strains of *Chlamydomonas*. During their London stopover, the Moewuses spent two days with Darlington before boarding a ship for Australia. On September 14, 1951, Moewus wrote to Sonneborn, describing his itinerary and stating that he was arranging for stocks of *Chlamydomonas eugametos* to be sent to him.

> I now only want to tell you that I have given one of my co-workers, Mr. H. W. Hagens, the task of sending you two sexually different stock of *Chlamydomonas eugametos* (absolutely pure cultures) in the course of October. The observation of group formation, the formation of couples, the zygote fusion and zygote reproduction is charming for every student to see, especially since the process can be initiated at any time. Only water and light are needed. My new address is:

Dr. F. Moewus
University of Sydney
Sydney, N.S.W.
Australia

In the next days a consignment of papers will be sent off also. I hope that we will continue to correspond and I will write to you again from Sydney. On 18.9 we leave Heidelberg, stay 6 days in London (of these, 2 days with Dr. Darlington) and after a 6 weeks' sea voyage, on the 4.11 we will then be in Sydney.

10. Great expectations: Moewus abroad

"He seemed like a very honest sort of individual to me. . . ."

Under Capricorn

In September 1951, the Moewuses set out for Australia. Franz Moewus was to be a Timbrol Research Fellow in the Botany Department of Sydney University. Moewus had a wide correspondence and, during the late 1940s, had mentioned to one of his friends in Australia that he would like to go abroad. At this time, the Commonwealth Scientific and Industrial Research Organisation (CSIRO) in conjunction with the universities was making a major effort to build up basic scientific research. During the war, England had neglected Australia, which had been heavily dependent on it for imports. After the war, it took a long time for England to develop again. The Australian Government decided to become as autonomous as possible from the mother country with respect to industrial development. But in order to do this, it realized that it needed to develop and strengthen fundamental research; it had to start from the bottom. The CSIRO received a great deal of support for its efforts to further scientific and industrial research, and it encouraged Timbrol to cooperate. Timbrol was an established chemical firm. The directing manager, Martin, and the research director, Professor Dr. Kaufmann who had immigrated to Australia after World War II, led the Moewuses to Sydney with the idea that the work on hormones in *Chlamydomonas* could possibly have some application for economically important problems in higher plants.

Moewus was able to take his stocks of *Chlamydomonas* with him to Sydney. One of his friends owned a suitcase factory and built a special case with racks for the cultures. The trip to Australia was paid for by the Australian Government. It was a one-class ship: All travelers had been engaged by the Australian Government; all were specialists in various fields. There were agricultural workers for the tropical Queensland sugar plantations, engineers for mining, road builders, surveyors, and so on. Most were English-speaking, and there were six Germans including the Moewuses. The 46-day boat trip to Australia was exhausting for

210

the Moewuses. They were held up for eight days in the Red Sea as it was evaporating. There had been a suicide on board. When the Moewuses finally arrived at Sydney, Alan Burges was at the harbor to greet them. All bags were sterilized upon arrival in Australia, but Burges somehow got their cultures in. Burges was an Australian citizen who received his Ph.D. in mycology at Cambridge before accepting a position in the Botany Department at Sydney University in 1947. Burges did not remain in Australia long. Apparently, his British wife was unhappy in Sydney, and he resigned in 1951 to return to England. During his stay in Australia, Burges did not publish any work. He was, however, extremely active in expanding the department and encouraging cooperative research with the CSIRO and with visiting scientists.

In November 1951, Franz and Liselotte Moewus arrived at the Botany Department of the University of Sydney. Franz Moewus joined the staff as a Timbrol Research Fellow to work on genetics studies in unicellular algae. By February of the next year, Liselotte Moewus also was made a Timbrol Research Fellow and began to exploit her expertise at culturing algae. Despite the trying trip, the Moewuses' experiences in Australia were generally happy ones. The Australian Government had been generous with travel expenses to Australia. The Moewuses were allowed to take any belongings they had to Sydney. They took their piano, dining room furniture, and boxes of books. They knew that if they wanted to stay in Australia, it would not be possible to have them sent later. The Timbrol firm sublet a beautiful house to them in Sydney; they integrated well and they worked hard. Franz Moewus lectured at various scientific institutions in Australia, and the Moewuses soon began to do novel collaborative research on a new and exciting biochemical genetic problem with the organic chemist Arthur Birch. This work would eventually generate a great deal of private controversy.

Arthur Birch had received his appointment at Sydney University in 1952, but like Burges soon left Australia for England. In 1955, he was appointed to the chair of organic chemistry in Manchester. In 1967, Birch returned to Australia as professor of organic chemistry at the Australian National University. He is well known today for his work on the synthesis of natural compounds. Among these natural products were flavonoid and anthocyanin pigments in plants. He was working on this problem when Moewus arrived in Sydney. Birch's central hypothesis on the biogenesis of natural products was what he called the "polyketide hypothesis," which today is considered to be entirely correct, and now explains the structure of 6,000 natural products.

When Moewus met Birch at Sydney, Moewus mentioned to him that he believed that flavonoids in rather small quantities acted like a hormone to influence mobility and sexuality in *Chlamydomonas engametos*. For example, Moewus held that quercetin, a flavonoid pigment, inhibited motility in *Chlamydomonas eugametos*. Moewus further claimed to have a number of mutants that could become motile when fed precursors of flavonoids. Birch thought this was "an absolutely ideal method of testing" the "polyketide hypothesis" he was then developing (Birch, interview, October 23, 1984).

Flavonoid compounds occupied a prominent position in the study of the biogenesis of plant pigments, largely because of their economic importance. Certain flavonoid compounds were among the earliest known natural dyestuffs. The conspicuous colors that anthocyanins impart to flowers, fruits, and leaves made them objects of great interest and speculation to scientists. The importance of flavonoid compounds in the tanning of leather, the fermentation of tea, the manufacture of cocoa, and the flavor qualities of foodstuffs led to intensive investigation of their chemical formation. Throughout the first half of the twentieth century, the biogenesis of flavonoids and anthocyanin pigments in plants was an object of great attention. It was hoped that an understanding of the manner in which flavonoid compounds were formed in nature would soon be achieved, and this further study could eventually help to elucidate the biological mechanism of one of the most common natural synthetic processes.

Indeed, despite much work on the genetics of flower color throughout the first half of the century, few conclusions had been reached regarding the biogenesis of flavonoids and anthocyanins, except for the demonstration that the processes were gene-controlled, and that flavonoids and anthocyanins were derived from the same precursors, if not in sequence. The major difficulties with genetic work on higher plants were the extreme complexity of both the genetic factors and pigment mixtures. In order to provide evidence of much value, Birch reasoned that most of the biochemical genetic work would have to be repeated using modern chemical techniques.

The unicellular alga *Chlamydomonas eugametos* was a much more amenable organism for the biochemical study of flavonoid biogenesis, since Moewus (1950) had reported that it used a flavonoid and an anthocyanin as sex hormones. Between 1939 and 1951, Moewus and his collaborators had carried out a large number of mutation experiments, using x-rays, radium, ultrasound, chemical compounds, and temperature shocks to induce mutations. On the basis of these experiments, Moewus claimed to have produced 37 mutants that were genetically female but could not copulate with male gametes. However, they did so when they were fed traces of the flavonoid isorhamnetin. Of these mutants, 26 were rendered capable of copulation when they were given quercetin. The remaining 11 mutants were held to contain quercetin (Kuhn and Löw, 1949). From these results, it seemed clear that the 26 mutants could obviously convert quercetin to isorhamnetin; the other 11 could not.

The question that interested Birch was, What steps led to the synthesis of the flavonoid quercetin? In other words, what substances, when added to the 26 mutants, could produce quercetin? The idea was then to do substitution experiments with the 26 mutants. The mutants would be incubated for four hours in a solution of the amino acids to be tested, and then a drop from the mutant solution was mixed with normal male gametes to see if they had become sexually activated.

There was, however, one immediate hitch. What Birch hypothesized was that flavonoid synthesis starts with acetic acid, and eventually the pathways branch out in various directions to form flavonoids and anthocyanin pigments. What was really new in his hypothesis was the role of acetic acid. Moewus, on the other hand, held that his mutants used a different precursor, *meso*-inositol, for flavonoid biogenesis, rather than acetic acid as Birch hypothesized. Birch "was absolutely certain that the polyketide hypothesis using acetate was correct." Nonetheless, as he recalls, "I had great doubts about Ockham's razor in biology; so I thought that it may be they operate by a different mechanism. So I asked Moewus to collaborate" (Birch, interview, October 23, 1984).

At this stage, all Birch knew of Moewus was that "he had collaborated with Richard Kuhn, who had gotten a Nobel Prize, and that a lot of Moewus's work was obviously heavily involved in this." In December 1952, Birch applied to the Rockefeller Foundation for additional research funds for work on flavonoid biogenesis. A grant for $6000 was approved, and Birch hired Liselotte Moewus as a research assistant, to be responsible for preparing cultures employed in the work. When Franz Moewus's Timbrol Research Fellowship expired that month, he became a temporary lecturer in botany.

The Moewuses began to collaborate with Birch and one of his doctoral students, F. W. Donovan. Birch gave Moewus the chemicals, and Moewus began to test his different mutant cultures to see which ones would respond by becoming sexually activated. As Mrs. Kobb recalls, everyone was excited to see the copulation under the microscope. How male and female *Chlamydomonas* cells react when brought together in a water drop on a microscope slide was an exciting sight. The team were also able to produce the quercetin reactions by using *meso*-inositol instead of acetate as a precursor. However, Birch soon ran into some anomalous biochemical results which, he recalls, puzzled him. They concerned some unexpected spots on paper chromatography. (Paper chromatography is a technique for separating the components of a complex mixture of organic chemicals, such as the soluble constituents of a plant. A few drops of a concentrated extract are placed on a corner of a sheet of filter paper, and the edge of the paper is dipped in a suitable organic solvent. The solvent moves through the paper by capillary action, and different substances in the drop move with the moving organic solvent at different rates, depending on their relative solubility, etc. The result is that the components of the mixture are separated into a row of discrete spots.) Birch recalls:

> So we started collaborating. Moewus had various tubes of green stuff and I gave him compounds which he fed in and he claimed that this would produce the quercetin reactions. So I said, "O.K., but your mutants should be accumulating – or probably accumulating – intermediates where the process is stopped, so let's look at the solutions." So, he gave me the solution, and a student of mine analyzed them by a method then available, which was paper chromatography, and this is where I first ran into something that puzzled me greatly – because phenylalanine is a really primitive precursor. We got a spot of phenylalanine which was partly two spots. Now paper is optically active – but this is typical of optically inactive phenylalanine. So, I said, "I simply do not believe. . . . There is a great puzzle here. I don't believe nature produces optically inactive phenylalanine for any purposes whatsoever. It's an article of faith like anything else but here you had optically inactive phenylalanine." (Birch, interview, October 23, 1984)

If one believed that Moewus was simply a charlatan, then the solution to the riddle might appear to be simple: He must have doped his cultures with the appropriate chemicals to obtain the expected results. Birch now suspects that Moewus must have put synthetic phenylalanine in his solutions. "It probably said 'phenylalanine' on the bottle, and Moewus didn't know enough to recognize that it should have said 'L-phenylalanine,' or else he didn't know that phenylalanine spots separate on paper – but it does. Nature doesn't work on DL-phenylalanine" (Birch, interview, October 23, 1984). Birch never made these charges directly to Franz or Liselotte. Mrs. Kobb first learned of them from me. Although Birch attempted to wash his hands of any responsibility for the results, Mrs. Kobb recalls (interview, May 26, 1987) that Birch himself had been there to see the copulation tests more than once. Birch did not offer an account of the procedures by which Moewus might execute the deception. But Mrs. Kobb did (interview, May 29, 1988), when commenting on my earlier draft of this story. On her account, there is no room for Moewus to have "doped" cultures:

> Franz had isolated the mutant from the zygote, because the strain had been transferred to Sydney as a zygote suspension. After he himself had obtained a clonal culture, he did a pre-check with cells under the microscope in order to make sure that the mutant character was still maintained. He then handed over this small spot-culture cell material to me. I extended it over several steps to finally 12 very large testtube slants. Whenever I made a transfer from agar to larger agar cultures, Mr. Donovan was invited to sit next to me in the sterile room. Finally – again in his presence – I washed the agar slants with distilled water and handed him the green suspension, a few 100 mls of it. And this material Donovan took to the chemistry department. From this, the paper chromatography was made. I must state again that from the moment Franz had given me the minute amount of cells from the mutant clone, he did not touch this material again. And it must have been obvious to Dr. Birch and Donovan, and others, that I did not know enough of biochemistry to handle any material of pigments, etc. Who should have "doped" the culture? And when Birch had been already suspicious, why in God's name did he not say so?

But Birch did not reach the conclusion that the results were somehow faked until much later. In fact, while the Moewuses were in Sydney, he believed the results. Scientists do not respond directly to nature; they respond to each other. At the time of the anomalous results, Birch knew only that Moewus's work along these lines was intimately related to that of the Nobel Prize-winning chemist, Richard Kuhn. Just after the problems with the paper chromatography and phenylalanine, Birch was in Adelaide and talked to the geneticist David Catcheside, from whom he first learned of any doubts about Moewus. Catcheside, Birch recalled (interview, October 23, 1984), "said that Haldane had calculated that he [Moewus] would have had to run one observation every second for twelve months without stopping." Actually, Ursula Philip was the first author of the criticism in question, not Haldane. Nonetheless, Haldane's name held significance for geneticists as Kuhn's did for chemists. Birch, of course, did not understand the background to this at all but thought it was very peculiar indeed. He questioned Moewus directly. Birch remembered that Moewus responded by "saying something about Mendel's original work" and "pooh-poohed" the criticisms, saying "it was a bunch of nonsense."

The question of Moewus's motives and an assessment of his personality played critical roles in influencing Birch's judgments of the validity of his results. Moewus was not a bombastic figure. He was very quiet and reserved. Birch offers the following description of Moewus's character:

> He was a very enthusiastic lecturer and a good lecturer and a good demonstrator. . . . He was very quiet, very reserved, quietly spoken, very confident. . . . The only thing I thought was a bit curious was that if I had been accused of fraud, I would have gone through the roof. But he didn't; he would just sort of say, "Well, those people just don't know how to do it. I am the world expert. I spent twenty years learning how to do these things. If those people spent twenty years, then I would say they had the qualifications to contradict me" . . . He was totally unperturbable. He showed absolutely no signs of guilt, anxiety, or anything else. His wife, on the other hand, was extremely jittery. There is no question that she was very upset by the whole affair – so much so, that I never tried to talk directly to her. (Birch, interview, October 23, 1984)

Birch later became very suspicious because of the anomalous result with phenylalanine. But at this stage, he was still convinced of the general authenticity of Moewus's work. In fact, he compromised his own theory of flavonoid biogenesis: He had previously believed that acetic acid was the main route; but in *Chlamydomonas eugametos,* at least, flavonoid biogenesis used a different precursor (*meso*-inositol), as claimed by Moewus. As Birch emphasized: "I was sufficiently convinced to publish a note in *Nature* that I had to withdraw eventually. I withdrew that in a subsequent article somewhere, saying there were doubts about the experimental work and that this should be neglected until any doubts were resolved" (Birch, interview, October 23, 1984).

Later, in 1957, Birch wrote a disclaimer to his work with Moewus (see Chapter 13). However, it certainly did not receive the attention that the *Nature* paper (Birch, Donovan, and Moewus, 1953) received. Indeed, the collaboration of Moewus with Birch was significant and was perceived by some readers as a corroboration of Moewus's biochemical genetic work. The exchange of credibility had come full circle: Birch was convinced because of Kuhn's involvement; now others were convinced because of the involvement of Birch. As late as 1962, V. J. Chapman, in his text *The Algae,* stated:

> [I]t is of importance to note that the chemistry of rutin and isorhamnetin in relation to this algae has recently been repeated at the University of Sydney. This is significant because all the previous biological and genetic work comes from a single investigator, whilst the biochemical work also

was performed at one place only. Many workers have therefore felt that the novelty of the results demanded repetition. (Chapman, 1962: 420)

Chapman's text was reprinted until 1969. In 1973, when a second edition appeared, all mention of Moewus disappeared.

Birch was not the only leading scientist in Australia who had faith in Moewus's work. Moewus was well received by the Jewish, Austrian-born geneticist Otto Frankel. Frankel was a student of the well-known German geneticist Erwin Baur during the mid-1920s. A few years after completing his thesis in Berlin in 1925, Frankel migrated to New Zealand, where he worked as a wheat geneticist until 1951, when he was appointed chief of the Division of Plant Industry of the CSIRO in Canberra. During the early 1950s, genetics research was almost nonexistent in Australia; Frankel played a key role in fostering its development. He had visited the Moewuses in Sydney and witnessed the copulation reactions with *Chlamydomonas*. According to Mrs. Kobb (interview, May 25, 1986), he was deeply impressed, and in the summer of 1953 he invited Moewus to Canberra for a week to give a lecture and a demonstration. The laboratories in Canberra were set up in many separate wooden barracks; there were greenhouses but with no cooling or proper maintenance. Frankel was looking for young geneticists to help build up his program and offered Moewus a job, informing him that he could have one of the barracks and a greenhouse to himself. Mrs Kobb recalls that her husband turned down the offer. He believed he would not be able to work on *Chlamydomonas* for several years under the existing conditions.

The second day of their visit in Canberra, there was to be a demonstration following Moewus's lecture. Moewus was to show how his mutant *Chlamydomonas* would become sexually active when fed quercetin. The demonstration failed. Mrs. Kobb believed the failed demonstration might have been due to the cells having been overilluminated. Before the demonstration, they had left their cultures overnight in one of the greenhouses. It had the brightest light Liselotte had ever seen. She recalls (interview, May 26, 1987):

> This was the first time the dear *Chlamydomonas* let us down entirely. He put two droplets together – no clumping – which worked for 30 years, flagella were formed. The flagella were sticking to the glass and they were vibrating. What could we do? Say, Come to Sydney, it works? Frankel knew; he had been to Sydney; he knew it worked. We had no explanation, except that it must have been this overillumination. It wasn't good for his reputation.

The Moewuses returned to Sydney to continue the work they had begun with Birch. The paper by Birch, Donovan, and Moewus appeared in *Nature* in November 1953. Four months earlier, both Franz and Liselotte Moewus resigned from their positions at Sydney University. Just when their work with Birch was in full swing, Moewus was given and accepted an invitation to be a visiting researcher at Columbia University to repeat some of his work under the observation of Francis Ryan. As will be discussed in more detail momentarily, Ryan was one of the first geneticists to work in Beadle's laboratory and learn the new techniques of using microorganisms for biochemical genetics. He played an important role in building up a program of bacterial genetics at Columbia.

John A. Moore was immediately responsible for linking up Moewus and Francis Ryan. However, biochemical genetics was not Moore's specialty. He was trained in embryology at Columbia University in the 1930s. In June of 1952, he, his wife Betty, and his daughter arrived in Australia. John Moore had a Fulbright Fellowship to work in the Zoology Department of the University of Sydney to study problems of evolution in frogs. Moore (interview, May 1, 1986) recalls:

> At that time, there were frogs that were called the same species that occurred in southeastern Australia . . . and in the southwest and, of course, there were some in between. And it seemed strange that you could have two biological species that presumably had been isolated for many many thousands of years. And, so what I wanted to do was go there and do some very crude experiments and collect specimens from the west and the east and hybridize them and to see what happened. And it turned out that there were two or three pairs of those species. . . . This was one of those times when research went extremely rapidly. I arrived in Sydney in winter there, and the rains were out and the frogs were out and I got a lot of material. . . . Things went so well that I had an answer to my question in three months.

Charles Birch made it possible for Moore to get a Fulbright Fellowship at the University of Sydney. Charles Birch was trained in entomology at the University of Adelaide. After working in Chicago and Oxford, he went to the Zoology Department at Sydney University in 1948. At that time he was more interested in teaching than doing full-time research. In 1954, Birch initiated and taught the first course in animal ecology ever taught in Australia. He was an outspoken critic of the mechanistic view of the world and wrote widely on population ecology, evolutionary biology, and the philosophy of biology. Moore had first met him at a zoological meeting at the University of Chicago and later at Columbia.

Birch had the idea that American biologists going to Australia should, as an obligation, do systematics of one group of organisms in Australia. Moore recalls that for Birch the problem was simple: Biologists did not know what species they had in Australia. As we have seen, classification was prerequisite for reporting experimental research on organisms. Moore quickly came to recognize the reality of the problem of classification of organisms in Australia. Only a week after he arrived in Sydney, some people from the veterinary school who were working on parasites in frogs came over to ask him if he knew what species the frogs were. They complained that they could not do any research since they did not know what names to put on their frogs.

Moore subsequently spent a great deal of time at the Museum of Australia, collecting frogs and classifying them. He eventually wrote a large monograph on the frogs of Australia which was published by the American Museum of Natural History. Up to that time, there was very little work on Australian amphibians; most of it had been done by people at the British Museum. Moore succeeded in helping to establish a major research program on the classification of frogs. As a result, Moore happily mentions, his own monograph became quickly out of date. He claims today that "one can almost make the statement that the frogs of Australia are probably better known than those from any other continent" (Moore, interview, May 1, 1986).

Moore did no active research in biochemical genetics of microorganisms. He was interested in what frogs were doing – specifically, their geographical distribution and the various species. Nonetheless, he was very aware of new developments in microbial genetics and appreciated the possible significance of Moewus's work. A brief review of the social relations in the Zoology Department at Columbia University at the time will help the reader understand Moore's interest in having Moewus go to New York.

During the previous three years, while still in his early thirties, Moore had the difficult task of being chairperson of the Zoology Department at Columbia. The Department of Zoology at Columbia had an illustrious history, beginning with one of its most celebrated chairpersons, E. B. Wilson. Wilson's major book, *The Cell in Development and Heredity*, written first in the mid-1890s and reprinted and revised until the mid-1920s, remains a classic to this day. It is well regarded as an attempt to synthesize, weigh, and digest the diverse and often conflicting theories of, and approaches to, heredity and development since the emergence of experimental embryology in the 1880s. Wilson was also instrumental in fostering the development of the famous *Drosophila* school led by T. H.

Morgan in the second decade of the century. When Morgan and A. H. Sturtevant moved to the California Institute of Technology in 1928, Columbia was still able to sustain its prestige. During the 1940s and 1950s, outstanding geneticists such as L. C. Dunn, Theodosius Dobzhansky, and Francis Ryan kept the department lively and in public view. However, the social relations in the department were not pleasant; Moore had been appointed chairperson at a very early age, with the difficult task of trying to establish harmony within the department. As Moore recalls,

> The Fulbright was an especially happy event for me because at a very young age I had been made the chairman of the department at Columbia – when I was in my early thirties. And this astonishing situation was the result [of the fact] that the department was an exceedingly strong department. It consisted only of birds of paradise, prima donnas that were outstanding throughout the world. They had reached the point of hardly being able to deal with one another. So who should we have as chairman. . . .? At any rate it was quite a trying three years. (Moore, interview, May 1, 1986)

At that time, Moore's closest friend at Columbia was Francis Ryan. They were both the same age; they had been in graduate school together, and they maintained a close friendship all of their adult lives. Ryan had obtained his Ph.D. at Columbia in experimental embryology and, in the early 1940s went to Stanford as a postdoctoral fellow. At that time, he was interested in investigating the effects of temperature on the rate of development of frog embryos. This was, as Moore puts it, "not a very swinging problem." But very early in his arrival at Stanford, he became interested in Beadle's work on *Neurospora*. The *Neurospora* story, of course, was just opening up then, and Ryan sat in at its beginning. Ryan subsequently returned from Stanford as an instructor, still in the early 1940s. He very quickly started a laboratory to work on *Neurospora*. However, during World War II, Ryan switched to working on gangrene, a medical problem of importance to the armed forces. After the war, Ryan returned to *Neurospora* and began to develop bacterial genetics and sought to establish seminars with microbial genetics groups at Cold Spring Harbor. Ryan died young at the age of 47, in 1963.

The shift from higher organisms to microorganisms was not an easy transition for geneticists. As one research domain receives attention, others feel the pinch. Many of the older geneticists such as Dunn and Dobzhansky, who worked on higher organisms, were far from sympathetic to the new developments in microorganisms. As Moore recalls:

> There was a very sad personal, political situation there. . . . The dominant people in genetics were L. C. Dunn and Theodosius Dobzhansky. They were much older than we were, world famous and so on, but somehow they, perhaps they felt threatened by the developments after World War II when people were working on *E. coli*, because that wasn't the way things happened in mice; that wasn't the way things happened in *Drosophila*. And Francis Ryan organized this seminar, and he brought all the people who were developing the field at that time; they had come from Cold Spring Harbor; they had come from wherever they were. So that the entire beginnings of what is now molecular biology unfolded in the lectures. Neither Dobzhansky nor Dunn ever attended a single one of these. . . . (Moore, interview, May 1, 1986)

Young biologists such as Moore, who did not work in microbial genetics, did, however, generally go to these seminars. And, as Moore recalls, at one of the seminars what was of public concern was that Moewus was describing all these exciting things. But people had real questions about the validity of Moewus's work. By the time Moore had gone to Australia, he "had the feeling this was very suspicious stuff." But as Moore explains, his doubts were "not on the basis of anything I had known. I had never read his papers, never heard him talk, but just the sort of scuttlebutt that you hear in the hallways or over a beer or whatever" (Moore, interview, May 1, 1986).

Very soon after Moore was in Australia, he became aware that Moewus was in Sydney and received a notice that Moewus was going to give a lecture. He and his wife, Betty, who had an undergraduate degree in biology, went to the lecture. Moore recalls:

> I was impressed. I mean, he seemed a very honest sort of individual to me. I guess one of the things I remember, back then there was that . . . the two [Franz and Liselotte] always seemed to be together; they seemed very close and that was wonderful, because they worked in the lab together; they would walk together across the campus, they would come to meetings together and so on. (Moore, interview, May 1, 1986)

There is no question that Moewus left a good impression on Moore. As Moore explained in a letter to me (February 24, 1986), it seemed to him to be entirely possible that the difficulties others were having in repeating his work were due to technical complications:

> I knew, of course, of the problem Moewus was encountering in having his experiments and conclusions accepted. Some, many perhaps, simply did not believe him and suspected fraud. My first impression was that Moewus was entirely honest and might well have done experiments that for one reason or another, others could not repeat.

Moore recalled speaking frankly to Moewus about the questions concerning this work, and asking him about the possibility of going to Columbia and working in Ryan's laboratory so "that people could see, and that would be that. He seemed to think that was a good idea" (Moore, interview, May 1, 1986). Moore sent a telegram to Ryan, suggesting that he invite Moewus to come to Columbia and have him repeat some questionable experiments. And Ryan cabled back. "Francis," Moore writes, "was a careful and highly respected scientist. I remember suggesting to him that such a resolution would be important in the rapidly developing field that was to become molecular biology" (Moore, letter to the author, February 24, 1986). Ryan had special grants for inviting visiting researchers. He agreed to invite Moewus.

Moewus seemed to be only delighted to have this opportunity to go to New York. His wife, however, was more than reluctant to leave Australia. Mrs. Kobb recalls that they had worked very hard in Australia, they had all their belongings sent to Australia, they had had the long exhaustive trip from Europe to Australia just two years earlier. Moreover, international travel for the Moewuses involved more than the usual amount of interrogation. When they left Germany 21 months earlier, they had to get a permit from the Australian Military Occupation Office in Bonn, which involved submitting police statements from all places where they had ever lived, being fingerprinted, and waiting for four months for an exit visa. In Sydney, they learned that, as German citizens, they were not allowed to transfer directly from Sydney to New York. Instead they first had to return to Germany and again undergo the lengthy ordeal of getting an exit visa to the United States. Liselotte Moewus had a nervous breakdown because she was so much against leaving. But the biochemical genetics of microorganisms was nonexistent in Australia apart from the Moewuses' work. From her husband's point of view, staying in Australia amounted to watching from the sidelines as the scientific world passed him by (Mrs. Kobb, interview, May 26, 1987). Eventually, the Moewuses were on their way to New York via Germany.

Cautious celebrations in the United States

The Moewuses arrived in Ryan's laboratory in the middle of January 1954, and they remained situated there until June of the following year. Francis Ryan and his wife, Elizabeth, made every attempt to help them settle in at Columbia University. The Ryans remained on friendly terms with the Moewuses and saw them socially frequently (Elizabeth Ryan, interview, May 6, 1986). But so did another couple, the Kaplans. Reinhard Kaplan's recollections give some indication of the attitude toward Moewus in Germany before the war and at Columbia during the mid-1950s. Kaplan received his doctorate in botany at Leipzig in 1937. He subsequently worked on x-ray-induced mutations in higher plants, as a scientific assistant at the Kaiser-Wilhelm Institut für Zuchtungsforschung in Muncheberg (later renamed the Erwin-Baur-Institut). Kaplan first met Moewus in Berlin in 1939 when Moewus gave a lecture on his work to a large audience. At that time, Kaplan had no reason to question Moewus's work; he knew only that it fitted his own thinking about the nature of the gene and gene action and that it was associated with the great Max Hartmann: "M's findings interested us as [they] fitted our view, derived from biophysical work, of the gene as a molecular entity with a specific action on cellular processes. I had no reason for doubt as M. was sponsored by the "great Hartmann" (letter to the author, July 5, 1986).

During the war, Kaplan was enlisted as an ambulance soldier and consequently heard nothing about Moewus. While he was a soldier, he developed plans to use bacteria to study spontaneous and induced mutations. Bacteria, of course, reproduced faster, and Kaplan believed they could also be used to develop more refined studies. His plans were realized in 1945, when he showed that bacterial mutations had the same biophysical characteristics as "true" mutations of higher organisms, Like Lederberg in America, Kaplan argued that bacteria possessed genes like other organisms. Around 1949, he heard rumors that bacterial genetics had also been initiated in the United States by Lederberg, Tatum, Luria, Francis Ryan, and others.

Kaplan reported his work on bacterial mutagenesis in 1951 at the International Congress of Microbiology held in Copenhagen where he first met Francis Ryan. He met Ryan again the next year in Rome, and Ryan invited him and his wife to continue their genetics work on bacteria in his laboratory in New York. They stayed at Columbia until 1955, when Reinhard Kaplan became professor of microbiology at Frankfurt. When he arrived in New York in 1953, Ryan mentioned to him that he had invited Moewus to give him the opportunity of settling the controversy around his work. As Kaplan recalls:

> I was only superficially informed on Moewus' work since it was distant from my field of biophysical mutation genetics and since after the war contact between scientists was still rather weak in Germany. I knew nearly

nothing about the doubts against M. and became simply curious whether they were right or not. I think Francis felt the same. (Letter to the author, July 5, 1986)

The research agenda for the Moewuses at Columbia University had been set up in Germany. When the Moewuses left Australia, they were held up in Germany for several weeks before going on to the United States. Franz Moewus visited Kuhn's laboratory in Heidelberg and Hartmann's new laboratory in Tübingen. By the time he arrived in Germany, Hartmann had given two of his assistants the job of repeating some of Moewus's experiments. The work concerned sterility mutants in *Chlamydomonas*. In one of his publications with Kuhn and Löw (Kuhn, Moewus, and Löw, 1944), Moewus claimed that he had isolated a sterile mutant, filtrates of which could prevent copulation of normally sexual male and female *Chlamydomonas eugametos*. The mutant was held to produce a sterility hormone which was identified as rutin (see also Moewus and Deulofeu, 1954). However, Hartmann's assistants, Herbert Förster and Lütz Wiese (see Hartmann et al., 1954a,b), could not confirm Moewus's claims. In their hands, rutin, obtained from filtrates of certain sterile mutants, did not prevent the copulation of male and female *Chlamydomonas eugametos*.

When Moewus arrived in Tübingen, he was immediately faced with the problem that Förster and Wiese were not able to confirm his results with rutin; rutin failed to prevent copulation. But were they following the proper procedures? Mrs. Kobb was the culture expert. She says they did not:

> I saw how he [Förster] did it. He smeared the palmelloid. He had an ultraviolet lamp on, day and night in his room. So I said, "That's not the way we grow it." And Hartmann knew it. So it was decided that Förster would come for three weeks to Heidelberg and see how our stuff worked. We showed him our way here. . . . It was quite obviously our method of growing the things. We went to the U.S.A. with the understanding between Hartmann, Kuhn and Moewus that this mysterious thing would be the first project to be done in the U.S.A. We started on it, and it did not take long. [Jacques Monod] came over for a lecture. He talked about shunts – that mutants could reverse using different enzymatic pathways. So we thought it must be something like this. . . . Over four months he [Moewus] wrote reports to Kuhn and Hartmann; it was all done in public. Ryan knew about it. (Mrs. Kobb, interview, May 26, 1987)

When the Moewuses arrived in Columbia in January 1954, they established their laboratory in a very large room in Ryan's area of the Zoology Department, and they became very active. Moewus gave several seminars and demonstrations of the basic mating reaction of *Chlamydomonas*, to various people in Ryan's laboratory. During the spring, he performed experiments with *Chlamydomonas eugametos* that were designed to illuminate some discrepancies between his previously published work and the work done by Förster and Wiese in Hartmann's laboratory – that is, why, in Förster and Wiese's hands, rutin did not prevent the copulation of male and female *Chlamydomonas eugametos*. Moewus reported his results in 1954. He claimed that his own attempts to reproduce the results in question were successful, and he offered explanations to account for the failure of the experiments of Förster and Wiese. One could account for their failure, he suggested, in terms of a biological difference in the strain they used. Moewus proposed that the failure of Förster and Wiese to confirm his results was due to the appearance of an enzyme in the sterile strain they used, which converted rutin into a harmless substance. Moewus reaffirmed that rutin could be isolated from the sexually sterile strain he employed and that "if rutin is offered to normal sexually active cells, it sterilizes them completely." However, Moewus was careful to add some procedural qualifications:

> These results are only valid if the strains are grown under our usual standard methods. Their essential point is that the experimental material is obtained by spreading a suspension of motile cells on agar, where they undergo a log-phase of growth and develop within 2–3 weeks a monocellular film. Different results are obtained, when *eugametos* and *agametos* [the sexually sterile strain] are grown by spreading thick layers of *non-motile cells* on agar, where they develop a slow growth. After one or more passages on agar (each time transferring the whole mass of non-motile cells onto the next plate), the physiological behavior of *eugametos* and *agametos* is changed. They are characterized by heavy production of mucilage: they have become palmelloid. Palmelloid *agametos* shows now normal fertility, palmelloid *eugametos* is now rutin-resistant. . . . In *agametos* the naturally produced rutin, in *eugametos* the offered rutin is converted to a harmless ombuoside. The appearance of this rutin resistance is probably not due to a spontaneous mutation. Experiments are in progress in order to explain this adaptive character. (Moewus, 1954: 293)

Ryan himself, however, apparently never saw these positive experimental results. Kaplan recalls that when he and Ryan came to observe the experiments, "the claimed effects were always lacking." Moewus "made excuses saying, e.g., that just before we came he saw them but meanwhile the cells were out of the right state" (letter to the author, July 5, 1986). Were these "excuses," or was there more to the phenomenon that the Moewuses had investigated than Kaplan recognized? Again we can call upon Mrs. Kobb for an alternative explanation. Mrs.

Kobb persistently emphasizes that she and her husband always saw *Chlamydomonas* as a plant, with its different phases influenced by light, temperature, and so on. Kaplan and others, she argued, did not appreciate the complexities – that when cells were transformed into the gamete state, they remained gametes only for a few hours, depending on when the cultures were exposed to light. Therefore, one could not simply walk into the laboratory at any time and ask for the experiments to be repeated. Mrs. Kobb (notes to the author, May 29, 1988) draws an analogy with a flowering plant: "After the bloom is over, who would ever recognize a color mutant if there were no flowers left?" This same issue of "excuses" versus temporality of the phenomenon arose again later in the heat of considerable controversy when Moewus gave some critical demonstrations at Woods Hole (see Chapter 11).

In the meantime, Ryan decided to repeat Moewus's experiments himself. But he also failed to confirm Moewus's results on the effects of rutin. As will be discussed in more detail in Chapter 12, Ryan later wrote a report of his attempts to reproduce Moewus's work, which was signed by Kaplan (Ryan, unpublished report, 1955). He recorded his failed attempt to repeat Moewus's work on the effects of rutin as follows:

> In May, when Moewus' experiments on this subject had been completed, I attempted to repeat them with material prepared by Moewus and the same conditions as he had used. The sexual behavior of various untreated cultures of *eugametos* varied from poor to excellent. Nonetheless, rutin, secured from Prof. Ray Dawson, did not prevent copulation even of cultures "grown under [his] usual standard condition – by spreading a suspension of *motile* cells on agar where they undergo a log-phase of growth and develop, within 2–3 weeks, a monocellular film." Frequent experiments since that time, using rutin secured from Moewus, yielded the same negative results with cultures grown under all of the conditions he had been able to achieve during the period; vigorous clumping and pairing of male and female *eugametos* occurred amidst crystals of rutin. In one of these experiments Dr. David Bonner was a participant. In addition to the failure of rutin to act in these experiments, "*Chlamydomonas agametos*" did not behave as reported in Moewus' communication; non-palmelloid "*agametos*" were fertile and their supernatants did not sterilize cultures of *eugametos*. (Ryan, unpublished report, 1955: 1)

Further attempts at repetition in Ryan's laboratory were postponed until December. In the meantime, Franz Moewus was engaged in a very active program outside of Ryan's laboratory.

Moewus's arrival in the United States caused a great deal of excitement, controversy, and anxiety. Plans were quickly devised for him to visit several universities and scientific institutions across the country – Purdue, Vanderbilt, U.C.L.A., and Cornell, among others. Before Moewus actually arrived in New York, plans were also made for Moewus to give a series of lectures during the summer of 1954 at the Marine Biological Laboratory in Woods Hole, Cape Cod. The Marine Biological Laboratory attracted scientists from all over the world, who rented laboratory space, gave lectures, and frequently demonstrations of their work.

On November 6, 1953, Philip B. Armstrong, director of the Marine Biological Laboratory, wrote to Sonneborn asking if he could write a letter in support of Moewus's candidature to be sponsored as a visiting foreign scientist at Woods Hole:

> There is some possibility that the Josiah Macy, Jr. Foundation may support one or more visiting foreign scientists at Woods Hole this next summer.
> I have written to Dr. Frank Fremont-Smith suggesting Dr. Moewus as one candidate for this support.
> Dr. Bold had previously written me suggesting that Moewus might be included in his staff. This would probably mean increased work for Dr. Bold so I felt the above might be used to get Moewus to Woods Hole and then Dr. Bold would be free to appoint someone to his staff who would better fill his teaching needs.
> I thought it might have a salutary effect if you wrote Dr. Fremont-Smith in support of Moewus. You are better acquainted with his achievements than I and certainly made a very good case for him in your letter to me.

With Sonneborn being such a leading and well-respected geneticist, his public support for Moewus was well known in the American biological community. However, Sonneborn himself was very cautious and avoided giving unqualified support to Moewus until the doubts about him were resolved in Moewus's favor. Sonneborn (November 13, 1953) wrote back to Armstrong, stating that he would write the letter, but he qualified it, warning Armstrong:

> If you would like me to write about Moewus, I could do it; however, I would be forced to mention the fact that his work is highly controversial. The people who are really acquainted with his work think that he is really great – either a great faker or a great scientist. Most people feel the former, but I am not yet ready to discard the latter.
> If you do not want me to write, just do not answer. I presume Dr. Bold can handle the matter and there will be no need for me to intervene one way or the other:

The application was successful. The Moewuses were going to spend the summer of 1954 at Woods Hole. Franz Moewus seemed to be only delighted to be in New York and at the center of attention in the American biological community. Shortly after his arrival in New York, he wrote to Sonneborn (January 20, 1954), describing how much he looked forward to the itinerary that was being made for him. (While in the United States, Moewus wrote all his letters to his American colleagues in English, which steadily improved.)

> At last, we have arrived safely at Columbia University, and I am just about to settle down in the nice and spacy lab we got at our disposal. My stay here will probably last only about 9 months, and this time seems to be short for all plans which are on the program. Dr. Bold proposed to come to Nashville [on] March 22–30, a date which gives me time to get settled here and to get some work started. You may know that Dr. Bold has arranged for me to go to Woods Hole to give a course about my own field of work. I am very delighted to have this wonderful chance, not only to see this famous place but also to teach "Chlamydomonadology" to a group of young and interested people.
>
> I should like very much to combine my trip to Nashville with a short (?) visit at Bloomington. Will you, please, let me know whether it can be arranged that I can come from Nashville to your place.

There was a great deal of interest in Moewus's work at Indiana University. Besides Sonneborn, affiliated with the Zoology Department, both Harold Brodie and Richard Starr in the Botany Department were anxious to have Moewus come to Indiana. Starr had begun to work on the genetics of alga and eventually established a major culture collection of algae at Indiana University (see Starr, 1964). Already, in November 1953, before Moewus arrived in the United States, the biologists at Indiana were discussing ways of bringing Moewus to their departments. Funds for visiting lecturers at Indiana were low and already committed for 1954. However, there was the possibility of bringing Moewus there through the auspices of the Semicentennial Fund of Indiana University. The purpose of the fund was to bring unquestionably topnotch people to speak at the university.

On November 11, 1953, Harold Brodie wrote to Sonneborn suggesting the idea:

> Starr tells me that the idea of bringing Moewus has been discussed. Probably neither of our departments has the means and it occurred to me that it could be financed through the Semi-cent. of the Grad. School. There is to be a budget for just such things.
>
> I will put it to Winther's Committee if you think it a good idea. If they agree, then someone would have to contact Moewus (he's at Columbia, is he not?) and find out what date would suit him.
>
> Would you make the arrangements if I get the Committee's OK?

Again, Sonneborn was cautious; Moewus's eminent stature was still in doubt, and Sonneborn felt it inappropriate to use the Semicentennial funds. He gave his reasons in an interdepartmental communication to Brodie:

> Starr and I have discussed the possibility of bringing Moewus here, and there are no funds available in either department for this. We had considered applying to the Semicentennial Fund but did not feel justified in this case for making such an appeal. The purpose of this fund, as I understand it, is to have unquestionably top-notch people come here to talk. There is so much doubt and confusion about Moewus that I felt it was not appropriate to apply to the Semicentennial Fund for his backing. In a sense it gives him a stamp of approval, which I think we all would hesitate to do. Moreover, the content of his communication would be of interest primarily to biochemists, geneticists, and algologists. It was for that reason that I also hesitated to ask Sigma Xi.
>
> What I propose to do is to drop him a note telling him I would like him to come here as my guest for as long as he wishes to stay. I think he would like to come and I think we could arrange for him to stop by when he is going to be near Bloomington anyway. If it is necessary to involve any money for his visit, I will try to arrange it some other way. I would be glad to hear your reactions to this. (Sonneborn to Brodie, November 13, 1953)

Although Sonneborn did not want to give unqualified support to Moewus, there is no question that he did want Moewus to come to Indiana. He wrote to Moewus (November 13, 1953):

> I am so sorry we did not have more chance to talk to each other in Tubingen. I am hoping very much [that] you will find it possible to visit me here at your convenience. Certainly I would like you to stop by here if you are anywhere in this neighborhood. We know that the funds for visiting lecturers are low and tied up this year, but somehow we would try to arrange for your expenses from some point nearby.
>
> I hope, when you do come, you will be willing to stay here long enough to have time to discuss your work and ours. Drs. Starr, Cleland, and Brodie would also like to have conferences with you.
>
> Please let us know when you plan to come this way.

The biologists at Indiana University were persistent and urged the Graduate School to find some funds to have Moewus come to Indiana. In a letter to Dean O. Winther, Harold Brodie (January 14, 1954) justified bringing Moewus to Indiana in the most glamorous terms:

> Some weeks ago, the Biological Seminar Group discussed the possibility of bringing Dr. Franz Moewus to the campus under the auspices of and with the support of the Graduate School Semicentennial Celebration program.

Dr. Moewus was not finally nominated because it was felt by some that the content of his address might not be of sufficiently wide interest.

However, Dr. Moewus is one of the outstanding biologists in the world. He was associated with the Max Planck Institute at Heidelberg and is now in the United States as Visiting Professor of Zoology at Columbia University. Vanderbilt University and Purdue are both trying to arrange that Dr. Moewus should visit them, and the biologists of Indiana would be missing a rare opportunity not to have him come to Bloomington during this tour.

It has been suggested that, if the Semicentennial Committee would recommend bringing Dr. Moewus and make this recommendation to Dr. Ralph Collins, serious consideration would be given to making a special appropriation for the purpose.

Dr. Moewus' lecture would *not* be listed with the regular Semicentennial program and the funds would *not* come from that budget.

Dr. Sonneborn and Dr. Starr are primarily interested, but we are all very much in favor of such an action.

It is suggested that a flat fee of one hundred dollars be offered to Dr. Moewus.

If this meets with your approval, could you refer the matter to other committee members and make your recommendation to Dr. Collins.

There is some urgency in the matter as it will be necessary to contact Dr. Moewus at the earliest possible date. From his present program, we would judge that he might come here during the last week in March.

The application was successful. The Moewuses were scheduled to arrive in Bloomington in late March and planned to give a lecture in Sonneborn's department. Moewus wrote to Sonneborn (March 25, 1954), joking about how he would "compose" some unpublished results specially tailored to suit Sonneborn's interest in nucleocytoplasmic relations.

Now everything is arranged for our trip to which we are looking forward eagerly. I am just about to "compose" some unpublished material for the talk in your laboratory. The best title would be perhaps "Interrelations of Gene, Enzyme, and Chloroplast Demonstrated by the ru^+ Gene in *Chlamydomonas*."

Mrs. Kobb recalls that they had a wonderful time in Bloomington. They were greeted by the president of the university after Moewus gave one of his beautiful talks to a large audience. Moewus did not run into any major difficulties during his stay at Bloomington. He reaffirmed his great respect for Sonneborn and his research on *Paramecium*. Upon returning to New York, Moewus wrote to Sonneborn (April 20, 1954), stressing the painstaking nature of the research in Sonneborn's laboratory and how *Paramecium* genetics seemed to be clearly related to his work on *Chlamydomonas*:

Being back from our trip and having recovered from all this talking-talking I want to thank you for the interesting time you gave us at Indiana University. I am sure this visit will be the spotlight of our American period. You will understand how much it meant to me to meet you and your group who are so well informed about my own field and who are carrying out similar tedious and intensive experiments on a field which seems to be so closely related in many aspects to the *Chlamydomonas* story.

Spots before their eyes

Moewus did run into difficulties when giving lectures elsewhere in the United States, however. One such occasion was when he gave a lecture in the Botany Department of the University of Pennsylvania. The lecture was similar to the one that he gave at Indiana University; it concerned the interrelations of genes and chloroplasts in *Chlamydomonas*. The difficulty centered on paper chromatography, and, similar to the anomalous result encountered by Arthur Birch in Sydney, it involved a particular spot on one of the chromatogram slides displayed by Moewus. It will be remembered that when Moewus worked with Birch on flavonoid biogenesis, anomalous chromatographic results were produced. It later occurred to Birch that Moewus might have been doping his cultures with synthetic chemicals. However, that interpretation was only one possibility. As discussed earlier, Birch believed Moewus, and he was willing to believe that a new biochemical pathway was discovered. The authenticity of Moewus's chromatogram slide was challenged by William Stepka, then assistant professor in the Botany Department of the University of Pennsylvania. Herbert Stern, who spent the academic year 1954–1955 in the Department of Botany at the University of Pennsylvania, recalled the event:

> Stepka had worked with Calvin and was considered to be an expert in paper chromatography, particularly in connection with chloroplast metabolism. After Moewus finished his talk (or, perhaps, during it), Stepka asked a question about a particular spot on one of the chromatogram slides displayed. That spot was a key point in the inferences drawn by Moewus I do not recall whether Stepka claimed it to be a fake at the seminar or immediately after it. If the former, then he also displayed his anger at the fakery afterwards. . . . Clearly, I am not so sure as to whether Moewus was confronted with the charge. The faculty had the highest regard for Stepka's integrity and scientific reliability; so do I. I do not recall anyone challenging Stepka's charge. (Stern, letter to the author, October 31, 1986)

The difficulty of inexplicable spots on paper chromatography occurred again when Bernard Davis invited Moewus to present a seminar before his research group located at the U.S. Public Health Service Tuberculosis Research Laboratory housed at Cornell University Medical College. At that time, Davis was one of the few researchers working on the biosynthetic pathway of the aromatic amino acids. Over a number of years, Davis's group had worked out a number of steps in the common pathway, using auxotrophic mutants of *E. coli* (mutants that require growth substances not needed by the normal wild type). They were then planning to move on, from the common pathway into the branches to the individual amino acids. Davis recalls (letter to the author, May 22, 1985), "I was amazed to read Moewus' report that he had managed, in a very short time, to verify with mutants of *Chlamydomonas* many of our steps and to add more on the specific paths to phenylalanine and tyrosine." However, the idea that the compounds were being added to the cultures rather than being accumulated by the mutants did not occur to Davis and Birch until the next summer which Moewus spent in Woods Hole.

Davis recalls that when Moewus gave his presentation to the group at Cornell, he was

> impressed by the elegant results, in which he used paper chromatography to demonstrate spots of various compounds accumulated by various mutants. One thing was very odd, but I stupidly discarded it too quickly from my thoughts: the accumulated tyrosine was a single spot, but the accumulated phenylalanine was present as two spots very closely together. The explanation, which came to me later, is that the standard commercial products at that time were L-tyrosine but DL-phenylalanine; and I did not immediately realize that the cellulose of the paper, being optically active, would interact differently with the two isomers of phenylalanine. For *Chlamydomonas* to make a racemic mixture of phenylalanines would be contrary to all we know about amino acid biosynthesis. (Davis, letter to the author, May 22, 1985)

Although Davis, like Birch, had doubts about the chromatographic results, he did not express them directly to Moewus. The first and most widely known public confrontations with Moewus occurred in the summer of 1954 at the Marine Biological Laboratory (MBL) at Woods Hole, Cape Cod, Massachusetts.

11. The Woods Hole trials

> I was there the summer Ruth Sager exposed him. He was demanding one coincidence after another. He was fudging something over Ruth Sager's eyes. I don't remember the details. (Lederberg, interview, May 26, 1986)

> I was really never interested in what Moewus was doing. Whereas there are people who are sort of dilettantes of science who like to read about controversial things and then pass judgment on them without knowing what the hell they're talking about – and I am not one of those. And we found quite a few of them before the Moewus thing was over. But as I was basically not interested in that problem, I hoped it would go away. (Sager, interview, May 15, 1986)

Summers get hot in New York City, and there were no air-conditioners in the laboratories at Columbia University in the mid-1950s. When the temperature began to soar, many leading biologists sought relief in the mild ocean breezes around Woods Hole. Founded in 1888, the Marine Biological Laboratory at Woods Hole became a Mecca for experimental biologists. During the late nineteenth century, the heyday of experimental embryology, Woods Hole offered embryologists a place to do collaborative research on amphibians and echinoderms – the favored organisms for embryological research. During the twentieth century, it continued to offer laboratory space which scientists could rent, and an outstanding library for writing up their research papers and monographs. During the 1950s, Woods Hole offered biologists a relaxing haven by the sea where they could meet and discuss with their colleagues when they were relieved of teaching duties during the summers.

Each summer Woods Hole attracted biologists from all over the world. But perhaps no summer caused as much calamity, excitement, and tension as the summer of 1954 when Moewus was there. As mentioned in the last chapter, Moewus had been invited to give a series of lectures in Harold Bold's course at Woods Hole. Moewus had 10 or 12 days in which he lectured every day. Many important people were in Woods Hole in the summer of 1954, including Boris Ephrussi, Joshua Lederberg, James Watson, Bernard Davis, Reinhard Kaplan, and Francis Ryan (see Figure 9), and many of them came to the lectures. Watson had just returned from Cambridge, triumphantly celebrating his work with Francis Crick on the structure of DNA (see Watson and Crick, 1953). He was one of the few who never came to Moewus's lectures. Mrs. Kobb recalls:

233

Figure 9. Left to right: David Bonner, Francis Ryan, and Ralph Lewin. Lewin's house near Woods Hole, circa 1958. (Photo courtesy Ralph Lewin.)

Watson didn't come to the lectures. At this time, he was very convinced of himself – running around with wild hair-flying, teeshirt – at night on the beach playing the fiddle and all these outlandish things.

And my husband and I are sitting in this room composing these lectures for the next day – almost no time to go to the beach ever, because he was in so much demand for performing.

And everyone sitting and discussing him while he was doing the work.

The students were delightful, and Harold Bold . . . was a true friend. (Mrs. Kobb, interview, May 26, 1987)

It is fair to say that Watson and his double helix were not the only center of attraction. Moewus shared the limelight. In fact, Moewus gave one of the famous Friday evening lectures at Woods Hole in the summer of 1954. This was the star hour of every biologist. The Friday evening lectures are well attended to this day. Only the most important scientists in the world were invited to give one. And Moewus, Mrs. Kobb recalls, gave a marvelous lecture which was followed by a reception. However, difficulties began the next day. A colloquium was held where scientists talked for about 20 minutes followed by discussion. Moewus reported on the effects of rutin on sexuality (see Moewus, 1954). His talk was based on his work which attempted to illuminate the discrepancies between his previously published results and those of Förster and Wiese: how culture methods were influencing the sterility effect of rutin in *Chlamydomonas*. In the discussion, a German biologist, Karl Grell, stood up and denounced Moewus, declaring that Moewus's results had been refuted by those of Förster and Wiese, and that Hartmann had repudiated Moewus's work.

This, of course, was the very matter Moewus was addressing. But Grell's English was poor; he was not a geneticist, and, according to Mrs. Kobb, he had not been informed about what was going on – that Moewus, Hartmann, and Kuhn had been corresponding for four months trying to sort out the difficulties. However, Grell did know of the work of Förster and Wiese, at least superficially. Förster and Wiese's first published report had just appeared in April 1954, in the leading German scientific journal, *Zeitschrift für Naturforschung*. Their first paper was prefaced by a statement by Hartmann himself (Förster and Wiese, 1954a: 470). As emphasized throughout this story, the verification of experimental results is not a simple matter. It requires judgments that the experiments were done properly, on the right material, judgments about the competency and integrity of the individual, and testimony from authoritative scientists. Hartmann's prestige as a great protozoologist had lent strong support to the validity of Moewus's work. Hartmann may have been lax in his scrutiny of Moewus's work because it tended to confirm his theories of sexuality. He now stated that he had personally overseen the work of Förster and Wiese:

> The experiments in the paper before you concern the frequently doubted results of Moewus on *Chlamydomonas eugametos*. The experiments were carried out with the greatest care and I have made sure for myself that in all cases they were carried out in the way described here and can be reproduced. Since the fundamental experiments of Moewus could not be reproduced on the material that was available, it was not possible to test the dependence of the formation of male and female gamones on light and their change due to light. It could only be shown that filtrates of male and female gametes contain a particular substance that works differentially. About experiments on these filtrates we will report later.
>
> Max Hartmann

This was the first time Hartmann publicly disclaimed any of Moewus's work. However, it must be emphasized that Hartmann never made any public statements to the effect that Moewus had deliberately intended to deceive his colleagues.

Grell was visiting the United States as a Rockefeller fellow from Hartmann's laboratory in Tübingen. He had a long history of listening to criticisms of Moewus's work in Germany. Grell completed his Ph.D. in zoology at the University of Bonn in 1937. His thesis was on the anatomy and histology of the scorpion fly. The following year he was awarded a position as an assistant to teach and do research in the Botany Department at Bonn. He first heard of Moewus before the war when, in one of the seminars in the Botany Department, a student gave a summary of Moewus's work. At that time, Grell remembers, many people in the department were skeptical of Moewus's claims and had doubts about him. He did not hear any more about Moewus until after the war (Grell, interview, October 27, 1985).

In the spring of 1940, Grell was taken into the army as a trooper in France. But he was soon taken out of the army, along with several other zoologists and chemists, to work in the Sanitary Corps against malaria. However, during the war years he spent most of his time working in the Antiplague Institute in Roscoff or Bonn. Grell recalls that their central question was hypothetical: "What would happen if plague breaks out in the German troops – because we knew that the Russians had specialists for plagues in their headquarters, and nobody knew what they would do. It was more or less theoretical what we did" (Grell, interview, October 27, 1985).

After the war, Grell returned to the Zoology Department of the University of Bonn and through the writings of Max Hartmann became interested in protozoology. During the late 1940s, he learned of the work of Sonneborn and his collaborators on *Paramecium* and began to write a book on Protozoa and their genetics, introducing the American work to German biologists and integrating it with recent work done in Germany. At that time, Grell had some cordial correspondence with Moewus, who kept him up to date on some of his work to be included in his book. On November 3, 1950, Moewus wrote to Grell about his work and that of Sonneborn:

> Thanks very much for your letter of the 31.10. With the same mail I am sending you some reprints including the first summary of *Chlamydomonas*. Unfortunately, there are no photos in the reprints. If by chance you need something on the subject for your little book, please write me. Heartfelt thanks also for your work. My wife particularly liked the essay on *Microcosmus*. She is a pure botanist and wants to include it in a lecture on Lysenkoism and Sonneborn's work, which unfortunately is chief witness for the Russian genetics. So your descriptions are very valuable.
>
> I am very grateful to you for leaving the Naples *Chlamydomonas*.

The Institute at Bonn had been partially destroyed during the war, and the conditions for research were not good, to say the least. In 1951, Grell accepted the invitation of Max Hartmann to work in his laboratory at the Max Planck Institute for Biology, which had been moved from

Berlin to Tubingen after the war. He arrived in Tübingen just around the time when Hartmann became "convinced that Moewus' work is a hoax" (Grell, letter to the author, March 5, 1985). Grell recalls that Hartmann was well liked by his students and assistants. Moewus had been one of his favorites, so Hartmann was greatly shaken by the Moewus affair: "For Hartmann, it was a stroke of fate [*Schicksalsschlag* in German] comparable with the loss of his only son at the Nanga Parbat expedition. Hartmann's mistake was his credulity." Grell himself would in fact succeed Hartmann, in a position which, Mrs. Kobb remembers, had been promised to Moewus for many years. As she put it, "Grell had stolen Hartmann's heart" (Mrs. Kobb, interview, May 26, 1987).

When Grell arrived in the United States in 1954, he became virtually the official spokesman for Hartmann. In 1954, he was visiting the Scripps Institution of Oceanography and was invited by David Bonner, a member of the famous *Neurospora* school, to give a lecture to the group at the California Institute of Technology in Pasadena. Grell's lecture was on the development and sex determination of a parasitic worm. The worm was completely transparent, and Grell had made a film showing the whole of development. His talk was not on biochemical genetics. Grell had first heard of the work on *Neurospora* by Beadle and his colleagues in 1949, when the news first entered Germany. He suspected, quite rightly, that they were more interested in the news he brought from Hartmann's laboratory than in his own work, and that they would question him about Moewus.

When Grell spoke out against Moewus in Woods Hole, however, he was received with intensely mixed feelings. Some greeted his remarks enthusiastically; others deplored them. Mrs. Kobb recalls the episode vividly:

> During the colloquium, my husband reported on this very matter: how the two culture methods were influencing their sterility. He reported in English – very good English – beautifully worked out.
>
> But Mr. Grell – who was there – he was not informed. He had not known about this work. He didn't understand a word about it. He stood up and made a large accusation against Moewus.
>
> There was a murmur. Everybody said, "Who is that man?"
>
> The moderator said, "Dr. Moewus do you want to answer that man?" Franz said, "No."
>
> What could he have said without blaming Hartmann, Kuhn, or everybody? Should he have said, "I have corresponded for four months with these people in Germany, and now I am telling what we have done against it."

> Grell was isolated in the dining room. He was sitting by himself. At that time people found it outrageous. In America, things like this are not done – shooting somebody like so. It got around. . . . (Mrs. Kobb, interview, May 26, 1987)

Grell's recollections are similar: "I had a lot of trouble in Woods Hole. I had the feeling – I must say it. We tried to repeat it, it had been done and we showed it was a hoax." There were two kinds of criticisms against him:

> Some people said that you cannot say that in the United States. They said you must be a little more polite and not just say "it cannot be." Someone [a German emigré] said to me, "You shouldn't speak against other Germans," and I said, "I have completely the other opinion because, if I have the impression that this man is a swindler, if he is a German, I am obliged to say it." (Grell, interview, October 27, 1985)

This controversy just marked the beginning of the confrontations at Woods Hole concerning Moewus and his work. The most serious difficulties occurred when Moewus gave a demonstration of some sexual phenomena in *Chlamydomonas* he had been reporting for many years. But this confrontation was preceded by still other confrontations.

A case of trickery

Feelings ran high during the summer of 1954 in Woods Hole. Lederberg, Sager, Ephrussi, Davis, and Watson spent the summer at Woods Hole. Early in the summer, Arthur Birch arrived looking for Moewus. Birch was in the United States on some university business making plans for a new chemistry school in Sydney. The work Moewus had done with him in Sydney was incomplete and Birch wanted to test it further. He was not yet convinced that Moewus's work was entirely unreliable. Moewus, he claimed, had promised to leave some mutants in Sydney, but failed to do so. He took the opportunity to visit Ryan's department at Columbia and later contacted Ruth Sager at Woods Hole. Sager and Birch went to see Moewus about the mutants he promised to leave in Sydney. Birch recalls the confrontation:

> So Ruth and I went to Woods Hole, and we saw Moewus in the lab . . . and I said, "Which are those mutants?" . . . so he pointed to a row of test tubes containing green solutions in a rack. So Ruth Sager picked this up and walked out. Moewus shouted something about contacting the police . . . stealing his stuff and the rest of it. However, he didn't. I held him back until she disappeared (not physically of course). She then tested these mutants at Columbia . . . and as far as she could tell, they were

normal (cells). They were not mutants at all! . . . (Birch, interview, October 23, 1984)

Birch conferred with Bernard Davis about the biochemical aspects of Moewus's work, particularly about the question of the double spot of phenylalanine. They agreed that Moewus must have doped his cultures. Davis's recollection of Moewus's response was similar to that of Birch:

> The next summer I spent at the Marine Biological Laboratory at Woods Hole, where Moewus had been given the opportunity, I believe, in one of the regular courses, to demonstrate his dramatic results. There was a great deal of private discussion of the increasing suspicion that he was a fraud. And this culminated in a visit to his laboratory by Dr. Sonneborn, who was widely trusted as the most judicious and broadly knowledgeable of microbial geneticists. I accompanied him with . . . [Boris Ephrussi]. By that time discussions with other biochemists had provided me with the explanation for the double spot of phenylalanine, and I knew that we were dealing with a disturbed person and a fraud. Unfortunately I cannot recall the nature of the evidence or the arguments that we considered, but I recall being angry at Moewus's cool insistence on the reliability of his results. I do not see how we can consider sloppiness as an explanation for his doping a culture with synthetic material that obviously came out of a stock bottle. (B. Davis, letter to the author, May 22, 1985)

The most controversial and publicized case against Moewus at Woods Hole involved a demonstration he gave in the presence of Ruth Sager. Demonstrations at Woods Hole were not uncommon occurrences. If one had something exciting and wanted to show people, one could give such a presentation. In the early days at Woods Hole, when biology was less hectic, demonstrations were more common. Sonneborn, for example, as a young scientist gave a demonstration to T. H. Morgan and other leading geneticists and protozoologists in the late 1930s at Woods Hole when he first learned how to control mating in *Paramecium*. The main difference in the case of Moewus was that it was carried out under a cloud of suspicion.

The demonstration in question concerned Moewus's claim that he had identified a gene that controls the production of the motility hormone crocetin. Moewus attempted to "show" that his putative *cro* (crocetin) mutant strain of *Chlamydomonas* fails to become motile or to form flagella in distilled water, but develops flagella and becomes motile in crocetin solution and in filtrates from nonmutant motile cultures. The demonstration he gave when Ruth Sager was present was not successful. In Sager's view, Moewus had deliberately attempted to trick her, and she claimed to have caught him in the act. The details of the demonstration in question will be dealt with shortly. First, however, it will be helpful to review Sager's position in the controversy.

The summer of 1954 at Woods Hole recalls some of the most unpleasant memories for Sager. She had no intention of, or interest in, refuting Moewus or participating in the controversy. As discussed in Chapter 7, since the early 1950s, Sager was working systematically on chloroplast genetics using *Chlamydomonas reinhardi*. Her work was only remotely related to the main focus of Moewus's work, which centered on chromosomal genes, sexuality, and the biosynthesis of aromatic compounds. At that time, Sager was interested in chromosomal genetics only as a necessary background for studying the cytoplasm. Sager had been skeptical of Moewus's work since G. M. Smith's first reports in the early 1950s. It will be recalled that Smith could not confirm any of Moewus's results on relative sexuality. However, as discussed in the previous chapters, many investigators, including Smith himself, did not consider his results as a refutation since he worked on a different species of *Chlamydomonas*. Moreover, as Smith claimed, his lack of confirmation could also have been due to his lack of technical skill.

Sager worked on *Chlamydomonas reinhardi*, not on *Chlamydomonas eugametos*. If Moewus's results were authentic, Sager believed:

> One possibility was that there were some subgenus of species which behaved quite differently from the one I was working on. It was perfectly clear to me that the species which had been isolated by Gilbert Smith – that it did not behave that way. (Sager, interview, May 15, 1986)

Nonetheless, Sager was very skeptical of the authenticity of Moewus's results on the effect of carotenoid pigments on sexuality. She had been isolating chlorophyll mutants and in the course of that work she had isolated some mutants that did affect carotenoid pathways. But none of these mutants that she had isolated had anything to do with mating-type inheritance as suggested by Moewus. But, again as Sager emphasized, she was "not at all either intellectually or emotionally interested."

In Sager's view, the most striking aspect of the controversy around Moewus was that many people were making judgments about things they knew nothing about. Sager knew the organism, but none of the others involved in the controversy at Woods Hole did:

> Here they were, willing to make all these decisions and they knew nothing about the system – the organism and how it behaves in the life cycle. They didn't know what a gamete was or how it looked. So it did not increase my respect for any of those people.

Sager put Birch and Davis in a different category:

> They had knowledge of a different kind, but they could certainly evaluate
> what was going on in terms of the biochemistry that was being claimed
> and unsubstantiated. But really the behavior of the mutants was the key.
> (Sager, interview, May 15, 1986)

Sager may have known the organism, but in the early 1950s she did
not yet possess the authority that had been granted to many of the other
participants in the debate. Her work on chloroplast genetics was still in
its early stages. She did not achieve a university appointment until after
1960, when she obtained a position at Hunter College. In 1961, she, in
collaboration with Francis Ryan, published her well-known book, *Cell
Heredity*, subtitled *An Analysis of the Mechanisms of Heredity at the Cellular
Level*. This book was considered by many to be the first molecular genetics textbook. It attempted to provide a wholly new synthesis of the field,
discussing and ordering current concepts that were emerging from the
"biological revolution" that was in full swing at the time. Sager's pathbreaking work on chloroplast genetics culminated in 1972 with the publication of her celebrated text, *Cytoplasmic Genes and Organelles*.

Sager was well known among the community of geneticists and biochemists who played leading roles in the molecular genetics revolution.
Yet, no one had consulted her about Moewus. Everyone knew that she
was committed to working on *Chlamydomonas* and that she was investigating chloroplast genetics. Bernard Davis took van Niel's course at Stanford
the same year as Sager did. At the time when she was at the Rockefeller
Institute, Davis had a laboratory across the street, so they saw each other
frequently. Ryan also knew of Sager's work; yet he did not consult her
about the idea of inviting Moewus to Columbia. In fact, she had no idea
that he was at Ryan's laboratory until she met him several months after he
had arrived. Sager was also good friends with Sonneborn. She remarked,
"I think he [Sonneborn] was a brilliant man. He did a lot of good things
for me. He was just wrong" (Sager, interview, May 15, 1986).

In Sager's view, the episode at Woods Hole was more or less forced
on her. "I was not interested in disproving Moewus; it wasn't on my
agenda" (Sager, interview, May 15, 1986). Nonetheless, when Sager
claimed that Moewus attempted to trick her at Woods Hole, her character and credibility as a scientist were questioned and came under great
scrutiny. As she suggests, perhaps it was a question of sexism:

> It was ridiculous. Here were all these people who considered themselves
> so important, busy making judgments when they knew absolutely noth-

ing about the system. Well, I knew them all [Lederberg, Watson, Davis etc.], and they all knew me, so it wasn't a question of being a hidden character; I certainly was not. But . . . they really did not take me seriously as a scientist. . . . Maybe it was prejudice against women, which was very very rampant in those days. Jim Watson is still very antifemale, and on that I am happy to be quoted. But I was very mad about it. I felt hurt. But when it really came down to the end, I was the one who made the decision. (Sager, interview, May 15, 1986)

Undoubtedly sexism was playing its usual role in the allocation of recognition. However, the issue of Sager's lack of credibility cannot be reduced simply to sexism, which was especially rampant in science of the 1950s. Sager went further in her speculations about the attitudes toward her and suggested that, "Perhaps they thought I had some ulterior motive." Indeed, one might raise the possibility of two ulterior motives lurking behind Sager's attitude toward Moewus. One might emerge out of the struggle for recognition in science. Sager may have been seen as competing for prestige with Moewus. If Sager felt she was the most qualified to make judgments, she was also one of the most suspect. This suspicion was not necessarily due to any peculiarities about her as an individual; it was the result of an intrinsic feature of scientific activity itself. A scientist is able to enhance her or his position in the social hierarchy of science, not just by receiving recognition, but also by granting recognition and credit to the work of other scientists that is most like his or her own and by withholding recognition and credit from work that is least like his or her own. As a result, there can be no impartial judgments in science because there are no judges who are not party to the case. The same characteristics that qualify an individual to serve as a judge in any particular case also necessarily endow that individual with an interest in the outcome of that case.

In addition to this particular bias resulting from the intrinsic internal politics of science, another issue needs to be discerned: the question of anti-Germanic sentiment after the war and its influence on Sager's attitude toward Moewus. Regarding anti-German sentiment, Sager commented:

It couldn't help but play some part. But I would say it's very hard to evaluate. . . . It's a real hornets' nest because . . . all the anti-German sentiment, antisemitism, anti-antisemitism. There are just gobs of emotion. . . . (Sager, interview, May 15, 1986)

Sager is of Jewish ancestry, and so is Joshua Lederberg who emphasized:

I think you can't ignore the political context. If Moewus had not been a Nazi then everything might have been looked at differently. . . . That was everybody's attribution about him. How could he have had all these facilities and so on during the war if he hadn't been a supporter of the regime? . . . The willingness or desire to lean over backward against all the other adverse aspects of assigning him credit – here is someone who has a source of disreputability on one other ground – becomes impossible. (Lederberg, interview, May 6, 1986)

Lederberg recalls that there was a great deal of anti-German sentiment after the war, which he believes is not altogether gone from American science. In Lederberg's view, any scientist who stayed in Germany would be suspect. However, as Sager suggests, it is difficult to weigh the extent of the influence of anti-German sentiment in the Moewus affair. If it helped to shape Sager's attitude toward Moewus, it was completely absent in Moewus's chief apologist in America. Sonneborn was also of Jewish ancestry, and, as I have discussed earlier, he had experienced considerable antisemitism in American science during World War II. In the case of Moewus, Sonneborn wanted the "internal" sociointellectual stakes in genetics research to take precedence over larger political attitudes.

Sonneborn arrived in Woods Hole late the very evening of the day of Moewus's controversial demonstration that was part of Moewus's student course on the genetics and physiology of algae. He quickly learned of the gossip that was circulating. Moewus complained that Ruth Sager was making propaganda against him. Sonneborn immediately established an informal committee to investigate the story – which included himself, Bernie Davis, and Boris Ephrussi. Sonneborn, Davis, and Ephrussi made it their business to get the story directly from Moewus. In a letter to Moewus, dated August 18, 1954, which Sonneborn intended to publish, he described in detail the events that took place during Moewus's demonstration. The charges against Moewus concerned allegations that he attempted to rig a demonstration by doping his cultures with a dilute iodine solution. Iodine was generally used at that time for making flagella visible under the microscope. Depending on the dose, the cells may be killed when exposed to it for some time. Sager's own recollections of the details of the demonstration are hazy. Sonneborn's report (August 18, 1954) remains the sole documentation of the episode; in all essentials, he claimed that it agreed with what Sager had said in the summer of 1954:

Dear Dr. Moewus:
    As you know, I have for 20 years followed with interest and care your remarkable publications on *Chlamydomonas* and other algae. Although in

1940 I published a critique of certain aspects of your work, through the course of the years I have not allowed my criticisms of others to close my eyes to what still seemed to me important accomplishments recorded in your publications. Because I was of the opinion that you had made great contributions, regardless of whether your critics were correct, and because this opinion was not generally shared by my colleagues, I published in 1951 an attempt to evaluate your position in the history of the development of several fields of current interest: the genetics of microorganisms, biochemical genetics, sexuality, and a number of special but theoretically important genetic problems. I have been told that this paper played a part in the invitation, which you received and accepted, to come to work for a year in this country. The fact that I have publicly participated in attempts to arrive at a just evaluation of your work, together with the potentially very great significance of that work, lays upon me a special obligation not to withold information in my possession which may be important for other biologists in their attempts to evaluate, understand, repeat or extend your work. From my previous correspondence and conversations with you, there can be no reasonable doubt, and I feel sure you have had none, of my persistent open-mindedness about your work, my friendliness towards you, and my desire to see that your accomplishments be properly recognized and correctly evaluated. This letter is consistent with that position. I shall limit this public letter to a statement of the facts, leaving to those who read it the task of drawing the conclusions that may seem to them warranted. I also wish you to be fully informed of my intention to publish this so that you may, if you wish, submit for concurrent publication a reply.

Before proceeding to a statement of the facts, let me make very clear that in themselves they do not directly bear upon the validity of any specific observation, experiment or idea set forth in your publications. For final judgment of this, we all recognise that independent confirmation is the final test. In biology, however, perhaps more so than in some other sciences, exact repetition of experiments with precisely the same material and under effectively identical conditions may be beset with great difficulties and independent confirmation may often require years of work. For short-term, tentative judgments, therefore, confidence in the investigator – for ill or good – assumes importance. The facts here set forth bear only on the question of confidence.

I shall limit myself to the facts as you presented them in discussion with Boris Ephrussi, Bernard Davis and myself on Aug. 13, 1954, at the MBL. In all essentials they agree with reports by other people involved. Finally, it should be made clear that this meeting took place in response to your expressed desire for a full and frank discussion, after I had told you that apparently serious questions concerning you were being discussed by others and you replied that you felt it was only just that the questions be set forth to you directly. This the three of us undertook to do. From this discussion the following facts emerged:

1. On June 6, 1954, an agar plate culture of a single clone of a male *cro* mutant of *Chlamydomonas* was subcultured onto three agar slants. Likewise these subcultures were made of a single clone plate of a female *cro* mutant. One male and one female subculture [were] left as a reserve in your laboratory; the other two of each sex were brought to the Marine Biological Laboratory for use in class demonstrations.

2. One male and one female subculture were used for class demonstration. The day before the demonstration, you personally verified that the cultures behaved in the fashion reported by you for *cro* mutants, viz: the cells failed to become motile or to form flagella in distilled water, but developed flagella and motility in crocin-solution and in filtrates from non-mutant motile cultures. These relations were then demonstrated to the class. The slants providing cells for these demonstrations were used up and are therefore no longer available.

3. A short time later (specify time), one of the two remaining cultures was similarly used for a demonstration to two investigators with research interests related to your own. This demonstration failed. It was marked by the following events.

   a. A sample was placed in a drop presumed to be distilled water. Contrary to expectation, the cells developed flagella and became motile. You suggested that the beaker of distilled water might have been contaminated with crocin, since crocin was being used in the laboratory and you had reported it to be very active in induction of flagella and motility. You therefore proposed to try again with distilled water from another container. For this purpose you used what was announced as pure crocin-free distilled water to set up a second trial with fresh cells. Briefly observing these under low magnification you saw they were non-motile, as *cro* mutants should be in distilled water. However, you were able to see – even under low magnification – that the cells had flagella. Although *cro* mutants should not have flagella, you made no comment. The two observers then looked. After examination under higher magnification, they concluded that the cells were dead. One of them therefore performed a simple experiment: two drops of a normal (non-mutant) motile culture were placed upon a slide. The coverslip that had been used in the second trial was placed upon one drop. The cells in this preparation also died while those in the other drop did not. The flask used in the second trial contained a dilute iodine solution – it was so dilute as to appear quite colorless.

   b. After a third trial, with a still different source of water, you agreed that flagellated motile cells were appearing in the putative *cro* mutant and, following the suggestion of one of the observers, entertained the hypothesis that back-mutation from *cro* to normal might have occurred. To see whether flagellated cells were present already on the agar slant culture, you proposed to transfer some cells directly from the agar slant to a drop of iodine solution. This revealed a low percentage of flagellated cells. You then graciously gave a sample of the culture to one of the observers to see if normal and *cro* mutants could be isolated.

c. When the three of us came to talk to you, though not earlier, you said that the fluid used in the second trial was dilute iodine solution, not distilled water. You identified the flask in question for us by means of a crack in it. It was not labeled. Neither was there a label on a flask of the same size, said to contain distilled water lying beside it on the same shelf. Although the fluid used in the second trial had revealed flagella plainly, you obtained a different (stronger) flask of iodine solution for the final direct observation for flagellated cells on the agar slant.

I asked you forthrightly if you could understand that the manner of your demonstration to the two investigators was such as to raise with them the question of whether you had deliberately intended to deceive them. I emphasized that your second trial, made with the announced intention of substituting crocin-free water for water suspected of being contaminated with crocin, was performed without comment from you; that they had at first seen that the organisms were non-motile, as required for a successful demonstration, and had then proceeded to discover not only that the organisms were dead but that the coverslips used in this trial carried material which killed other normal motile cells. You admitted that the impression of intent to deceive could well have been obtained, but denied such intention. You said, on the contrary, that you expected the facts to be obvious to them and that it is your practice to allow observers to find out the facts for themselves rather than be told by you; but that in this case it was a mistake in judgment, to allow matters to proceed in a way that could be otherwise interpreted.

When we inquired as to whether it was your regular practice to have superficially identical flasks of apparently clear water remaining side by side unlabeled, when one contained distilled water and the other dilute iodine solution, you replied that this was not practiced in your research laboratory, but that at the MBL there were four people using the laboratory, including yourself and your wife, and implied you had no control over what others might do. From your account, you did not know the flask employed in the important second trial contained iodinated water until after you used it.

As I said above, I do not wish to draw conclusions, but to limit myself strictly to a factual report, the facts being set forth as you stated to the three of us. One purpose of this letter is to freeze the facts before they can be distorted by repeated oral transfers. It is not my purpose to interpret the facts or express an opinion about them. Like myself, each interested reader may do this for himself. I must however inform you that, although the facts are susceptible of more than one interpretation, I now wish to dissociate myself from the task of trying to analyse and evaluate your work and shall content myself to wait for independent confirmation. Your eagerly expressed willingness to assist Dr. Davis [in] carrying out independent studies of your biochemical mutants in which he is interested leads me to hope that the wait may be short on certain critical points.

No matter what interpretation of the events at Woods Hole we might make, as Sonneborn wrote in the introduction of his letter to Moewus of August 18, 1954, Moewus's inability to reproduce his results in Woods Hole "did not bear directly upon the validity of any specific observation, experiment or idea set forth" in his publications. As Sonneborn emphasized, "In biology . . . perhaps more so than in some other sciences exact repetition of experiments with precisely the same material and under effectively similar conditions may be beset with great difficulties." In biology, it is not uncommon for a researcher to encounter difficulties in reproducing his or her own results. As machine-like as organisms are constructed to be, they cannot always be forced to provide consistent and neatly packaged results that fit into the tables, charts, and graphs of short articles. Organisms may not behave faithfully so as to facilitate technical precision. The behavior of the organism under observation may be controlled by a complexity of factors that cannot be rigorously controlled. In Sonneborn's view, "Independent confirmation may often require years of work." In the short term, at least, we are left with judgment about the competency and motives of the individual researcher.

When appreciating Sonneborn's decision to dissociate himself from Moewus and his work, one has to consider all the other criticisms made against the validity of Moewus's work, and the amount of time and effort Sonneborn had invested in it. As he wrote to Moewus after he defended his work at Cold Spring Harbor in 1951, he had taken considerable risk in speaking out in Moewus's defense. However, the events in Woods Hole had greatly shaken Sonneborn's confidence in Moewus. He no longer felt confident to raise the ante any further in Moewus's favor. The risks were simply too high to stay in the game. Although he could not be sure, it appeared as if Moewus had been bluffing all along, and now he was caught in the very act of rigging his experiments. It was now better for Sonneborn to cut his losses and bow out of the contest rather than risk losing his own hard-earned credibility.

12. Closing the controversy

The Moewus question seems to have been pretty settled in this country, too, what with Ryan's statement and all the rest.
(Elly Hinreiner to Ralph Lewin, November 11, 1955)

Moewus returned to Ryan's laboratory, dismayed by his experiences at Woods Hole. He complained to Ryan that his stocks were in poor condition; further attempts at repetition were postponed until December 1954 (see Ryan, unpublished report, 1955: 2). In the meantime, Moewus continued to do a great deal of lecturing in the United States. During this period, he wrote and lectured on flavonoid biosynthesis. It will be recalled that Moewus and his co-workers in Germany had claimed that flavonoid pigments acted as sex hormones in *Chlamydomonas*, while in Sydney, Birch, Donovan, and Moewus had published detailed descriptions of the biochemical path leading to flavonoid synthesis. Despite Birch's revelation at Woods Hole that Moewus was probably doping his cultures with compounds out of stock bottles, Moewus insisted that his work with Birch was entirely sound. In fact, according to Birch, Moewus wrote him, after the Woods Hole episodes, insisting that he publish a full account of their work. Birch refused (Birch, interview, October 23, 1984).

In November, Moewus visited G. M. Smith and others who were working on *Chlamydomonas* at the University of California at Los Angeles and at the University of California at Davis. His lectures caused a great deal of excitement, but there were no major attacks made on the credibility of his work, at least to his face. Moewus elaborated on the critical conditions that he claimed were necessary to produce the positive results with his mutant strains of *Chlamydomonas*. He also promised to send stocks of his putative flavonoid-requiring mutants to researchers at the University of California at Davis. His *Chlamydomonas* cultures arrived a month later with three pages of instructions. The following letter from Elly Hinreiner (December 20, 1954) at Davis to Ralph Lewin at the Maritime Regional Laboratory of the Canadian National Research Council in Halifax, Nova Scotia, describes the atmosphere and attitude toward Moewus during his visit.

248

Dear Ralph,

Many thanks for the *Chlamydomonas parvula* concentrate which arrived yesterday; I certainly appreciate your interest. A few preliminary spot tests indicate that the stuff almost certainly is not a flavonoid, and with the amount available and lack of any other hunches as to its nature, I doubt that I'll be able to do much toward identifying it. I'll let you know if I have any bright ideas though.

Dr. Moewus visited Davis and gave us two seminars about a month ago. He stirred up a lot of discussion and most people I talked with seemed unwilling to believe him, but had no really conclusive reasons for doubting his work. Even Dr. Geissman at U.C.L.A. [where Moewus also lectured] said he was on the fence until he heard that Kuhn had tried unsuccessfully to repeat the earlier work and is planning to publish a refutation. What I got out of it mainly was that the conditions of light, nutrition, etc. under which the organism produces these substances are extremely critical, and it is therefore easy for Moewus to claim that people who could not reproduce his work did not culture the *Chlamydomonas* properly. It's hard for me to understand though, why, if the correct conditions are so important, Moewus did not publish his methods in greater detail.

I was about to write you that so far the promised cultures had not arrived, when Mort Starr phoned to say that they came in this morning along with three pages of directions. So perhaps we shall see!

Very best wishes to you and Joyce for Christmas and the New Year. I certainly hope that your projected visit to California materializes soon.

Hinreiner and her colleague, Mortimer Starr, did not know exactly what the *Chlamydomonas* cultures were, whether they were mutants or normal wild type. As Hinreiner wrote to Ralph Lewin (January 18, 1955):

Dear Ralph,

Enclosed is a copy of Moewus' directions for *Chlamydomonas* culture. There seem to be a few typing errors, but I think they're all obvious ones. The cultures themselves (so far we have only one pair of *C. eugametos*, but have been promised others when we learn the culture techniques) seem to be thriving. Mort hasn't tried anything beyond vegetative growth yet though, since provisions for controlled illumination and temperature haven't been completed.

I appreciate your offer to send me the concentrates you have made from *Chlamydomonas* cultures. I'd like very much to try some chemical tests on them if I can ever get ahead of the work that keeps piling up here, but that doesn't seem very likely for the near future. I hope you'll save the solutions for me though.

We missed you at the A.A.A.S. meeting, but hope that you and Joyce will find some other excuse to come out our way soon. Perhaps the Federation meetings in April?

It seemed to Hinreiner that Moewus was being very cooperative, sending stocks and detailed procedural instructions to repeat his work. Lewin, however, was not impressed. He, like Arthur Birch, Ruth Sager, and Bernard Davis, doubted that Moewus had any flavonoid-requiring mutants of *Chlamydomonas*. Lewin tried to keep up to date with gossip and published literature pertaining to Moewus's work, but all he saw was an abundance of clear evidence of disconfirmations. He maintained these conclusions often irrespectively of those who did the original work. Some recent work by D. Lewis and his collaborators at the John Innes Horticultural Institution in England is a case in point. They failed to confirm other work of Moewus – the role of rutin and a related compound, quercetin, which Moewus (1950) claimed inhibited pollen germination in the ornamental flowering shrub, *Forsythia intermedia*. In Lewin's view, their inability to confirm Moewus's results represented more testimony to the claim that Moewus was systematically faking data.

But did Lewis's experiments actually refute Moewus's work on *Forsythia?* Lewis himself did not think so. Instead, he offered an alternative possibility for the conflicting results. Although Lewis (1954: 266) and his collaborators could find no quercetin in *Forsythia*, they maintained that their analysis was incomplete since they examined only one of the parental species used in their crossbreeding experiments. Lewis (1954: 266) reasoned: "One of these species may contain quercetin; this, if we assume segregation in later generations of the hybrid, would explain the presence of quercetin in the one plant examined by Moewus."

Indeed, if one considered the work of Lewis and his collaborators on its own merits (which D. Lewis did), it would be very easy to account for their failure to confirm Moewus's work in terms of their own incomplete experiments or in terms of technical differences. However, when Ralph Lewin considered them in the light of several other failed confirmations of Moewus's reports, he interpreted them in terms of another *refutation* of Moewus's claims. Thus, although Lewis did not interpret his results as a refutation of those of Moewus, to Ralph Lewin and others (see Raper, 1957) they represented an integral part of the whole pattern of results and judgments concerning the failure to confirm Moewus's reports: Moewus was publishing fake data.

However, as discussed throughout this story, in order to maintain this second alternative interpretation, one had to deny that the difficulties others were having in repeating the work were due to some intrinsic complexity of Moewus's original experimental procedures. Lewin took this view. He had been working on *Chlamydomonas (moewusii)* for five or six years. To him, those who remained open with regard to the validity of Moewus's work were simply naive. In his view, Moewus could not vindicate himself by gestures such as sending stocks and detailed descriptions of technical procedures. Indeed, Lewin denied the complexities of the technical procedures that Moewus claimed were necessary for doing genetics work on *Chlamydomonas*. As he wrote to Hinreiner (January 23, 1955):

> Dear Elly,
> Thanks very much indeed for your letter of 18th Jan., in which you enclosed Moewus' instructions on *Chlamydomonas* culture. But I confess myself baffled as to what all that is for. I gather that you aren't primarily interested in getting mating: his instructions, which are essentially the same as those of his original thesis, back in 1933, seem to be designed chiefly for this end. Actually, all those precautions aren't necessary: using the strains he sent me and my usual culture techniques (continuous illumination at 200–500 ft.c. [foot-candles], temperature about 23°C, etc.) I can get as good mating with his strains as with my own. But I thought you were interested in a particular strain producing flavonoids? From what I now understand, Moewus has sent you a pair of strains which mate – probably the same ones that everybody has been getting from him for the last 3–4 years – and has promised to send you the flavonoid mutants if and when you are good children and prove yourselves proficient in handling the wild-type! Is this so, or have you actually received a living culture of a flavonoid-producing mutant? Because that seems to be one of the critical things in the whole business, particularly since neither Esser & Straub (*Biol. Zentralbl.* 73:449–55, 1954) nor Lewis (*Adv. Genet.* 6:235–84, 1954) has been able to confirm the *Forsythia* story in any essential feature.
> Reading this through, I see that I have expressed myself in a somewhat incoherent fashion. But that's how I feel at the moment.

Constructing facts: Lewin versus Hartmann

Lewin had no faith in the reality of Moewus's published results on sexuality. In short, he believed that Moewus was merely reporting results expected by theories of his "father professor," Max Hartmann. Indeed, Lewin saw a direct relationship between Moewus's results and Hartmann's theory of relative sexuality. If Moewus was fabricating data, as Lewin believed, he was not simply doing it in a haphazard or irrational way. Nor was he doing it as a joke to prey on scientists' gullibilities. In Lewin's view, there was a method behind this madness. Moewus strongly believed in the ultimate truth of Hartmann's theory of sexuality. Moewus himself was publishing brilliant and sophisticated results and interpretations which followed logically from his premises based on Hartmann's theory.

In Lewin's view, if Moewus's published results were false (i.e., if his data were merely theoretical interpretations without any empirical foundation), it was because Hartmann's theory of sexuality which guided his work was also false. If Moewus published results that could not be confirmed, it was because they did not correspond to reality. And if Moewus's published results did not correspond to reality, it was because Hartmann's theory did not correspond to reality. Hartmann, in Lewin's view, had been willing to believe Moewus's results only because they supported his theories. Up to this point in the controversy, no one in the United States had publicly challenged Hartmann's credibility as a reliable and thorough researcher. On the contrary, Hartmann's authority had been a major source of credibility for Moewus's reports. Lewin took on the task of challenging Hartmann. He stated his views clearly in a paper on "Sex in Unicellular Algae" (Lewin, 1954), which quickly led him into direct confrontation with Hartmann himself.

Lewin's published report was based on a paper he gave at a symposium on "Sex in Microorganisms" presented at a meeting of the American Association for the Advancement of Science held in Philadelphia on December 30, 1951. The main objective of the symposium was "to present the evidence for 'sex' in the principal groups of microorganisms, from the viruses through bacteria, fungi, and unicellular algae to the protozoa" (Wenrich, 1954: iii). Biologists had given a great deal of attention to the nature, origin, and evolution of sex; most of it had been based upon conditions in higher organisms. But sex itself remained an elusive concept for biologists; it seemed to escape all attempts to define it conclusively. In their well-known book on the evolution of sex, Geddes and Thomson (1889: 112) made the following statement:

> The number of speculations as to the nature of sex has well-nigh doubled since Drelincourt, in the last century, brought together two hundred and sixty-two "groundless hypotheses," and since Blumenbach quaintly remarked that nothing was more certain than that Drelincourt's own theory formed the two hundred and sixty-third.

Hartmann's attempt to give a unified theory of sexuality in his *Die Sexualität* (1943) was regarded as one of the more notable books of the twentieth century on sexuality. However, many contemporary investigators of sex realized that it was difficult, if not impossible, to give an all-embracing definition of sex or sexuality because of the variations in the process and the differences of opinion about the subject. Hartmann's definition of fertilization and sex was that two sexually different haploid cells unite to form a diploid one. Others defined sex as the change from a haploid to a diploid phase – a definition that did not involve the necessity of sexually differentiated types (see Wenrich, 1954: 335). Hartmann's theory of sexuality was mentioned in several of the papers offered at the symposium on sex in microorganisms, but it received its most thorough and most critical treatment in Lewin's paper when he discussed the work of Moewus.

Lewin qualified his review by pointing out that he did not intend to discuss the literature on sexuality in algae exhaustively. Instead, he wrote, "Particular attention will be given here to studies of cultures and experiments under controlled conditions, where a measure of reproducibility can be expected in the results" (Lewin, 1954: 101). Taking this as his point of departure, Lewin did not discuss Moewus's published results in great technical detail. Instead of assessing Moewus's work solely in terms of detailed laboratory procedures and elaborate faulty results, as others had done, Lewin adopted an alternative strategy; he attempted to reconstruct Moewus's work by placing it in a larger historical context of a heated debate over concepts of sexuality which raged during the 1920s and 1930s between biologists at Berlin led by Hartmann, and others at Prague led by Mainx and Czurda. In particular, the debate centered on explanations of homothallism – "the condition in which a complete sexual cycle can take place within a single clone." Homothallic organisms are contrasted with the better known heterothallic organisms. Heterothallism is "the state in which two haploid clones of different genotype – different genetic mating type – are required for sexual reproduction" (Lewin, 1954: 104).

"Hartmann (1943)," Lewin (1954: 104) wrote,

> adopted the theoretical anthropomorphic concept, borrowed perhaps from Aristotle, that there can be no sexual union without sexual differentiation. If compatible gametes are morphologically and, as in homothallic species, genetically identical, then, he postulated, there must nevertheless exist some invisible physiological difference between them, and the strain is said to possess "bisexual potency."

While Lewin dismissed Hartmann's theory as being a speculative anthropomorphic concept, with no experimental foundation, he suggested that the opposing view of Mainx and Czurda was supported by experimental observation:

> On the other hand, Mainx (1933) and Czurda (1933) saw no reason for adopting this hypothesis, for which they found no corroboration in their

experimental observations on algae and other organisms, and they freely accepted the fusion of identical cells in syngamy just as it occurs in the hyphal anastomoses of fungi, or the formation of plasmodia in Myxomycetes. (Lewin, 1954: 104)

Moewus, Lewin suggested, had come to Hartmann's rescue by reporting experimental evidence to support his claims. Lewin continued:

> And then, in the early 1930s, Franz Moewus, a student of Kniep and Hartmann at the Kaiser Wilhelm Institute, reported experimental support for the theory of bisexual potency from his investigations of unicellular algae such as *Protosiphon* (1933[b], 1935a), *Polytoma* (1937), and *Chlamydomonas eugametos synoica* (1938a). (Lewin, 1954: 104)

Lewin pointed out that Moewus claimed that "male" and "female" gametes are produced in a clonal culture of the above-mentioned homothallic organisms. When mating takes place, there would usually be an excess of one of the sexes. One could take the "residual gametes" of the supernumerary sex and test them for mating type by mixing them with heterothallic test stocks of known mating types and observing whether or not mating occurs. Moewus, following Hartmann, thus claimed that there existed physiological sexual differences in homothallic organisms. Sexuality in these organisms was claimed to be largely controlled by the presence of soluble "sex substances" in the nutrient medium (see Chapter 3).

But these results of Moewus had been attacked by other workers in Prague, who failed to confirm them. Lewin continued: "Pringsheim and Ondratschek (1939) in Prague sought in vain to confirm Moewus' assertions about residual gamete behavior in *Polytoma* and *Protosiphon*, thereby lending further weight to the initial objections of Czurda (1933)." The biologists at Prague not only offered experimental evidence to refute Moewus's evidence of sexual differences in homothallic organisms, but, as Lewin pointed out, Pringsheim and Ondratschek had also argued on evolutionary grounds against Hartmann's theory and the theoretical implausibility of Moewus's results (Lewin, 1954: 104).

Lewin then extended his criticisms of Moewus's work to embrace Hartmann's theory of relative sexuality. His discursive strategy was to reduce the evidence for Hartmann's theory of relative sexuality to that provided by Moewus and then emphasize the fact that it had been criticized by many workers.

> Another concept inherent in Hartmann's theory of sexual differentiation is that of relative sexuality. Several examples of such behavior in different genera of algae, such as *Chlamydomonas*, *Polytoma* and *Protosiphon*, have

been reported by Moewus, and in certain cases viable zygotes have been obtained by mating "strong" gametes with "weak" gametes of the same sex. . . . Relative sexuality has apparently not been established in any unicellular alga investigated by other workers. No report of sex reversal in alga has come to the notice of the present author. (Lewin, 1954: 105)

Lewin (1954: 126) concluded his attack by pointing out that Moewus's genetic results had been subjected to severe criticisms and that "Moewus's replies to these criticisms (1941, 1943) are far from satisfactory."

By placing Moewus's work in a general theoretical structure erected by Hartmann, Lewin was, in effect, boldly challenging Hartmann's theories and his scientific authority. In this context, the legitimacy or "truth" of Moewus's reports (i.e., the willingness to believe them) could not rest on Moewus's shoulders alone. If Moewus's work was considered false, it was because many biologists outside of Hartmann's laboratory considered it to be false. If it were considered true, it was only because Hartmann had lent credibility to it because it supported his theories. The implication was that Hartmann had allowed his own biases to interfere with the rigorous scrutiny of Moewus's published work.

Hartmann took great objection to Lewin's review. In a paper published both in German (1955a) and in English (1955b), Hartmann went to great lengths to defend and distinguish his theory of selective sexuality and the existence of sex substances from all reliance upon the work of Moewus. His paper was entitled "Sex Problems in Algae, Fungi and Protozoa: A Critical Account Following the Review of R. A. Lewin." His primary strategy was to trivialize the arguments put forward against Moewus's work by the biologists at Prague, and to enlist a series of investigations by other biologists, that were not considered by Lewin and that supported his theories. Hartmann began his rebuttal by claiming that Lewin was wrong to reject his theory of sexuality as an "anthropomorphic concept" borrowed from Aristotle. He dismissed the criticisms of the school at Prague and claimed that the statements of Czurda, Mainx, and others had "already been refuted in detail," and that the "evidence for the existence of physiologically diverse gametes" was based on experimental methods that did not originate with Moewus. Hartmann continued:

> Furthermore the indication for sex-specific agents (gamones) in gametes of algae is questioned, obviously since Lewin seems to believe that sex substances in unicellular algae have been mentioned only by Moewus "whose published results contain a number of discrepant features." For the same reason the existence of relative sexuality is doubted. Actually, the term "gamone" for sex-specific agents did not originate with Moewus,

as Lewin seems to assume, but was proposed by Hartmann and Scharta (1940) [*sic;* 1939] after consultation with Prof. R. Kuhn. (Hartmann, 1955b: 321)

Hartmann also defended himself against the charge that he had not given Moewus's work sufficient scrutiny. When Lewin wrote his review in 1951, the work of Förster and Wiese (1954a,b) carried out in Hartmann's laboratory had not yet appeared. It will be recalled that they reported a failure to confirm many essential features of Moewus's biochemical reports concerning sexual processes in *Chlamydomonas*. Förster and Wiese had stated that their study was carried out "at the wish of Prof. Hartmann and Prof. Kuhn." But Hartmann (1955b: 321–322) himself took full credit for initiating their work and testified to its validity. Hartmann realized full well that his authority and integrity were being questioned with the implication that he was willing to accept Moewus's results without scrutiny because it confirmed his "false" theories. He took a moment to point out his senior position in the field of algal research (based on decades of work as opposed to Lewin's few years). Appropriating the work of Förster and Wiese, he claimed that the investigation of Moewus's work, which he initiated, was the "most careful and thorough examination yet made":

> I claim to be entitled to make such a criticism of Lewin's report not only on account of my own investigations, which extend over decades, and those of my collaborators and students with regard to sex in algae, but also by the fact that my present collaborators, Förster and Wiese, have, at my suggestion, accomplished the most careful and thorough examination yet made of the statements by Moewus on sexuality in *Chlamydomonas eugametos* and on its gamones in particular. In agreement with American investigators, they could not verify the assertion of Moewus on the carotenoid nature of the gamones in this species. Moreover, they showed that the cited statements of Moewus are wrong and why. They proved the existence of gamones in *Chlamydomonas eugametos* and *C. reinhardi*, determined the specific effect of the gamones, and showed them to be proteins with sugar compounds.

Hartmann then attacked Lewin for biased reporting and biased interpretations of the controversy that had raged between the biologists at Prague and those at Berlin. He argued:

> Lewin . . . lays greater stress on the accounts of Pringsheim and Ondratscheck [Ondratschek] (1939) (who could not confirm the findings of Moewus on monoecious strains of *Polytoma* and *Protosiphon* with regard to the behavior of residual gametes) than on the positive statements of Haemmerling, Rosenberg, Moewus and Lerche concerning other organisms. (Hartmann, 1955b: 324)

He further dismissed the evolutionary argument used by Pringsheim and Ondratschek to deny the likelihood of the existence of nongenetic mating-type differentiation, which, according to Hartmann and Moewus, was largely controlled by the presence of sex substances in the nutrient medium. First, Hartmann (1955b: 324) claimed that the "reasoning of Pringsheim and Ondratscheck is of a purely hypothetical and teleologic nature and thus cannot prove any facts."

Hartmann also offered additional arguments to support the now familiar argument that the inability to confirm some of Moewus's claims through replication was not "decisive" in disproving the reality of his claims. By this point, the argument that the failure to confirm Moewus's results might be due to different techniques employed was ubiquitous in the controversy. According to this argument, technical inconsistencies in carrying out the experiment were responsible for the inconsistent results obtained. Hartmann, on the other hand, raised another possibility. Like Sonneborn, he suggested that Nature herself was too inconsistent to be an objective judge of the reality of scientific observations. In other words, it was not just the experimenters that might be inconsistent. It was not just their procedural differences, their theoretical biases, and their expectations – their seeing "what they want to see" – that might lead to different experimental results. One also had to take into consideration the uncertain behavior of the organism under observation. The organism itself may behave differently at different times and under different conditions. The variables acting on the behavior of the organism may be so complex as to inhibit technical precision. If this were true, different observers, using identical procedures, could legitimately make different observations. The conflicting observations of a replicator could not refute those of another, nor could two conflicting observations cancel each other out. Thus Hartmann offered another reason why replication or experimentation could not definitively decide the issue. In the end, he suggested, one had to rely on the reliability of the investigator. In Hartmann's words:

> That they [Pringsheim and Ondratschek] also failed to find two sexual types in some cases is neither remarkable nor decisive. It goes without saying that in some cases the residual gamete method will fail and it will be impossible to demonstrate the presence of sexually differentiated gametes for some monoecious isogamous algae. This would hold especially for cases in which the gametes are not released all at once but emerge only successively in the course of one day or of several days, or if a chemotactic or phototactic isolation of the residual gametes cannot be achieved. The negative results of Pringsheim and Ondratschek might have been due to

such conditions. In any case a few negative findings cannot affect the conclusive force of positive results by acknowledged reliable investigators.
(Hartmann, 1955b: 324)

After reducing the reality of biological phenomena to the reliability of
investigators, Hartmann then claimed that Lewin gave an "insufficient"
account of the evidence in favor of relative sexuality. Lewin discussed
only the dubious work of Moewus on *Chlamydomonas eugametos*. Hart-
tbmann conceded that Moewus's reports of relative sexuality in *Chlamydomonas eugametos* could not be confirmed and were dubious. However,
he believed that similar results of Moewus concerning other species of
Chlamydomonas were correct:

> Förster and Wiese (unpublished) could not confirm Moewus on the occur-
> rence of relative sexuality in strong and weak mutants of *Chlamydomonas
> eugametos* (put at one's disposal by Moewus). I convinced myself of the
> correctness of their observations. . . . Nevertheless I still believe that his
> described cases of relative sexuality in crosses of *Chlamydomonas paradoxa*
> with *Chl. pseudoparadoxa* are correct. But being so far unconfirmed, they
> are left out of consideration here. (Hartmann, 1955b: 328, 329)

Hartmann (1955b: 329, 330) claimed that he himself had obtained "the
first proof" of relative sexuality in algae in 1914. He explained, however,
that on account of the outbreak of World War I, the material perished.
In 1924, he resumed his studies of relative sexuality at Naples and found
that his colleague, Victor Jollos (1926), had "demonstrated" the occurrence of another case. Jollos's results were discussed in detail in the
books of Kniep (1928) and Hartmann (1943). Hartmann further defended the independence of the evidence for the existence of sex substances – gamones and termones – from reliance on the work of Moewus. He claimed that the occurrence of gamones had been established
beyond doubt in other genera by Jollos (1926), himself, and others in
Germany.

Hartmann added to his defense the recent work of Förster and Wiese,
which, he claimed, furnished proof of the existence of gamones in *Chlamydomonas eugametos*. He maintained, however, that although gamones
did exist in this species, they were not carotenoids as Moewus claimed:

> Förster and Wiese could prove without a doubt by chemical investigations,
> especially by paper chromatography, that the gamones of *Chlamydomonas
> eugametos* in filtrates consist of high molecular weight proteins with a sugar
> component (so-called glycoproteids). (Hartmann, 1955b: 341)

Hartmann (1955b: 344) concluded his review by insisting that he had
"furnished proof that the phenomena of general *bipolar sexuality*, of *bi-*

sexual potency, of *relative sexuality* and the *presence of gamones* should not be looked upon as a theoretical anthropomorphical concept (Lewin)." On the contrary, he claimed, "They are well founded on facts which [he felt compelled to say] are of decisive significance for a general theory of fertilization and sexuality."

The Moewus controversy was a tar baby; as Lewin knew well, anyone who got caught in the controversy would become suspect by it. He now got himself caught in the thick of it. By publicly dragging Hartmann and his theories to the fore of the controversy and challenging him, Lewin had boldly put his own credibility on the line. He was quick to respond to the charges leveled against him by Hartmann that his account was biased, incomplete, and so on. He published his reply the next year in the "Letters to the Editors" section of *The American Naturalist*, where the English version of Hartmann's paper had appeared. Lewin stated that he had not intended his paper to be a comprehensive review of relative sexuality. It was confined to a treatment of the literature on sex in algae. This disqualified the charge that he did not mention the work of Jollos and others who worked on other organisms. Nonetheless, he continued to emphasize his differences of opinion concerning the evidence for Hartmann's theory of sexuality. And then, he charged Hartmann with taking his words out of context and distorting his original meaning concerning his discussion of sex substances:

> If one of Hartmann's basic theses is that physiological differentiation is an *essential* condition for syngamy [union of gametes in fertilization], then I feel I must continue to disagree. I know of no unequivocal evidence for intraclonal sex differentiations in any homothallic . . . unicellular alga, although Lerche's unpublished experiments with *Haematococcus* appear indicative. Even if such a condition were confirmed in one case, it might be rash to generalize. . . .
>
> Finally, I might point out that an incomplete quotation from my article . . . distorts the original meaning and that, in his discussion of sex substances, Hartmann . . . appears to have confused sex-substance activity of types 2 and 3. I make these corrections merely in the hope that further misunderstandings may not creep into the literature. (Lewin, 1956: 331)

In an attempt to disarm Hartmann, Lewin wrote to Hartmann (February 2, 1956), insisting that he did not mean to attack his theories. He suggested that "difficulties of language and the complexities of translation" may have falsely caused Hartmann to become alarmed. The remark about Aristotle should not be taken seriously, and there was no need to get defensive:

Dear Dr. Hartmann,

I feel somewhat honoured that you should have deigned to take so much notice of my article on "Sex in Unicellular Algae"; but I also feel most apologetic that I should have caused you to feel that this was an attack upon your theories. Much of the misunderstanding, I feel sure, can be attributed to the difficulties of language and the complexities of translation. I am sending a brief explanation, or reply, to the *American Naturalist*, and I thought I would send you a copy at the same time for your information.

The recent work of Förster and Wiese in your laboratory is extremely interesting, and I am sorry that the information was not available at the time I wrote my article. I look forward to reading their results with the glucoprotein factors which may be active sex substances; this might add significance to my own studies on extracellular polysaccharides of green algae (*Phycol. News Bull.* 25: 6, 1955).

To deal with a few further points in your criticism, I might say at once that my reference to Aristotle was not to be taken too seriously. I did not attribute the method of residual gametes *originally* to Moewus; I merely stated that he claimed to have used it. Likewise I did not think that Moewus had originated the term "gamone"; I just stated that he had used the term, not always consistently.

Finally, I am at a loss to understand why you refer to *Acetabularia* as haplomonoecious. I gather that Dr. Haemmerling will be unable to come here next month; this is regrettable, especially since I should have liked to discuss this point further with him.

My colleague, Michael Bernhard, wishes to be remembered to you.

Despite Hartmann's attempts to defend and distinguish his theories from the reliability of Moewus's reports, there is no question that the criticisms of Moewus tended to discredit the validity of Hartmann's theory, the possible existence of relative sexuality, and its importance as a scientific problem. In his book, *A History of Genetics* (1965: 119), A. H. Sturtevant wrote: "There is no doubt about the reality of the phenomenon of relative sexuality, and it is certainly something that needs study and analysis, but the papers of Moewus have unfortunately given it unpleasant associations."

Time running out: persuasive arguments not enough

While Lewin and Hartmann were engaged in polemics over the evidence provided by Moewus for the existence of sex substances and relative sexuality, Moewus was busy lecturing and publishing more papers on the genetics and biochemistry of sex-determining hormones in algae. He continued to behave as if his results were positively sound, supported by actual theory and solid empirical evidence. In 1955, he published another account of the "Biogenesis of the Flavonoids" in the *Annals of the New York Academy of Science*. He supported his claims with the biochemical work of Birch and Donovan, and with other recent biochemical work carried out on flowering plants at the University of California at Davis. Thus Moewus (1955: 663) wrote:

> The biological part of this investigation was supported by chemical identification of quercetin and its precursors from algal material. This work was carried out by Birch, Donovan and Moewus. Using standard methods of paper chromatography, they could identify inositol from the PHLO-mutant, phenalalanine from the TY-mutant, tyrosine from the DIPHA-mutant, 3,4-dihydroxyphenylalanine from the DIPRO-mutant, 3,4-dihydrophenyl-propionic acid from the DICIN-mutant. The X-mutant material contained, indeed, phloroglucinol and 3,4-dihydroxycinnamic acid as well.
>
> Just recently, Giesmann and Harbine reported the occurrence of cinnamic acid derivatives in a white flowering mutant of *Antirrhinum*, while the normal wild type is pigmented by flavonoids. Thus, he got the same blockage at the place of our X- or QU-mutant. Moreover, he found in three other plants that the flowers contain 3,4-dihydroxycinnamic acid esters and flavonoid pigments hydroxylated in 3;4 positions. Also, peach leaves contain Kaempferol and quercetin along with esters of both 4-hydroxycinnamic and 3,4-dihydroxycinnamic acid. This part of our biosynthesis seems to be a rather common pattern in green plants.

Lewin and others were busy enlisting scientists to provide evidence that Moewus's work was unsound and indeed fraudulent. Moewus, on the other hand, was busy enlisting other scientists to support his claims. He continued to argue that the difficulties others were having in repeating his work were due to technical complications. As discussed previously, when Moewus visited California, he emphasized the critical techniques necessary to repeat his work on *Chlamydomonas*. During his visit to California, Moewus thought that he had been able to clear up some misunderstandings concerning the physiological part of his *Chlamydomonas* research. As mentioned earlier, in 1955, Hartmann reported that Förster and Wiese, working in his laboratory, claimed that the sex substances in *Chlamydomonas* were not carotenoid pigments, but "high-molecular-weight proteins with a sugar component." However, in 1955, Moewus, perhaps not knowing about Hartmann's report, continued to insist that the sex substances were carotenoids. In fact, he claimed that the carotenoid part of his work was being confirmed indirectly through some recent work by G. M. Smith and his collaborators at Stanford and by others in Japan. He disqualified contradictory results of Lewin, based on different light conditions he used in his experimental procedures. He wrote to Sonneborn (March 19, 1955), reporting the good news. He argued that he needed more time to do detailed studies of light and temperature regimes which had to be worked out in order to systematize the technical procedures for studying growth and sexuality in algae. However, Ryan had received a Fulbright Fellowship to go to Japan in June as visiting professor at the University of Tokyo. Moewus's time at Columbia University was running out. Moewus stated that he would be returning to his native Germany, but suggested that he would like to visit Sonneborn before he left the United States:

> You have not heard from us since we met at Woods Hole. I should like to give you a short report on what we have done during these months. The hot climate and the high room temperatures during September and October handicapped completely our experimental work.
>
> During November I was out in California and had a most interesting and stimulating time. I did not meet with any difficulties in discussions after my talks; on the contrary I had the feeling that I could create some understanding for the physiological part of *Chlamydomonas* research. You will be interested in hearing that G. M. Smith, together with Dr. French from the Carnegie Institution, have found that the action spectrum for sexuality covers the region between 300 and 480 nm which corresponds very well to the absorption of carotenoids. Smith's cultures are kept on agar and are illuminated by natural daylight and some additional fluorescent lamps which comes very close to our Heidelberg setup, while R. A. Lewin's permanent "daylight" illumination of 400 foot candles is a completely different light condition, and so his results cannot be compared with Smith's results. Another confirmation of the carotenoid part of my work has come from Japan. I had sent some crocin and some instructions how to use it to Dr. Tsubo (Kobe), and he could show that crocin induces formation of flagella in the dark in his own sexual strain, which does not have flagella, if grown on agar. In this species, like *eugametos*, flagella formation normally is induced by flooding and illuminating or by crocin in the dark. I do not know when Dr. Tsubo will publish his results.
>
> In December we started a series of investigations on the influence of environmental conditions on the copulation behavior. We found out that for our autotrophic organism the light source (wavelengths, intensities) are the first factors which have to be standardized. Besides we have studied the characters of the sexual population of our agar cultures (work plates). We had at our disposal only one light cabinet with constant temperature; thus, we could not set up different illumination conditions at the same time, but only one by one. That's why we needed four months and probably we need another one, until we can sum up our results.
>
> On account of the methodological difficulties of studying the interrela-

tions between growth and sexuality in an autotrophic organism, we tackled the problem by using *Polytoma*. We had isolated a highly sexual homothallic strain at Woods Hole. First we developed methods of getting our cultures strictly synchronized so that the total population entered the sexual phase at the same time, which is actually the same method we use in our *Chlamydomonas* work. These synchronized cultures were submitted to all kinds of nutritional and starvation conditions, comparing the wild type and a sterile mutant [which] we had obtained from it by UV-irradiation. Although we got a sex determining influence with *eugametos* + and − filtrates and with isorhamnetin, we have not yet done experiments using *Polytoma* filtrates of sexual populations on nonsexual populations. In order to get active and powerful filtrates we first had to develop a method of growing *Polytoma* on agar plates. We hope we can do these experiments during the next weeks.

You will know that F. Ryan is leaving for Japan at the beginning of June and that the microbiological group more or less will be "Gone with the Wind." As far as I know, Dr. Wittkens from Cold Spring Harbor will replace him more or less pro forma, since she obviously does not plan to spend all her time here at Columbia. Thus, at the end of June we will finish up our work here and we will leave for Germany.

Do you take part in the meeting of the American bacteriologists held [in] New York in May? I would like so very much to talk over with you a few things, f.i. in what form I could publish best our recent results. In case you do not come to New York and because I really want to see you once more before I leave the country, I would like to come to Bloomington for a short private visit of one or two days. Could you let me know whether this plan would suit you and what time at the beginning of April would be the best?

Sonneborn did not refuse to see Moewus. He wrote Moewus (March 22, 1955), pointing out that he would be in Philadelphia around the end of the month and suggesting that they might meet there. Sonneborn's wife, Ruth, would try to make the arrangements.

Although Moewus said that he intended to return to Germany, he was not eager to do so. In fact, there was no position for Moewus to return to in Germany. As discussed in Chapter 9, the faculty at the University of Heidelberg where Moewus taught had long discussed the difficulties with his work. In 1949, the dean, Ludwig, advised Moewus to ensure that his stocks would be sent to leading researchers so that his work could be examined by well-known scientists. Because of the doubts about Moewus's work, the faculty had refused to ask for Moewus's professorship when it was considered in 1950 and 1951. Moewus had been granted a special leave of absence from his teaching duties at the University of Heidelberg until May 1955. However, by January of that year, the failure of workers in Hartmann's laboratory to confirm Moewus's work was well-known in Germany. In November 1954, when Moewus's difficulties were again discussed, the dean of Natural Sciences consulted Hartmann.

Hartmann (December 3, 1954) informed the faculty that Förster and Wiese could confirm the existence of gamones affecting sexuality. However, the sex substances were probably not what Moewus held them to be. According to Hartmann, "Their results make it likely that the substances, which according to Moewus's reports would appear under the influence of light, are not to be found." Hartmann would not recommend that Moewus be promoted to professor: "On the basis of the experiments of my pupils, I do not think it is justified that you should ask to give Dr. Moewus the title" (Hartmann to the dean of Natural Sciences, University of Heidelberg, December 3, 1954). On January 19, 1955 the dean of Natural Sciences, Jensen, wrote to Moewus, pointing out the new papers that appeared, which could not confirm some of his work. These papers, Jensen claimed, seemed to corroborate the doubts about Moewus's work which long existed among faculty members. The dean claimed that Moewus created obstacles for those who wished to test his results. (Presumably, the dean is referring to Moewus's alleged reluctance to supply André Lwoff with cultures he requested; see Chapter 6.) He asked Moewus to respond immediately. Moewus did not respond. In May 1955, he was taken out of the faculty register.

Moewus had lost his important allies in Germany. He was now homeless, and he was constantly on the lookout for a scientific institution in the United States that might accept him. This proved to be a difficult task. One of the first problems he encountered was trying to find influential scientists to write recommendations on his behalf. He tried to obtain support from Sonneborn. After meeting with Sonneborn in Philadelphia late in March, Moewus gave a general lecture at Fordham University on biochemical genetics in *Chlamydomonas*. Following his lecture, he was invited by one of the directors of the Navy Research Office to make an application for a research grant. The application required that Moewus be located in a university or other research institution. Moewus had in mind the University of California at Los Angeles. He believed that several investigators, including G. M. Smith at Stanford, would support his application for financial support. But he also knew that Sonneborn's support might clinch the deal. On April 27, 1955, Moewus wrote to Sonneborn, describing his situation, asking for advice, and if he would write a letter in support of his application:

Today I should like to give you a short report on my activities concerning my planned stay here in the States. Recently when I gave a lecture at Fordham University on "Problems and Aspects of Biochemical genetics in *Chlamydomonas*," Dr. Flynnt, head of Navy Research Division, New York, was present. Since he was interested in the subject and since he knew about my work from his days as editor of *Biological Abstracts*, he proposed to me to make an application to his organization. As far as I can overlook, a contract with the NRO would be the best chance I can expect at the moment. There are two difficulties which have to be clarified immediately. In the application I have to propose a place where I am going to carry out the proposed research program. The second point is that the financial year of the Navy budget begins with July, but the money would probably not be available before October/November. The latter difficulty might be overcome if Dr. Buzzati-Travserso, who invited me last fall for a longer stay at La Jolla, has still this invitation in his mind. I have written to him, whether we could spend the intermediate period at La Jolla. However, more important is the decision at what university we should settle down as "paying guests." The somewhat unfortunate work conditions at Columbia have resulted in an urgent necessity to find a place, where we would have a better start. The program I am going to propose would be: "Function and Biosynthesis of Pigments" and "Interrelation Between Growth and Sexuality in Microorganisms." That university would be good for my own purpose where I could find microbiologists interested in genetics and physiology as well, and where I could collaborate with biochemists interested in pigment chemistry. Don't you think Los Angeles would be a good place in this respect? – having a good zoological department, having botanists who are working in the related field of photoperiodism, and with the best expert in flavonoid chemistry.

I have asked Prof. Jahn, whether he would consider to take us in his department. Moreover, I have written to Prof. Geissman to support this suggestion. Would you be so kind as to write also a short letter to Dr. Jahn in case you think it a reasonable idea that we should go to UCLA. Or would you know of another place? The application has to be sent in as soon as possible.

Last week Prof. G. M. Smith spent 5 days with us in our lab. One day we took him down to Dr. Huttner and Dr. Provasoli, where – as I believe – we could crack somewhat the ice! And just by chance R. Lewin dropped in, when Smith was here. So we had quite a gathering of Chlamydomonadologists and could talk over many points. How could I ever have such personal contacts if I would go back to Germany! Smith, Goddard, Gyorgy (Philadelphia) and Szent-Györgyi (Woods Hole) have told me, they would support my application to the Navy. Probably your own word would count even more in this matter.

I am sorry, that I am bothering you with my own affairs. You will imagine how important it is for me to get over the uncertainty and unpleasantness of not knowing where to go.

Sonneborn was not willing to help. The rhetoric of enlisting scientists for and against the validity of Moewus's work could go on forever. In Sonneborn's view, persuasive arguments had run their course; it was now time for definitive experiments. But in order for Moewus to do the definitive experiments, and clear up what he claimed to be the technical difficulties of obtaining consistent results, he needed to find a scientific institution that would take him. Sonneborn refused to recommend him for a position in the United States. He could only suggest that Moewus should return to Germany, as Moewus himself originally suggested. Sonneborn wrote back to Moewus (May 2, 1955), stating his position in no uncertain terms:

> Your letter of the 27th came as a real surprise since you told me not very long ago that you were definitely going back to Heidelberg where you had the best of all possible facilities for carrying on your work. That seems to me much more important at the present time than proximity to others who may be interested in discussing your work. What is needed now is decisive experiments, not persuasive arguments. I don't see how you could possibly expect to have anything like as satisfactory a laboratory setup anywhere in this country.
>
> For that reason and because, as I have told you several times since last summer, I want to remain strictly out of this whole affair until the most important dubious points are settled – I don't see how I can support your application to the Navy or write a letter in your support to Dr. Jahn. I am quite sure that the various people you mentioned as supporting your application will be more than adequate. Some of them at least are much closer to the problem than I am and their word should be more decisive. I regret than I have to withdraw from association with your project in any way but I think no one has exerted more efforts than I in the past to see that you get a fair hearing. Subsequent developments, as you know, have greatly shaken my confidence so that in good conscience I can no longer go on record in support of your findings. I trust that others now can take over where I leave off.

There is no doubt that Sonneborn had done as much as anyone could ever expect to ensure that Moewus got a fair hearing in the United States. As he stated in his letter to Moewus, if others wanted to continue where he left off, that was their business, but he himself had extended his generosity to the limit. As much as he may have liked Moewus and appreciated his technical breakthroughs and insightful ideas, he could no longer in "good conscience" support his results. By 1955, Moewus had lost another one of his most important allies. That year he would become more and more marginalized, as many biologists simply refused to associate with him professionally. Indeed, although Moewus pleaded that it was important to be closely associated with the lively group who worked on *Chlamydomonas* in the United States, what he may not have known was that some of them were simply unwilling to associate with him. The attitude of Ralph Lewin is exemplary.

In 1955, Lewin was invited to contribute a paper to a symposium on "Sexuality and Genetics of Algae," organized by the Phycological Society of America. Studies of sexuality in algae had developed into a flourishing research front by 1955, and the organizer had attracted nine speakers who took different experimental approaches to the various aspects of the problem. Lewin suggested that he might speak on the "Control of Sexual Activity of *Chlamydomonas* by Light and Darkness" (Lewin to G. W. Prescott, May 15, 1955). This problem, of course, had occupied the attention of Moewus since the early 1930s. As we have seen, Moewus's work had originally been thought to provide a point of departure for illuminating various poorly understood phenomena such as heterothallism and homothallism, the relations of genes, hormones, and environment in determining sexuality. At the 1955 conference, Sager was to speak on "Mating Type Mutation and Its Consequences in *Chlamydomonas*." G. M. Smith had been invited to talk on some aspect of his work on *Chlamydomonas*, and several others were lined up to give papers on sexuality in various other genera of algae.

Lewin was interested in participating in the conference until he saw Moewus's name on the final list of speakers. Moewus was to speak on "The Manifestation of Homothallic Behavior in Algae." In dismay, Lewin withdrew his name from the symposium program. He wrote to Prescott (June 13, 1955) that when he received his invitation to speak, he "considered it an honour, and was pleased to accept." However, Lewin explained:

> I did so on the understanding that my associates would be scientists of undisputed integrity, and that our combined contributions would reflect favourably on the stature of the Society.
>
> I was not then aware that Dr. F. Moewus would be one of the invited contributors. This, I feel – for reasons which you will probably understand – is liable to prejudice the success of the symposium as a serious contribution to science, and I have qualms about the advisability of my remaining on the programme in its present form.
>
> In the current circumstances, perhaps you would forgive me if I withdraw my name and title from the symposium programme.

That month, an event occurred that led Lewin's attitude toward Moewus to be shared by the biological world at large.

Ryan's report (June 1955)

The event that struck the final blow to Moewus's reliability and the authenticity of his published results was the publishing of the following short note by Francis Ryan, in *Science* in June 1955:

> During the stay of Franz Moewus and his wife in this laboratory, sympathetic and conscientious efforts were made by members of our group to repeat a number of his experiments. These involved monoecious and dioecious strains of *Chlamydomonas eugametos*, some mutants of the latter and a monoecious strain of *Polytoma uyella*. Culture fluids of male and female *eugametos* and the compounds phenylalanine, rutin, isorhamnetin and paeonin [peonin] were utilized; for all of these, specific activities had been reported. The material and conditions were those of Moewus, the experiments were of his choice and he participated in some of them. Since after sixteen months no substantial confirmation of his claims is at hand, even for experiments originally performed by Moewus and his wife in this laboratory [F. Moewus, *Biol. Bull.* 107: 293 (1954).] our attempts at repetition have been discontinued. A mimeographed description of the experiments that were done can be obtained from the present author.

It will be recalled that before Moewus went to Woods Hole, Moewus had begun to set up, with Ryan's help, experimental conditions to repeat some of his experiments. The specific point of concern centered on Moewus's claim that the carotenoid derivative rutin acted as a sex-inhibiting substance. More specifically, Moewus was supposed to try to account for the conflicting results of Förster and Wiese in Hartmann's laboratory: Why did rutin not prevent the copulation of male and female *Chlamydomonas eugametos* when Förster and Wiese performed the experiment? Moewus insisted that rutin did prevent copulation and suggested that the failure of Förster and Wiese could be due to the appearance of an enzyme that converted rutin into a harmless substance in the stocks they used. However, when Ryan tried his hand at the experiments, using Moewus's stocks and procedures, he also failed to obtain positive results with rutin. Further attempts at repetition were postponed until the winter. It will be recalled that when Moewus returned from Woods Hole, he complained that his stocks were in poor condition. Ryan's (1955) longer mimeographed descriptions of the experiments performed in his laboratory highlight the difficulties he faced when attempting to evaluate Moewus's work:

> Attempts at repetition were postponed until Dec. 1954, because of Moewus' belief, upon his return from a summer at Woods Hole, that his stocks were in poor condition (and this despite the fact that he had kept duplicate

sets at the Haskins Lab., at Columbia, and with him at Woods Hole). He reported a continued inability to grow material (even of his supposedly flagellaless *cro* mutant) that was in part not flagellated on agar and that would not become motile in the dark. Hence, experiments of his choice, with material prepared at his hands, were performed with his collaboration at this time. Notable among these was a set with what he designated as *"Chlamydomonas synoica."* These experiments involved the participation of Dr. Reinhard Kaplan, Phyllis Fried, and Miriam Schwartz. When plates of *"synoica"* were flooded with water, copulation occurred within the suspension, as had been reported (Moewus, F. 1938 . . .). No increased copulation was observed when this suspension was mixed with male *eugametos*. When it was mixed with female *eugametos*, however, a noticeable increase in the frequency of pairs was observed, as though there were a residue of cells of male sex. Cultures of *"synoica"* flooded with the supernatant of a female *eugametos* culture or with a solution of isorhamnetin, secured from Moewus, in a concentration of 10^{-5} gram per ml. copulated only with female *eugametos* cells, as though the monoecious *"synoica"* culture had been converted into a female one (Moewus, F. 1940[e] . . .; [Kuhn, R.,] Moewus, F. and I. Löw 1944 . . .). On the other hand, when cells of *"synoica"* were flooded with the supernatant of a male *eugametos* culture they copulated among themselves but also with male *eugametos* as though they had not been affected by the supernatant but contained an excess of female cells. In the course of this experiment all participants deliberately left Moewus with the experiments in his laboratory for a period of approximately two hours; all observed copulations occurred after our return. In short, our results in general confirmed those which had been reported earlier in various papers by Moewus.

However, the results of the experiments reported in the preceding paragraph could not be obtained when the experiments were repeated. Cultures of *"synoica"*, whether they became motile upon flooding with water or only partly so, did not show internal copulation. Nor did supernatants of male or female *eugametos* nor isorhamnetin nor peonin, secured from Prof. Richard Kuhn, influence their sexuality (Moewus, F. 1950 . . .). From that time cultures of *"synoica"* behaved as female and copulated only with male *eugametos*. These further attempts at repetition which failed were done without Moewus' participation. Throughout this period plates were incubated in a refrigerated, illuminated box in Moewus' laboratory, but they were labelled so that Moewus could not possibly know the nature of their contents.

Furthermore, on the same day that the first experiments on *"synoica"* were performed with Moewus, a "termoneless" and two "gamoneless mutants" were studied. The first, T31, had been reported to copulate with male *eugametos* when flooded with isorhamnetin or phenylalanine, but not when flooded with water (Birch, A. J., Donovan, F. W. and F. Moewus 1953 . . .). No copulation was observed when water-flooded cells were mixed with male or female *eugametos*; when flooded with isorhamnetin copulation only with male *eugametos* took place; but when flooded with phenylalanine (10^{-6} gram per ml) copulation occurred only with female *eugametos*. In the second instance, male and female "gathe mutants" copulated, respectively, with normal female and male *eugametos* only when flooded with water and not at all when flooded with supernatants of male and female *eugametos* (Kuhn, R. and F. Moewus 1940 . . .). These observations were also made after our deliberate absence from the laboratory.

In May 1955 attempts to repeat Moewus' work with *Chlamydomonas* were discontinued, and the repetition of other relevant experiments was deemed impossible at the time, because of the fact that most cultures, even when prepared by Moewus, did not become motile upon flooding in light and because all the palmelloid cultures were reported to have died. The conditions of light, temperature and medium were varied to rectify the situation. Only a few cultures grown at room temperature in light from a north window behaved as they previously had by becoming motile and when these cells, which included "*synoica*," were tested they reacted as reported above; that is, they behaved as female rather than monoecious and failed to respond to isorhamnetin, peonin or supernatants of male or female *eugametos*. The *eugametos* so grown were also insensitive to rutin.

Because of an inability to germinate zygotes by Moewus' recommended procedure, fresh stocks were not available for experiments from suspensions of zygotes maintained by Moewus or produced during the first experiments performed with him. Nor did Moewus secure fluorescent lamps and chemicals from Heidelberg as he had promised to do. When Prof. Gilbert Smith visited Moewus in our laboratory in April 1955, after repeatedly announcing his intention for nine months, no material was ready for a demonstration of the effects of filtrates on *eugametos* cells grown in the dark which was the experiment he had come to see.

During the winter and spring of 1955 attempts were also made to repeat some of Moewus' experiments with a monoecious strain of *Polytoma uvella* on which he was currently working. To date he has not published this work so that our repetitions were of what might be considered preliminary claims. One of these was that supernatants of male and female *Chlamydomonas eugametos* and the compound isorhamnetin prevented copulation within a culture of monoecious *Polytoma* and determined the sex of cells accordingly. In a first experiment which, as mentioned above, deliberately was left by us for two hours, this seemed to be the case; cultures treated with isorhamnetin or female supernatant [failed to copulate] while control cultures copulated internally. However, numerous repetitions of this experiment, conducted without Moewus' participation, failed completely. Finally Moewus joined us in a repetition that was carefully scrutinized throughout. Although Moewus claimed that the isorhamnetin was converting the monoecious *Polytoma* into females, he did not convince Phyllis Fried, Miriam Schwartz nor myself that the conclusion could be drawn from the cultures under observation, because sexuality in all cultures was so weak as to make the identification of copulation in any one uncertain. When Dr. Vincent Cirillo visited our laboratory this experiment was repeated again. It was concluded that isorhamnetin had no demonstrable effect on the sex of *Polytoma*.

Furthermore, before and during Dr. Cirillo's visit, two mutants of *Polytoma* isolated by Moewus from cultures that had been treated with ultraviolet light were examined for sexual behavior. Both by themselves clumped, paired, and fused; one did not form zygotes and the pairs eventually disappeared, the other formed "thin walled" zygotes. Mixtures of the two mutants showed behavior no different [from that] when they were alone. It was concluded that at least one of these strains was self sterile but that they were not reciprocally hetero-phallic as Moewus had reported them to be.

The sum of the confirmable work performed by Moewus during his sixteen months in this laboratory is reported in *Microbial Genetics Bulletin* 12:17–18 (1955). *Polytoma uvella* was shown to have a diphasic growth curve under certain conditions; at the onset of the second phase the population becomes sexual. These observations have been substantially reproduced by ourselves and by Dr. Cirillo.

It should be emphasized that the experiments reported here had as their object a repetition of those performed by Moewus. It was felt that with his assistance after sixteen months some confirmation would be found, at least under the conditions originally used by him in these laboratories to perform the work he published from here (Moewus F. 1954 . . .). Friendly relations with the Moewuses were honestly maintained because all collaborating members of this laboratory initially believed that the inability of other workers to confirm Moewus might have had a basis in procedural differences. The failure of our attempted repetitions should not be taken, however, as indicative of some excessive intrinsic complexity of original research with these microorganisms. Further details of these experiments will be provided to anyone who requests them.

Dr. Moewus read the whole of this statement before it was mimeographed and in the presence of Dr. Reinhard Kaplan and myself stated that he could not disagree with it.

Ryan's report was seen by many American investigators as a final confirmation of Moewus's intention to deceive his colleagues. All the difficulties in reproducing Moewus's work could be explained under a single theory: Moewus had simply "fabricated" crucial aspects of his results. He was publishing theory *as* data. Elly Hinreiner wrote to Ralph Lewin on November 11, 1955, claiming that the stocks she and Starr had received from Moewus were not flavonoid-producing mutants; that Lewin was right; and that Ryan's statement definitively settled the controversy over Moewus:

Dear Ralph,
I'm appalled at myself for the length of time that has elapsed since I received your last letter and the collection of culture filtrates you generously sent. There wasn't anything new to report at the time though (and actually there still isn't much).
None of the filtrates I looked at (a few were smashed in transit) gave

any of the typical reactions of flavonoid compounds (no response to magnesium and hydrochloric acid, neutral lead acetate, ferric chloride, nor any marked intensification of the yellow color in alkali), but I realize that you didn't particularly expect them to. The copious white precipitate that forms upon the addition of base is, I presume, due to inorganic constituents of the medium. Some of the tests had to be made on the neutralized filtrate from this alkaline treatment, rather than on the original solution, because of interference of the precipitate. However, I noticed that the alcoholic extract of the *Chlamydomonas parvula* supernatant that you sent me earlier also gives a small precipitate. This could also be due to small traces of inorganic cations remaining in the alcoholic solution, but I'm wondering if the pigment itself is precipitated by alkali. It appears to be, but it's hard to tell in such a dilute solution. What is the composition of the medium you use to grow the algae?

The Moewus question seems to have been pretty well settled in this country too what with Ryan's statement and all the rest. Apparently you were right about the cultures he deigned to send us, i.e. they were just a pair which mate, for us to practice with, not flavonoid producers. Mort didn't seem to know exactly what they were when I wrote before.

Mort has no less than the daughter of R. Kuhn working with him (as a graduate student) on the structure of the blue pigment of *Corynebacterium insidiosum*. It will be interesting to hear what Professor Kuhn has to say about the *Chlamydomonas* work when he visits next spring.

I hope your work is prospering as usual. My very best personal wishes to you and Joyce.

Sincerely yours,
Elly

There is no doubt that Ryan's short note in *Science* was decisive in closing the controversy surrounding Moewus's work. Throughout this controversy one could reduce the disconfirmation of Moewus's results to the status of personal opinions by appealing to three central arguments: (1) the intrinsic technical complexities of Moewus's experiments and the lack of technical skill of others, (2) the different nature of various strains of *Chlamydomonas*, and (3) the authority of Max Hartmann and of Richard Kuhn who had collaborated with Moewus. Correspondingly, the failure of the attempts to reproduce Moewus's work by Ryan and his collaborators with Moewus's participation proved devastating for Moewus on all three accounts.

In his longer mimeographed report, Ryan had been careful to emphasize his view that the failure of the attempted repetitions should not be taken as "indicative of some extrinsic complexity of original research with these organisms." It was clear from Ryan's report that he believed Moewus was bluffing, that Moewus was able to repeat experiments when he was left alone but consistently failed to do so when witnesses were present. Yet, Moewus himself still maintained the original authenticity of his results. Could Ryan's report really be taken as a definitive refutation of Moewus's claims? Certainly, in a social sense it could be, and in fact it was. Was it really true that Moewus never had mutant stocks which behaved in the way he described? Or, had something gone wrong with Moewus's strains while at Columbia that killed the mutants, or made them ineffective? Mrs. Kobb suggests that this second possibility was indeed the case. She explains that the problem may well have been the conditions of culturing used in Columbia:

In Heidelberg, Moewus had his strains of *Chlamydomonas* cultured in a specially designed cabinet. The cabinet was about one square meter, illuminated with so-called "cold" light bulbs. When the Moewuses arrived in Columbia, in January 1954, there was no problem; they simply put their petri dishes in a north window. But before the Moewuses left for Woods Hole, it got hot, around 28°C, and condensation droplets began to appear on the cover of the petri dishes. Mrs. Kobb recalls that Ryan was generous; he was willing to build a special cabinet with four mirror-lined drawers.

The cabinet was built, but they changed the type of neon bulbs; they had a different spectrum. At that time, industry was trying to make neon bulbs colder and colder, and that was achieved by omiting as much of the warm part of the spectrum as possible. Liselotte was largely responsible for the culturing. She recalls that they were not able to stabilize the cultures during all the months after they returned from Woods Hole. Mrs. Kobb believes that the change in the type of neon bulb changed the behavior of the mutants (interview, May 26, 1987):

> I worked for six or eight months getting lights from all different firms – Sylvania, Westinghouse, whatever. Nobody would tell me exactly the spectrum. But I found out that what these lamps do, in order to get cold light, is to cut out the red wavelengths – the warm ones.
>
> Obviously that was the reason these cultures turned to a growing state – a slumbering state. In order to wake them up to make the gametes again – not just floating around – you had to test them three or four times quicker than the others and then they would work again.
>
> I was in the middle of this when we left Columbia. . . . They all knew we were working on that problem.

This, in Mrs. Kobb's view, accounted for the inability of Moewus to repeat his own work on the effect of light on the sexuality of *Chlamydomonas*.

Mrs. Kobb also offers an alternative account of the inability of Ryan to confirm Moewus's results on *Polytoma*. It will be recalled (Ryan, un-published report, 1955) that Ryan and his collaborators could not repeat Moewus's result – that isorhamnetin was affecting the sexuality of *Polytoma*. As Ryan (unpublished report, 1955) wrote:

> Although Moewus claimed that the isorhamnetin was converting the monoecious *Polytoma* into females, he did not convince Phyllis Fried, Miriam Schwartz, nor myself that the conclusion could be drawn from the cultures under observation, because sexuality in all cultures was so weak as to make identification of copulation in any one uncertain.

Was copulation "so weak," or were Ryan and his collaborators simply untrained so as not to be able to detect it when it occurred? Mrs. Kobb believes the latter to be the case. Mrs. Kobb recalls that Ryan's attempt to repeat this work marked the final break with him. When he could not do it, he simply gave up in frustration:

> When we had come back from Woods Hole, he [Ryan] had heard about this stuff [the demonstration for Sager]. He was still willing to believe our side. It was not a matter of deciding today or tomorrow. The experiments had to go on, it may take a year to finish.
> So Franz had found *Polytoma* over there [in Woods Hole] which turned out to be a marvelous sexual strain.

In order to test the effect of isorhamnetin on the sexuality of *Polytoma*, one had to put it under the microscope and click (count) paired couples. However, Ryan and his assistants did not work on algae; they worked on bacteria, which required different standard experimental procedures. "Ryan," Mrs. Kobb recalls, "thought it was easy."

> So he said, "We are going to repeat it." So he put a microscope on the other side of the laboratory and had two technicians. [We were to] work at the same time. We took cultures, and they took half. They started to count. . . . Now to our great amazement they didn't know how to use the microscope.
> . . . We gave them the proper pipette the same as we had, but they didn't use it. Is it a pair; is it a single, is it a zygote? They couldn't judge. . . .
> Then Ryan came in. He took over. He couldn't do it either. It was pathetic. He got red in the face. He got up and said, "I won't be fooled!" or something like this and stormed out. Sure they were looking like fools. (Liselotte Kobb, interview, May 26, 1987)

In Mrs. Kobb's view, Ryan's report and final behavior were unjustified. With regard to the failure to repeat some of the work on *Chlamydom-*

onas, she claims that there were real technical failures in cultivating it. With regard to the repetitions of the work on *Polytoma,* she claims that Ryan himself simply did not have the technical skills to follow the experimental procedures that she and her husband used.

13. Burying the black box

In Memoriam. Seldom in the history of biology has a discovery called forth the perplexity and mental anguish that has been expended in the vain hope of achieving a tolerable rationale for the chemical behaviour and biological activity of *cis–trans*-crocetin-dimethylester. Those who have known it best will mourn it least. (Raper, 1957: 154)

Nailing the coffin

By the late 1950s, there had been, as we have seen, many reported failures to confirm various aspects of Moewus's published claims. However, the published literature criticizing Moewus's work still remained scattered in the journals of various specialties. Moreover, as we have seen, when each failed confirmation was taken on its own terms, it was not clear what it meant. One had a repertoire of possible interpretations to consider: Did it mean that the replicators had erred in some way in the experimental procedures? Was the behavior of the organism used in the experiment influenced by environmental factors that were yet to be satisfactorily controlled? Were the replicators working on a different organism? If one felt that none of these explanations were reasonable, then did the failed confirmations mean that Moewus had simply made mistakes? Or did they mean that Moewus, for unknown reasons, was reporting experiments and results that were fakes? The point of each failed confirmation was too soft and weak ever to nail a solid lid on the controversy. The fact that alternative explanations were always possible had kept the controversy alive.

However, by the mid-1950s, the collective point of the failed confirmations became stronger and sharper in the eyes of many. Moewus had lost important allies. The events at Woods Hole, the failed replications in Hartmann's laboratory, and the failure of Moewus himself to reproduce his results in Ryan's laboratory weighed heavily against him. Moreover, Moewus had become more and more marginalized, more and more vulnerable to attacks, and, as we shall see, he soon found himself in no-man's-land – an experimental scientist without a laboratory to defend himself. Many biologists who were active in the controversy, or who had followed the Moewus stories closely, came to believe that Moewus's reports on sex determination were fakes. The controversy had gone on long enough; perhaps it could go on forever if allowed. Many scientists had invested valuable time and effort into it. Their investments were not paying off, and they could not afford to go on any longer. A revolution in genetics had occurred, and there was other more rewarding work to be done; soon there would be Nobel prizes to be awarded. It was time to stop the controversy and, for better or for worse, make judgments, once and for all, to bury the controversy. Certainty about Moewus's methods and motives remained concealed in a black box that was now rapidly closing. Persuaded by the reports of Ryan and of Förster and Wiese, they brought together recent failed confirmations and old unpublished criticisms to hammer the last nail into the lid of the coffin of Moewus's fraud.

It will be useful to summarize the main points of "correction" to the coordinating mechanism of sex determination in *Chlamydomonas eugametos* that appeared in the published criticisms of Moewus's work in 1954 and 1955. In effect, this was done by John Raper in a critical review on "Hormones and Sexuality in Lower Plants," in 1957. It will be recalled that it was the identification of the specific compounds and their origins being pinpointed to precisely traced biochemical pathways that made Moewus's published data appear to be so unique to many investigators.

1. According to Moewus, motility of vegetative cells could be achieved only in light, or by the addition of the carotenoid, crocin. However, Förster and Wiese (1954a,b) claimed that light was not required for the indicators of motility; cells could achieve full motility in darkness. Not only was crocin *not* required for motility, Hartmann (1955b: 341) claimed that Förster and Wiese (unpublished) "could prove . . . the complete absence of crocin" in *Chlamydomonas eugametos.* Raper (1957: 153) argued that "the twin demonstration that neither light nor crocin is required for motility tears a gaping hole in the central fabric of Moewus' scheme."

2. Moewus claimed that light was necessary in the conversion of motile vegetative cells of both sexes into sexually active gametes. He attributed these effects to the release of cis and trans forms of crocetin dimethyl ester in specific and characteristic ratios for the several valences of sexual expression, ranging from strong female to strong male. Förster and Wiese (1954b) claimed, in effect, that Moewus's reports about the effects of light on sexuality were only half true. They claimed that males were apparently sexually activated only upon illumination but that light was not necessary for the sexual activation of the female. However, Förster and Wiese (1954b) reported that by performing the experiments indicated by Moewus, it was quite impossible to demonstrate the purported gamones or compounds of *cis*- and *trans*-crocetin dimethyl ester and their dependence on light. Instead, they offered a radical reinterpretation of the identity of the copulating substances (gamones). According to Hartmann (1955b: 341), "Förster and Wiese could prove without a doubt by chemical investigations, especially by paper chromatography, that the gamones of *Chlamydomonas eugametos* in filtrates consist of high-molecular-weight proteins with a sugar component" (so-called glyco-proteids). As Raper (1957: 153) pointed out, "Clumping as well as copulation now appear to be due to the activity of gamones, whereas originally these functions appeared distinct."

3. Moewus had claimed that rutin, a glycoside of the flavonoid quercetin, secreted by certain mutant strains, was a specific inhibitor of gamone production. Its presence supposedly resulted in the suppression of copulation between normal male and female gametes. However, Raper argued that both Förster and Wiese (1954a) and Ryan and his collaborators found rutin to be without effect upon the copulatory process.

4. Moewus said that isorhamnetin (a flavonal) and peonin (an anthocyanin), related to rutin through the common precursor quercetin, determined bisexual cells as female and male, respectively (termones). Ryan and his co-workers could not confirm these specific effects.

On the basis of the work of Ryan, and of Förster and Wiese, Raper (1957: 153) argued:

> When the account of sexual hormones in *Chlamydomonas eugametos* is "corrected" to the extent required by the results of these two independent investigations, performed in both cases with identical strains of alga used by Moewus, the original work loses much of its earlier impressiveness. When to these "corrections" are added those that necessarily follow – the excision of (a) the alternate set of termones, a *picrocrocin*-like substance (female) and 4-*oxy*-β-*cyclocitral* (male), both derived from a joint hydrolytic product with crocin from a common precursor, and (b) the cis and trans isomers of crocetin-dimethyl ester, the esterified hydrolytic product of crocin – nothing is left of the elaborate structure of the hormonal coordinating mechanism as originally described.

Raper also listed the failure of Lewis (1954) and Esser and Straub (1954) to confirm Moewus's work on pollen inhibition in *Forsythia*. He concluded his summary with what he considered to be the In Memoriam, stating: "Seldom in the history of biology has a discovery called forth the perplexity and mental anguish that has been expended in the vain hope of achieving a tolerable rationale for the chemical behaviour and biological activity of *cis–trans*-crocetin-dimethylester." For Raper, and many others, the controversy was now dead. The following year, George Beadle and Edward Tatum shared the Noble Prize with Joshua Lederberg. The next year, Raper's In Memoriam could be extended fully to embrace Moewus himself.

A scientist's no-man's-land

The Moewuses did not leave Columbia on good terms with Francis Ryan. The final outcome of their stay in Ryan's laboratory marked a tragic turning point in their lives. Problems began almost immediately when Moewus tried to work elsewhere in the United States. Before their time was up at Columbia, before Ryan's report in *Science*, Moewus received an offer from a firm called the National Drug Company, in Philadelphia. Agents from the company were very interested in the rutin story. It will be recalled that when Moewus was in Heidelberg during the war, he claimed to have isolated a sterile mutant of *Chlamydomonas*, filtrates of which prevented copulation of normal *Chlamydomonas eugametos*. The mutant was held to produce a sterility hormone, identified as rutin (Kuhn, Moewus, and Löw, 1944). However, as discussed in Chapter 10, Förster and Wiese, working in Hartmann's laboratory, could not confirm that rutin acted as a sterility hormone. Nonetheless, Moewus maintained that his results were sound. The National Drug Company had the intention of producing rutin as a pharmaceutical product, where it could be used medically to treat capillary narrowing and widening. Moewus was offered a grant but was left with the responsibility of finding a place to do the work. There were three possibilities: Woods Hole, La Jolla, or the University of Miami. After some consultation, the Moewuses decided upon the University of Miami; it was a new university; they thought they might have a real chance there.

They received their income tax refunds from working in Ryan's laboratory and bought a used car. The firm forwarded them $3000 to get settled in Miami. At this stage, they had to buy everything. The outlook for the Moewuses seemed fine until they reached Miami. In the meantime, news of Ryan's report in *Science* had reached the firm. As Mrs. Kobb put it, "These people in Philadelphia went dead." They did not answer the phone, nor did they respond to letters. The Moewuses worked in the laboratory for a little while until they had exhausted their money. In the meantime, Moewus wrote letters to all of his colleagues in the United States, but no one responded.

The Moewuses were in serious trouble. They had arrived in the United States on a research exchange visa, which meant that they could not do any other work except scientific work, so that American citizens would not lose jobs. Americans were very strict about this. They went to see a German-American lawyer to see if they could do anything about the company's having broken their contract. The lawyer informed them that as foreigners they did not have much of a chance to win. All the company had to do was drag out the trial over several months until their visas expired, and they would then be transported out of the country as undesirable aliens.

Mrs. Kobb (interview, May 26, 1987) remembers all too well the remaining tragic years of her life with her husband. She herself contemplated suicide:

> There was no legal way of protesting. We were condemned to die, and I wanted to die. We were humiliated – how long could it go on? Franz was out of work for two full years without a penny.
> I went to the unemployment office. I was sitting in line with drunkards, drunken sailors – hoping to get a job that would pay one dollar. I couldn't find anything.

Finally, Liselotte found an opening at the Chemistry Department of the university. She remembers:

> They all knew my husband, and they all knew about this tragedy – and the chemist was a very nice man. We looked like starvation. . . .
> Dr Sigel . . . a young man in his thirties – just setting up a department of virology within the Department of Microbiology of the medical school. He knew the story about Moewus; he saw how he looked. He was Jewish, born in Germany – had all the reason to be nasty, all the reason in the world. But he was not. He was the nicest man I ever met in my life. And he said, "We cannot do anything for Franz, because we do virology here. But you are good at the microscope."

The new laboratory was affiliated with the polio hospital in Miami. Liselotte was hired at $200 a month, and she worked while her husband could get no work. Finally, Franz Moewus got a job:

> He tried and tried – drove around. And then he heard of this little research laboratory near the airport. There was a little development – ten or so units built like warehouses and each one equal: a little office, a big room made of one or two sections. They did research work for clients. It was a commercial outfit. And they had a little microbiology laboratory – not really, it was a cubicle – windowless, dark. But once in a while they had a microbiological problem to solve and so they said, "You can work here, and when a problem comes in, you get paid for it." But at least he had a room.

Franz Moewus now had a working address again; he could apply for grants. Liselotte Moewus made $200 a month, and her husband got a few jobs. Then, in 1957, Franz Moewus applied for a few grants. One was to the National Cancer Institute. This was three years after they had left Ryan's laboratory in New York. "It was Christmas. He was so happy – that was in 1958." It was a small grant. Moewus put $8000 of it into the laboratory. The Moewuses took $700 out to go to Nassau so they could enter the country again on a permanent resident visa. Their life seemed to begin again. Moewus brought back the cultures from Columbia which were in bad shape, and he constructed a makeshift culture cabinet. His small cubicle had no windows, no air-conditioning. He could not afford to buy an incubator. Instead, he fitted an old ordinary refrigerator with a 200-watt bulb and a thermostat set for 12-hour rhythms of light and dark. But tragedy soon struck again. One day, after a long weekend when Moewus returned to his cubicle, he found that the thermostat had given out. The bulb had been on over three days - the cultures were practically "cooked" – including all of the 125 slant cultures. Mrs. Kobb recalls that her husband wrote to Richard Starr, who was in charge of the large *Chlamydomonas* culture collection. But Starr did not answer Moewus's letter and did not send cultures. In the early hours of May 30, 1959, Liselotte found her husband lying dead on the floor. As she put it, Franz died of "a broken heart."

Liselotte published her husband's last scientific paper posthumously in 1959 (see Moewus, 1959). She continued to work for five more years as a trained virologist, applying for and receiving research grants. In 1961, she married a retired lawyer, Joseph Kobb, and returned to her native Germany to live happily with him in Heidelberg.

Was there anything there?

The Moewus affair, in one sense, had ended, but the full truth of what happened can never be known. Did Moewus deliberately fake all of his results, none of them, or just some? These questions remain a matter of speculation and conjecture. Although Moewus's scientific data were considered dubious, and he was judged to have perpetrated a fraud, did this mean he made no contributions? Had all this time and energy, all this writing, all this discussion, all this marshaling of numbers and experimental data, all the elaborate laboratory procedures, all the social arrangements and public demonstrations, and all the anxiety been over nothing? Were scientists looking at clouds and seeing whatever they wanted to see? Was there nothing to hold onto? If the data Moewus reported were useless, what about his technical achievements and insightful ideas? Sonneborn, it will be recalled, had argued forcefully in 1951 that Moewus should receive the highest recognition regardless of the criticisms that had been made against his data. To Sonneborn, Moewus had been a pioneer in the development of microbial genetics, offering insightful ideas and new techniques. Moewus was one of the first to systematically investigate the genetics of microorganisms. He was also the first to show how to use microorganisms for biochemical genetics.

Even as late as 1955, Sonneborn continued to see great value in Moewus's work, even though he was convinced that the experimental results Moewus reported were dubious. His opinions on the merit of Moewus's work in 1955 are well stated in a letter to Vernon Bryson, program director for genetic and developmental biology of the National Science Foundation. In 1955, before leaving New York for Miami, Moewus applied to the National Science Foundation for support of an ambitious proposal. Bryson possessed a copy of Ryan's mimeograph about Moewus; but he also recalled Sonneborn's elaborate defense of Moewus at Cold Spring Harbor in 1951. On October 19, 1955, he wrote to Sonneborn to solicit his current opinion:

> Doctor Grobstein called me on the telephone today because he is very much disturbed about the implications of support for Doctor Moewus, from whom we have an ambitious proposal. Although it is not the business of the advisory panel to make policy, there are certain issues in the Moewus case that go beyond the consideration of merit, and may even qualify the merit rating. I am in possession of Dr. Ryan's mimeograph about Doctor Moewus, and have heard another person on his own initiative discuss the events last summer at Woods Hole. Further exploration is personally distasteful, but the issues are going to come up and we will have to be prepared.
>
> Both Doctor Grobstein and I recall a Symposium lecture given at Cold Spring Harbor in which you came to the defense of Doctor Moewus. Several years have intervened, providing time for you to weigh your original views. At the request of Doctor Grobstein, I am asking for your current opinion. If a letter seems inappropriate, you may prefer to telephone me collect.
>
> May I add vociferous approval to the letter of Doctor Waterman in which he thanks you for past services of inestimable value to the National Science Foundation.

Sonneborn's reply to Bryson (October 24, 1955) requires no introduction:

I have no reason to be unwilling to put my comments on Dr. Moewus in writing. As you know, I have tried for years, against much opposition, to see that Moewus got full credit for whatever he deserves. I still think that the more important things I said in my Cold Spring Harbor paper about him are justified, for Moewus certainly anticipated a great deal of important work, priority for which is usually given to others. The situation, however, is terribly complicated because the data that formed the basis for the important contributions in the way of methodology and ideas are subject to very serious questions. It is hard to disassociate the two aspects of Moewus, namely the ideas and methodologies which he clearly grasped and forcefully presented and which have been proved to be sound by the work of others; and, on the other hand, the experimental data relating to them which are, to say the least, quantitatively dubious.

The incidents at the Marine Biological Laboratories to which you refer are well known to me. Bernie Davis, Boris Ephrussi and I made it our business to go directly to Moewus and try to get from him the facts concerning the "demonstration" he gave for Ruth Sager. While it looked like a case of deception (and deliberate deception), Moewus was completely forthright and honest in his account of it to us and, no matter what impression we may have gotten, we could certainly not have incriminated Moewus before any impartial judge on the basis of the facts which were essentially agreed upon in his account and Ruth Sager's. What impressed me more than this was the laxity of laboratory technique indicated by failure to label flasks which looked alike and contained clear liquids – one contained redistilled water and the other a very dilute iodine solution. I am also exceedingly impressed by Ryan's failure to attain success in the experiments he tried to run.

In view of the whole picture, I have refused to recommend Moewus for grants and positions and have told him that I would no longer go to bat for him as I have in the past. I feel now that the situation is so far out of hand that it is quite essential for him to set the situation straight before he can expect further support either in the way of scientific friendship or financial aid. I would therefore be completely unwilling to back Moewus' request for support from N.S.F. and, in fact, would urge upon the panel to consider well that this man has not only failed to resolve the many difficulties which have arisen in the course of efforts to repeat his work in the past decade but that he has, on the contrary, raised repeatedly more and more questions in the minds of others about the reliability of his reports and experiments. I am afraid that I have to conclude that there is something very "fishy" about this man and/or his wife. The fact that he has gotten a job in this country is perhaps as much as he deserves at present. I went over the whole situation with him in detail last spring and was completely in favor of his decision to return to his Heidelberg laboratory where, as he said, conditions were ideal for carrying on the experiments that need to be done in order to analyse the reasons for the disagreements between others and himself. I was therefore surprised and disappointed when he asked me to support applications for positions in this country and I flatly refused to do so. In sum, I do not feel able to defend Moewus any longer or even to support his applications for positions or research aid. Nevertheless, I am still of the opinion that a great deal of what he has reported in his papers is probably correct in principle and of much value. I recognize, however, that his quantitative results are hardly ever to be taken seriously. With things in such a mess, Moewus' job is to straighten out the points of difference one by one and not go off into a dozen new paths of investigation which will raise some dozens of more questions when attempts are made to repeat them.

In the end, Sonneborn viewed the essence of the conflict over scientific priority and discovery as a debate between two principles of hierarchization in scientific practice: the debate between the principle of giving primacy to observation and data, with all that it entails, on the one hand, and the principle that privileges the establishment of techniques and theory and its correlative interest, on the other. The discoveries, as defined by Sonneborn, had two faces, but when they were placed on Moewus, one of them became lost to view in the dark shadow, all of the difficulties, caused by the other. Although Moewus insisted that he did not fabricate his results, he now passed from being celebrated by many as one of the outstanding biologists of the world and a "founding father" of modern microbial and biochemical genetics to being merely an aberration to be ignored and forgotten.

Apologies from Germany

It was not until 1960 that Richard Kuhn and Irmentraut Löw wrote a brief retraction pointing out some "corrections," qualifications, and doubts concerning the papers they coauthored with Moewus. Their paper, entitled "About Flavonal Glycosides of *Forsythia* and About Substances in *Chlamydomonas*," was published in *Chemischen Berichte*, the chemical journal in which several of their papers had appeared. They did not discuss the "purely physiological and genetic investigations of Moewus that had appeared in other periodicals; they referred to the writings of Mainx (1937), Raper (1952), Esser and Straub (1954), "Hartmann and co-workers" (Förster and Wiese, 1954a,b), Ryan (1955), and Renner (1958) who "have distinguished themselves particularly for the extensive testing or correction of them" (Kuhn and Löw, 1960: 1010). Instead, they concentrated on results concerning the chemical isolation of natural substances in the flowering plant *Forsythia* and the alga, *Chlamydomonas*.

First, they stated that they repeated some tests concerning their previous reports of the existence of quercetin and lactose in *Forsythia*. It will be recalled that Moewus claimed that quercetin acted as an inhibitor to pollen formation. Yet, Kuhn and Löw (1960: 1009) could find no quercetin when tests were made on the same plant from which Moewus claimed he originally obtained his material. They claimed that their failure to confirm their previous reports was not due to technical difficulties. Actually, they changed their techniques in identifying substances, and they used the most recent techniques of paper chromatography, which they claimed were more definitive: "Small fractions of the previously reported quantities of lactose and quercetin would have been certainly found with the help of the methods of paper chromatography now used" (Kuhn and Löw, 1960: 1009).

Kuhn and Löw emphasized that in their collaborative work with Moewus on *Chlamydomonas*, they had "always used plant material bred and prepared by F. Moewus." But they no longer had Moewus's mutant stocks of *Chlamydomonas*. It was impossible to say whether or not the *Chlamydomonas* strains they had used previously, naturally contained the substances they had previously reported. As they argued, "Without newly breeding the mutants concerned and without renewed chemical isolation it remains questionable whether [their previous tests had been made on] substances really contained in the algae." These included crocin and crocetin dimethyl ester, rutin, quercetin, isorhamnetin, and peonin. The existence of these substances in *Chlamydomonas* and their effects on the organism remained in doubt.

They offered no account of the activity that went on in their laboratory. They drew disciplinary boundaries and insisted that only the chemical results published by them were sound. After listing the various papers they published with Moewus, they concluded: "It should be noted, that the purely chemical results in these papers, that relate to the methods of isolation, crystallization and unraveling of the constituents, remain free of the reservations above" (Kuhn and Löw, 1960: 1010).

According to Birch (interview, October 23, 1984), many biochemists found this retraction to be unsatisfactory since without Moewus's work the biochemical work was simply routine. Moreover, if Kuhn's collaborative work with Moewus had been found to be correct and sound, there would be no question that he would share the recognition gained. Yet, in the reverse circumstances Kuhn simply washed his hands of all responsibility. Birch himself wrote a retraction in 1957 stating:

> Work on *Chlamydomonas eugametos* would appear to indicate that the phloroglucinol ring of quercetin is derived from *meso*-inositol. . . . This conclu-

sion was based on the effects of added and isolated substances on the sexual processes of the organism, observed by Dr. F. Moewus. However, a communication from Professor F. J. Ryan and his associates at Columbia University informs us that some fundamental experiments in this connection could not be repeated even with Moewus' assistance. Accordingly, we consider that the communication of Birch, Donovan and Moewus should be disregarded. (Birch, 1957: 206)

Undoubtedly, Moewus's notoriety threatened to damage seriously the credibility of German investigators and their theories. As discussed previously, Max Hartmann (1955a,b) went to great lengths to defend and distinguish his theory of relative sexuality from all reliance upon the work of Moewus. However, like Kuhn, Hartmann gave no account of what took place in his laboratory and how Moewus could have convinced him and so many other scientists for so long. Hartmann, like Kuhn, did not question the social and institutional structure of their science. Nor did he state conclusively that Moewus had deliberately intended to deceive others by publishing fictitious results and experiments.

The first and only published account in Germany, that contained explicit statements that Moewus's reports represented attempts at deliberate deception, was published by the celebrated botanist Otto Renner in 1958. Renner was director and professor of the Botany Institute of the University of Munich. His paper appeared in the central scientific journal *Zeitschrift für Naturforschung*. Renner not only condemned Moewus but also criticized German biologists in general for allowing the confusion sown by Moewus to get so out of hand.

Renner began his paper by mentioning his own failed attempts to confirm Moewus's observations pertaining to pollen formation in *Forsythia*. He claimed that his observations reinforced the failure of Esser and Straub (1954) to confirm Moewus's claims. Renner then turned to Moewus's work on *Chlamydomonas*, to argue that Moewus's reports did not simply result from poor observations and poor technique. He claimed that in several cases Moewus could not have done the experiment at all.

First, Renner referred to Moewus's reports on flagella formation in *Chlamydomonas*, said to be affected by crocin. After testing the technical procedures Moewus gave, Renner claimed that Moewus only pretended to have done the experiments. For example, in order to examine the effect of crocin on flagella formation, Moewus said that he made microscopic preparations by first fixing the cells in alcohol and subsequently counting the number of cells with or without flagella. Now, under these conditions, Renner claimed, the flagella become detached from all cells. "A 'mistake' here," Renner (1958: 399) argued, "was hardly in question. It was more likely a deception."

Renner made some additional remarks relating to Moewus's behavior in an attempt to establish his claim that Moewus's work represented a deliberate attempt to deceive. Renner wrote that Professor Seybold tried unsuccessfully, in 1949, to get Moewus to show the protocols for his genetic experiments on *Chlamydomonas*. According to Renner (1958: 400), Seybold "meritoriously drew serious conclusions from the refusal." When the tide of public opinion had swept away uncertainties about the validity of Moewus's work and judged Moewus to have largely fabricated important results, what at first was judged by some to be unscientific behavior (i.e., judging the validity of results without repeating and testing them) was now applauded.

Renner used still other arguments to persuade readers that Moewus had deliberately fabricated results. In Renner's (1958: 401) view, "Most judgers [of Moewus' results] started out from the trusting opinion that Moewus actually carried out the experiments on which he reported." However, Renner pointed to the "enormous numbers of constantly 100% successfully executed tetrad-analyses." Time alone, in his opinion, could not have permitted Meowus to carry out such extensive observations:

> When I put the question before him on January 14, 1958, that this means he allowed himself 1 second for 800 cells, he did not respond. . . . Moewus knows how to throw sand in the eyes of naive readers, through particularly exact numbers, or through special pointers to the scientific precision of his method. (Renner, 1958: 402)

On February 24, 1958, Renner wrote to Moewus, saying that he did not believe "any of his chemotaxis work at all apart from some trivialities," but that he "would much prefer if he could defend himself." Moewus did not respond. Renner also claimed that years earlier he wrote to Moewus, raising questions similar to those Raper later articulated and published. According to him, Moewus answered in "his primitive way." "I observe the experiment and ascertain the data. Any other particularities, as to what actually occurs in the drop, is beyond our knowledge." Renner's judgment in his last letter, unanswered by Moewus, was that "the connection of the chemotaxis experiments with the conception of the copulation determining [the] cis–trans-ester-system is illogical, and because of this, the observation cannot be right." Renner continued:

> As I now know the algae is not particularly chemotactically sensitive, and the method invented by Moewus, which is also intended to grasp the different speeds of reaction, as to be expected, is barely useful. But this sad drama does not depend on either of these. The authors who took this table seriously at first, must have approached it with a rock solid trust in the ethos of a scientist. (Renner, 1958: 403)

Now that Moewus's work was judged to be largely fictitious, Renner claimed that he had suspected this all along. But some might think that this only makes matters worse. The question arises, If this was so, why did he and other German biologists not publicly criticize Moewus and attempt to stop his prolific writing from getting out of hand from its earliest beginnings? After all, a published critique by him or other leading botanists and geneticists could have seriously damaged Moewus's credibility very early. Conversely, the fact that he and various other leading biologists in Germany did not publicly attack Moewus's results during the 1940s and early 1950s certainly lent credibility to Moewus's claims. Renner gave only the following sketchy account of some of the controversy surrounding Moewus in Germany during the war years and the late 1940s:

> In the last days of the year 1938, Moewus sent me a MS from Erlangen of a treatise which had been made in Heidelberg, "About the Chemotaxis of Algae Gametes" for publication in *Flora*. As publisher, I corresponded with the author about the manuscript, as I felt coresponsible for every sentence to be printed, and with various colleagues who worked with Moewus, until the end of February 1939; I rejected the acceptance of the MS at the end of January with the "urgent advice not to let it be printed elsewhere as it stands at the moment." A discussion between Moewus, and a few colleagues, and myself, had been arranged at Jena on the suggestion of F. v. Wettstein, but did not take place for various reasons. Had it taken place, much calamity would likely have been prevented.
>
> The opus then appeared . . . with various additions resulting from pointers in my correspondence, but without any substantial improvement. As it appeared in print, I could have demonstrated the doubtfulness of the most important series of experiments with the arguments which I give in the following, but in order not to just negate something to do with an object I had no acquaintance with, I suggested to my assistant, Dr. Helmut Doring, to test a few of the results. Mr. Moewus readily provided cultures of *Chlamydomonas* for use. Unfortunately, Doring did not get past his first experiments, which contradicted Moewus, . . . because he was conscripted into the army in the spring of 1940 and fell in Russia.
>
> After the war, I repeatedly asked Mr. Moewus for new material. Only

in the spring of 1950, through the intervention of Mr. Prof. R. Kuhn, did I obtain living agar cultures of *Chl. eugametos,* male and female. Miss Malvine Seyfferth, now Mrs. Hagen, made many interrupted experiments primarily on chemotaxis, on these clones, which are now completed. . . . Mrs. Hagen will shortly report on her results. (Renner, 1958: 400)

Seyfferth's paper appeared in *Planta* the following year (see Hagen-Seyfferth, 1959).

To understand the reasons for the reluctance of German biologists to criticize Moewus's papers earlier, one has to consider the system of scientific authority in Germany. An attack on Moewus's results reached the very pillars of German scientific authority. To challenge Moewus's reliability and credibility indirectly meant impugning the integrity of Max Hartmann, a powerful professor and famous protozoologist. It would also be perceived as a hostile gesture toward the celebrated chemist Richard Kuhn, one of the outstanding and most powerful scientists of Germany. Whatever the reasons, German biologists were as reluctant to criticize Moewus's work as non-German biologists (especially geneticists) were to accept it. As Renner (1958: 400) painfully concluded:

We German biologists have been amiss, not guarding more strenuously against the confusion sown by Moewus – geneticists above all others have not held back on their sharp criticism – and as a result have left it to [biologists] abroad to lay the ghost to rest belatedly. At the same time, no one could hardly have attempted working together with Moewus with more good will than did Prof. Ryan of Columbia University – an attempt that ended in ignominy. More than one of us thought of taking up the amazing experimenter in his institute and observing him at work, but it went no further than the intention. In his graveside speech [*In Memoriam*] on Moewus's sexual – physiological opus, Prof. Raper . . . of Harvard University stated with a few bitter words what this so productive researcher did to biology alone with this part of his production. . . . We [German] nationals are obliged to confess that the case of Moewus belongs to that chapter of the human comedy in which scientific satires like that of Piltdown Man are recorded.

Renner's paper was the only published account that even alluded to a critique of the social nature of science when discussing Moewus's work. But Renner's paper did not go unnoticed. Ralph Lewin was impressed by it and wrote to Renner (December 12, 1958), suggesting that a book be published on the Moewus controversy to illustrate its lessons for scientists:

Dear Dr. Renner,
    I was extremely interested to read your article *"Auch Etwas über F. Moewus u.s.w."* in *Naturforsch.* 13b:399, and look forward with interest to the

forthcoming publication on *Chlamydomonas* by Mrs. Hagen, whose husband I had the pleasure of meeting briefly this summer. I hope you will
be able to spare me a reprint of your paper.

The full story of Dr. Moewus would be well worth writing, and would
make fascinating reading. There are of course large numbers of anomalies
in the literature published by this author for which you had of course no
room to mention. There were S- and E-shaped *Chlamydomonas* species. The
long, early paper on *Dauermodifikationen* – a popular subject before the rise
of biochemical genetics and tetrad analysis – is full of surprises: The subtle
change that appeared over the years when the role of carotenoids shifted
to that of another group of yellow pigments, the flavonoids; this involved
the unfortunate Drs. A. J. Birch and F. W. Donovan in the remarkable
paper which they jointly published in *Nature* (172: 902, 1953), in which
tyrosine-less mutants could grow on mineral media, and other strange
features emerged. The sensitivity of the Kresswurzel test for auxin (F. M.,
Biol. Zbl. 68: 118, 1949) has already been criticized. Since my work has
been largely on *Chlamydomonas*, I have made a hobby of following the
Moewus stories; and, since he was only 50 years old last Sunday, perhaps
they may yet continue.

My reason for writing to you is seriously to suggest that a compilation
be made, similar to that of the small, brief fraud of the Piltdown Man; and
that it be prepared for publication. I am not prepared to do this myself,
but would be pleased to cooperate as far as possible. Whether circumstances in Germany in the middle 1930's were especially favorable for such
developments (e.g., Kuhnemann, also on *Chlamydomonas*; Ondratschek-
Reinhard, on various flagellates) I am not prepared to say. However there
is a profound lesson for us all in these stories, and it would be worth
emphasizing again.

Concealing the nails

During the 1960s and 1970s, there were no published attempts to understand what had happened in the Moewus affair. Scientists who had
been involved in the controversy returned to business as usual. They
did not question the nature of their enterprise, the difficulties they had
in assessing the validity of Moewus's claims, and how he came to be
considered as the perpetrator of a major fraud. Perhaps the successes
of genetics and molecular biology tended to overshadow, or appeared
to deny, the need for methodological reflection and close scrutiny of the
nature of their science. When geneticists wrote historical accounts of
their discipline, they focused on its triumphs, they celebrated its heroes,
and their achievements, punctuated and ornamented by various Nobel
prizes in physiology and medicine (see Table 2).

An exception to this rule was made in the case of the Lysenko affair,

Table 2. *Nobel Prize recipients for work in genetics, 1933–1969*

1933	Thomas Hunt Morgan (1866–1945), for research on formal chromosomal genetics
1946	Hermann Joseph Muller (1890–1967), for discovery of the induction of mutations by x-ray
1958	George Wells Beadle (1903–) and Edward Laurie Tatum (1909–1975), for contributions in biochemical genetics; and Joshua Lederberg (1925–), for discovery of sexual recombination in bacteria
1959	Arthur Kornberg (1918–) and Severo Ochoa (1905–), for studies of the chemistry of DNA and RNA
1962	James D. Watson (1928–), Francis H. C. Crick (1916–), and Maurice H. F. Wilkins (1916–), for elucidating the intimate structure of DNA
1965	François Jacob (1920–), André Lwoff (1902–), and Jacques Monod (1910–1976), for their discoveries concerning the genetic control of enzyme and virus synthesis
1968	Marshall W. Nirenberg (1927–), H. G. Khorana (1922–), and Robert W. Holley (1922–), for "cracking the genetic code" and elucidating the means by which a gene determines the sequence of amino acids in a protein
1969	Max Delbrück (1906–1981), Salvador Luria (1912–), and Alfred Hershey (1908–), for their work on bacteriophages

which was treated as a lesson about how dogmaticism, authoritarianism, and the abuse of state power can help create and sustain erroneous theories – about what can happen when untrained politicians interfere in scientific matters. In effect, the Lysenko affair is used to bolster the need for the "self-regulating" professional autonomy of science (see Joravsky, 1970; Medvedev, 1971). A critical examination of science itself as culture is left untreated. And, Lewin's call for a book describing the Moewus affair, as we have seen, necessarily entails this very thing.

One might have thought that historians of science would have paid more critical attention to the social organization and methodology of science. History and Philosophy of Science departments grew quickly during the 1960s and 1970s in the United States and England. The establishment of new departments and graduate schools devoted to the social study of science occurred as a result of the large sums of moneys that were poured into Western universities following Sputnik. In the 1960s the primary focus of history of science was the history of ideas. The social organization of science was not of primary importance. And, during the 1970s, those who did focus their attention on the social organization of science carried out their studies almost in isolation from the ideas and technical content of science. It was generally taken for granted that an overriding scientific method existed, but it was "seen" largely as the domain of philosophers; it was certainly seldom investigated by historians. During the 1960s and much of the 1970s, historians of science simply followed the precedents of scientists themselves. Their job was dedicated primarily to explaining scientific texts, speculating on how ideas came to the mind of great scientists, and describing the growth and progress of science. Historians focused on the victors: Failed discoveries and discussions of fraud were generally excluded from or downplayed in their accounts; Moewus was written out of history.

The only published account of Moewus's work written after 1960 appeared in 1976 in the Appendix of a text edited by Ralph Lewin on *The Genetics of Algae*. It was written by the algal geneticist C. Shields Gowans. Moewus published over 100 papers covering a broad range of problems. When the controversy came to a close in the late 1950s, it was not clear how much of Moewus's work was unsound. There were many loose ends that needed to be brought together and tidied up. Gowans's extensive review of the critiques developed against Moewus's papers represented an attempt finally to clear up the rubble left in the wake of the controversy, to see what, if anything, could be salvaged as useful to algal geneticists. Gowans began his review by asking the reader to be suspicious of all of Moewus's work. One could not use discovery language such as "has shown," "revealed," "demonstrated," etc., in relation to any of Moewus's papers. Gowans made the point explicit (1976: 311): "[A]ll sentences written in italics (and in the present tense) should be preceded by 'According to Moewus. . . .' Discussion by the present author (usually in the past tense) are printed in ordinary roman type."

In addition to summarizing the criticisms of others, Gowans pointed out that after 20 years of work, he and his collaborators failed to produce any evidence to support any aspect of Moewus's genetic work on *Chlamydomonas*. Gowans (1976: 327) was careful to point out that they

> used strains received directly from Moewus (through G. M. Smith) and strains having independent origins from Moewus (Strains 9 and 10) in the Indiana Culture Collection; see Starr, 1964). Similarly, Lewin . . ., working with isolates of *C. moewusii*, was unable to support any of Moewus's claims beyond some details of gametic behavior mentioned in Moewus' 1933 paper.

Gowans's (1976: 327) conclusion was very brief: "In view of the considerable amount of evidence presented and summarized above, it is considered advisable to discount any and all of the published work of Franz

Moewus unless such results have been repeated and confirmed independently."

Lewin himself wrote an epilogue to follow Gowans's review (Lewin, unpublished, 1976). It was meant to place the Moewus affair within a larger sociopolitical context. However, for reasons that will be discussed shortly, it was deleted from the book at the "proof" stage. Lewin wrote that Moewus was a bright young man who was

> quick to appreciate the importance of a variety of new developments in the biology of his time – *Dauermodifikationen*, microbial genetics, plant hormones, tetrad analysis, chemical mutagenesis, biochemical genetics, etc. – and he eventually contributed all too generously to the published literature on these and other subjects. He studied first under Kneip and later, under Hartmann, who – undoubtedly gratified by Moewus' apparently clear-cut experimental confirmation of his ideas of relative sexuality – helped the young man to prominence, first in Heidelberg, later in Berlin.

Hartmann, in Lewin's view, was Moewus's uncritical sponsor. However, Lewin did not stop with placing the responsibility on Hartmann. He considered two additional issues: First, the very synthetic nature and novelty of Moewus's work made it difficult to evaluate. Second, one had to consider the larger political context, which in Lewin's view was a major factor in accounting for what he considered to be Moewus's fraud:

> Perhaps the scientific climate of Nazi Germany was an unfavorable one for critical studies on the inheritance of racial characteristics, even those of algae, and we now know that a lot of data were falsified in that deplorable epoch. Evidently most of Moewus' contributions to the published literature must be regarded in this light. He adroitly trod the strips of no-man's land between taxonomy, physiology, biochemistry, cytology and genetics, in such a way that few had the breadth of background (or the professional courage) to question in print the authenticity of Moewus' claims. (Physical chemists do not usually read the phycological literature: conversely – at least in those days – few phycologists were confident about the mystiques of physical chemistry.) Furthermore, under the political pressures of the time, scientific criticism was repressed. Moewus published more extensively than ever between 1935 and 1945. Unfortunately, his results could not be confirmed or even made available to workers in the same field in his native Germany.
>
> After the Second World War, in the cold light of reason, more and more scientists began to question the validity of Moewus' interpretations, and even the objectivity of his published data. . . . The year in Ryan's laboratory at Columbia University, where Moewus was afforded every opportunity to demonstrate at least a few of his remarkable claims, and his summer in Woods Hole, where critical biologists abound as usual, ended in anticlimax. . . .

In the long run, were the publications of Moewus a help or a hindrance to the advances of alga genetics? Did they ultimately stimulate valuable work by others in the field, or did they discourage it? I think mostly the latter. It needs a brave, almost foolhardy, soul to set out in search of lands which have, reputedly, already been discovered, but for which the essential sailing directions are deliberately withheld.

Then was it all a big fraud? If so – as many of us now believe – it was indeed a huge one, extending longer in time and wider in scope than, say the better-known peccadillos associated with Kammerer's toads or the Piltdown skull.

The moral of course should be plain. We are all responsible, as scientists, for the good name of science. The publishers of books and reviewers of topics, no less than university teachers and the laboratory workers themselves, must maintain a spirit of healthy skepticism – especially when someone claims to have demonstrated a phenomenon at variance with established dogmata, and *especially* when the same scientist makes several unconnected hard-to-believe claims. He may be honest and brilliant – we should all hope he is – but this should never be naively taken for granted. We owe it to science and posterity to be wary of tricksters, even in our midst. Perhaps this is the main lesson to be learned from the Moewus saga. (Lewin, unpublished, 1976)

Lewin's account raises a number of issues that deserve close scrutiny. But before addressing them, it is first helpful to understand why Lewin's account was excluded from the book on *The Genetics of Algae*. Its omission helps us to understand the rule that scientists use to reach closure in their published accounts of scientific controversies. As discussed throughout this book, the anti-German and anti-Nazi sentiment surrounding World War II played a significant part in predisposing scientists outside of Germany to be extra critical toward Moewus and his work. In this sense, the anti-German sentiment helped to construct the fraud socially. Lewin himself understood the controversy in different terms: The Nazi context was responsible only for the biases of Moewus himself and of those scientists in Germany who were reluctant to criticize him publicly; it was not responsible for Lewin's own biases. Nonetheless, as Lewin later recognized, the social influences affecting judgments in the Moewus controversy could easily be turned around. Social issues of all kinds therefore had to be excluded. Lewin made the political bias explicit, and his account was therefore omitted. As Lewin explained (letter to the author, November 27, 1984):

I was advised to delete the little bit about politics because (a) the book was to be a purely scientific one, in which in the ordinary course of events, one would not include discussions of that sort, and (b) if I did include them, someone would be sure to suggest that, as a Jew by birth (if not by conviction), my anti-Nazi sentiments might have colored my judgments.

This case illustrates clearly how and why the social nature of science, including the judgments and the prejudices of the victors, is removed from scientists' accounts. Moewus's critics could win on purely "empirical" or rational grounds. If we were left only with Gowans's review, of 1976, we would be left with the impression that the factors influencing the victors' judgments were reduced only to questions of replication, confirmation, and disconfirmation. In this context, only the losers have biases that would require special explanation. A sociopolitical discussion of the context and stakes would only muddy the issue and raise doubts.

14. Where the truth lies

> The full story of Dr. Moewus would be well worth writing, and would make fascinating reading. . . . My reason for writing to you is seriously to suggest that a compilation be made, similar to that of the small, brief fraud of the Piltdown man; and that it be prepared for publication. I am not prepared to do this myself, but would be pleased to cooperate as far as possible. . . . However, there is a profound lesson for us in all these stories, and it would be worth emphasizing again.
>
> (Lewin to Renner, December 12, 1958)

Those readers who have followed me as I threaded my way through the maze of technical procedures and heated arguments in the controversy surrounding Franz Moewus may well ask at this stage, Well, what are the important lessons to be learned from this detailed study? Needless to say, any answer to this question relies on our understanding of what the story is about. Any suggestion that this story is about a swindler in the midst of scientific truth, about an ambitious psychopath who tried to fool an unsuspecting scientific community would be simplistic and naive. Such an interpretation of the Moewus controversy would be as misconceived as any attempt to claim that Moewus was the victim of anti-Nazi sentiment or of a mobbing by the scientific community. It is not a story of good guys and bad guys, of heroes and villains.

The plot of the Moewus controversy is much more complex than either of these alternatives. Indeed, it would be difficult to imagine a richer one. In this story we have the origins of molecular biology beginning with an enormous fraud controversy. As our setting we have a context which we all know interfered with good reason – Nazi Germany. In complete symmetry, we have most of Moewus's critics as Jewish scientists. But to make it difficult for them, we have our suspect's work sanctioned and often co-authored by a leading Nobel Prize-winning chemist. But this is all too predictable, cloak-and-dagger stuff. We need a twist in the plot. First, we have a leading Jewish geneticist defending him, and then we learn that the suspect is not a Nazi at all. The story becomes more complex as we add to it other interests – competition among research programs and conflicting theories – and then crown it with a priority dispute among leading geneticists in the United States.

This story is not about objectively deciding who acted properly and who did not. Certainly, if a set of institutional rules or methods existed, we should be able to superimpose it on our actors to distinguish the good scientists from the deviant ones. This perhaps is the first lesson highlighted in the Moewus controversy: There are no institutionalized rules or criteria for distinguishing among them. There are no deviant scientists; there are only scientists. There was a range of just about every different attitude and behavior possible in the Moewus controversy. All of them depended on the social, political, and technical interests of the disputants. Some scientists, such as Delbrück, refused to take Moewus's papers seriously. Others accepted them uncritically. Lwoff and Lederberg publicly criticized Moewus's work in a wholesale manner, and dismissed him early on. Sonneborn and Beadle criticized certain details of Moewus's results but kept an open mind about their general validity; they wanted to test them by having his experiments repeated. Others, such as Lewin and Nyborn, protested against having to repeat his experiments. Still others, such as Seybold and Renner, questioned Moewus's work in private but were reluctant to comment on it in press.

When the controversy was alive and vibrant, incorrect versus correct behavior was not black and white. To scientists such as Sonneborn, those who publicly attacked and condemned Moewus before his experiments were repeated were guilty of stone-throwing and spreading gossip. On the other hand, to scientists such as Lewin, who believed that Moewus's claims were dubious and that he was a swindler all along, those who publicly attacked or condemned Moewus early were the heroes, or perhaps those who chose to ignore his work acted in the right way. Those who did not challenge Moewus in press and dismiss him early lacked "professional courage."

But the controversy surrounding Moewus is not a matter of fraud and ideology, on the one hand, versus science and truth, on the other. It is not a matter of how a "criminal" slipped throught the "policing mechanisms" of science. On the contrary, it is a story about how a scientist's work came to be the object of a great deal of scrutiny and controversy. It is a story of how a biologist working on arcane problems of sexuality in microorganisms came to be at the center of protests and demonstrations over the foundations of molecular biology. It is about the generation and degeneration of a scientific controversy; about the strengths and weaknesses, strategies, and tactics of the contestants. In short, it is about the politics of scientific truth.

In the course of this investigation, we have seen where the truth lies on both sides of the controversy. Herein lies a second important lesson to be learned from the Moewus controversy. "Cooking" and "trimming" are not exclusively the activities of "deviant" scientists; they are intrinsic features of the scientific knowledge-making process. Fiction itself is a crucial and often inescapable part of the truth-telling mode. Those who postulate a science without fiction postulate a fictitious science. Scientific arguments are not based simply on an accumulation of the "facts" that are piled on top of one another. Instead, the facts themselves are often selected to suit theoretical expectations. Sometimes they are transformed slightly and shaped to fit the particular edifice a scientist constructs. Scientific facts do not speak for themselves; scientific reports cannot be taken too literally.

We have seen how scientific statements are manipulated and transformed as they pass on from one writer to another. Authors of scientific review papers often take creative liberties, superimposing their own interpretations on the work of others in order to make convincing arguments for their version of the truth. We saw this interpretative procedure at work as scientists used evidence from the work of others to attack the validity of Moewus's claims. We saw how Lederberg enlisted Sonneborn and Smith to claim that Moewus's work had been "challenged in every detail." By enlisting their names, one gives the impression that they agreed with this claim, but, as we have seen, they did not. We saw how Lewin and Raper enlisted the work of Lewis and his co-workers as another refutation of Moewus's work. Again, Lewis and his collaborators made claims to the contrary.

A second related form of this interpretative procedure in scientific "reviews" occurs when scientists selectively report only those studies that best fit their views and ignore those that do not. Scientists often proclaim that in presenting an argument, one should mention the existence of any results contradicting those put forward by the scientific author; not to do so would tend to deprive the writer of all reliability. Although we would all agree with the spirit of this statement, we must recognize that compiling evidence for and against an author's views is not a straightforward affair. What is to count as evidence itself is negotiable. One can shape the evidence to suit one's views by restricting or expanding the range of literature one wishes to consider. The debate between Lewin and Hartmann in the mid-1950s over the theory of relative sexuality is an example. Lewin considered only the evidence for relative sexuality in *algae* (his own specialty) and argued forcefully that it was faulty, whereas Hartmann broadened the literature to include all Protozoa to show that the evidence for relative sexuality was solid. Each of these individuals accused the other of misrepresenting the other's arguments and the evidence for relative sexuality.

In effect, Lewin and Hartmann accused each other of "cooking" or "finagling" – of telling only part of the truth. However, this form of "misrepresentation," this kind of rhetoric, has to be distinguished from what scientists label "fraud." Neither Lewin nor Hartmann would have considered each other as perpetrators of fraud. What counted as valid evidence for each of them depended upon their own unique perspectives. In such cases, ambiguities are common. What amounts to suppression of data cannot always be distinguished from the decisions authors of scientific papers routinely make in deciding not to publish data judged to be faulty in one respect or another. Again, what counts as good reason in the marshaling of data is not always black and white.

The "cooking" of evidence is pervasive in science, and there seems to be little escape from it. But it is most obvious in scientists' accounts of discovery when they superimpose new meaning on past scientific work but write in such a way so as to leave the impression that the discovery had always been there. We have seen this in its most striking form in Chapter 2, when examining the construction of the myth surrounding Garrod's "neglect" and subsequent "rediscovery" as "the founding father" of biochemical genetics, and as the originator of the theory of gene action which Beadle was proposing in the 1940s and 1950s. Beadle took creative liberties, superimposing his own interpretations on the work of Garrod in order to make convincing arguments for his version of the truth. In Babbage's terminology, Beadle had "cooked" the historical data by choosing only that which fit his own hypothesis best and discarding or ignoring the rest. Beadle and those who followed him reinforced the notion of a timeless truth that lies above and beyond scientific activity and interpretation.

These kinds of discovery accounts are part of the repertoire of methodological resources which scientists use to argue for their version of the truth. We saw this again in Chapter 5, when we were following the controversy over the meaning of Mendel's work. Mendel's "discovery" was not found in his writings. It was imposed on them through various historical interpretations, and it was won through social battles. We saw how Bateson, during the first decades of the century, appropriated Mendel's work to support his own non-Darwinian views. In his textbooks, Bateson told a story about Mendel's long neglect which best suited his own struggle against Darwinians, biometricians, and non-experimentalists. We saw how, in contrast, R. A. Fisher attempted to reconstruct Mendel as a good Darwinian to support his own views about the mechanism of evolution.

Scientists' accounts of the history of their domains are important aspects of their knowledge-making process. And the kinds of accounts they write vary greatly according to the truth they want to convey. Many scientific texts are written in terms of historical narratives, in which the past contributions of scientists are presented in a logical sequence, leading the reader to be convinced of the truth and solid evidence underlying established knowledge. In fact, the experimental results of science, and the contributions of past scientists, are left soaking and change their flavor in a soup of historical interpretations. But these accounts are often "trimmed" and "cooked" to convince the reader not only of the truth of specific discoveries but also of the objectivity of scientists themselves. By omitting frauds, by giving one-sided accounts of scientific controversies, by ignoring the import past scientists ascribed to their own findings, by simplifying them and adjusting them so as to give them contemporary relevance, scientists concoct the illusion of intellectual consensus and harmonious flow of "facts" in their field. As discussed in the previous chapter, it was through this truth-telling process that Moewus and the controversy surrounding him were written out of history.

There is no question that what scientists say about the history and nature of their activity, and how historians perceive it are often very disparate. If scientists were held responsible for bringing their understanding of how scientific theories are developed and accepted or rejected, and how facts are transformed to artifacts in line with that of the historian, a major revision in scientific pedagogy would be necessary. Over the past decade or so, many historians have deviated from portraying science as the objective workings of superrationalistic researchers working in a harmonious way, who accept or reject theories as they agree or fail to agree with scientific "facts." They have come to recognize that there is no universally efficacious scientific method for discovery and evaluation. Instead they have shown that "method talk" must be understood as important rhetoric that scientists use in conflicting ways in their knowledge-making and knowledge-breaking claims (see, e.g., Schuster and Yeo, 1986). Investigating the great scientists of the past, historians have shown that they frequently fall short of the idealized image of the "normal" scientist. These historians have challenged the power of experimentation to verify scientific knowledge and have argued that scientists often act in a subjective manner. They have emphasized that the domestic politics of science, competition, power, and authority play important roles in directing scientific work – the kinds of questions that are deemed important, the phenomena that are deemed interesting, the techniques that are deemed most suitable, and the answers that are deemed acceptable.

Those who still deny the importance of power and authority in directing science and in shaping the outcome of controversy, those who still may appeal to an underlying impersonal logic, or a timeless scientific method as a vital force directing the progress of science have little use for this literature. Many philosophers of science who claim that there is a right way to go about doing science now, and also in the future, have therefore avoided contemporary writings in the history of science. Instead, they have used fictitious examples when discussing scientific method. Herbert Feigl (1970: 3) frankly admitted that for some time philosophers of science "rather unashamedly 'made up' some phases of the history of science." And others who follow Lakatos, such as J. C. C. Smart, have suggested that it is quite legitimate to use fictionalized history of science to illustrate pronouncements on the correct scientific method. As Smart (1972: 268) put it, "Fictitious examples are as good as factual ones." Indeed, some philosophers of science seem prepared to legitimize *any method* by which they can illustrate that there exists a *single timeless correct method* in science. To those who postulate the existence of such a scientific method, fiction holds more truth than fact.

Scientists and historians of science have occasionally discussed the differences in the knowledge they produce and the merits of using the writings of historians of science in training science students. Sometimes the discussions have become quite heated. Much of the difference depends on conflicting notions of truth the historian and the scientist employ. Scientists often portray truth as something that lies out there in "reality," to be unveiled or "discovered." Scientists who look back on the writings of their predecessors often do so only to search for bits of this truth. To the historian, on the other hand, truth is something that is constructed and won in certain social, theoretical, and technical contexts; it is not discovered. Whereas the scientist is often engaged in decontextualizing knowledge, it is the historian's job to recontextualize it.

In part, the historian's doubts about scientific objectivity and reservations about "the Truth" are supported by the fact that scientific concepts constantly change, sometimes quite radically, despite new pronouncements about "the Truth" and "the facts" by each successive generation of scientists. However, it would be misleading to give the impression that the contextualist view of science has been accepted by all historians of science, or that the view of scientists as neutral "fact finders" is accepted by many scientists. Scientists themselves are often aware of the dubiousness of maintaining objective truth. As André Lwoff (1957: 249) put it, "Those who claim to possess the truth should remember that heterogenesis [spontaneous generation] was once considered 'the truth'. And the spectre of heterogenesis should be allowed to rest in the famed purple sheet where scientists shroud their dead gods."

Nonetheless, it has been suggested that since the 1950s, more and more scientists have turned away from considering the history of science as a legitimate part of the science curriculum. It has been argued that the history of science has subversive aspects that might be harmful to science students. The stakes in the dispute have been presented by Stephen Brush (1974) and are well captured in the title of his paper, "Should the History of Science Be Rated X?" As Brush (1974: 1170) argued:

> If science teachers want to use the history of science, and if they want to obtain their information and interpretations from contemporary writings by historians of science rather than from the myths and anecdotes handed down from one generation of textbook writers to the next, they cannot avoid being influenced by the kind of skepticism about objectivity which is now widespread. They will find it hard to resist the arguments of the historians, especially if they bother to check their original sources.

Indeed, writers of scientific textbooks seem to be able to take great liberties when dealing with the historical process of knowledge production. If scientists who write textbooks were responsible for contextualizing scientific knowledge – by examining the strategies and tactics through which controversies are won – the knowledge in their domains might appear to be much more open-ended. The extent to which scientists use contextualist history of science therefore depends on their own teaching strategies. Those who want their students to challenge orthodoxy, or want to convey some understanding of science as a social activity, have found some stimulation in recent writings in the history of science. On the other hand, the science teacher who wants to train students into accepted theory, indoctrinate them into the role of a "fact finder," and make pronouncements about "the correct scientific method" is better off avoiding contextualist history of science; it might only make students "deviants." They are better off using the "fictionalized" history of science textbooks. Thus, if scientists' accounts of the history of their domains are often "trimmed, cooked, and forged" to suit theoretical expectations, it is precisely because fiction often holds more "truth" than does fact.

The truth that lies in scientists' accounts of discovery and controversies reported in scientific review papers and textbooks is, in principle, no different from that found in reports of original data. According to the testimony of the leading scientists mentioned in Chapter 5, the reporting of results biased by the experimenter's expectancy is ubiquitous in genetics. There seems to be little escape from the kind of biases represented by "cooked" results. However, it is not one's theoretical beliefs alone that lead to the reporting of what others might consider to be faulty or biased results. A scientist also has to learn what others expect of him or her when producing convincing arguments. There is no universal efficacious scientific method to appeal to. If such a method existed, one might expect scientists today to be able to do science better than they could two or three hundred years ago. One should be able to convey this method and, in each generation, train scientists who are better than their predecessors. But no such method exists for the scientist any more than it exists for the fiction writer. Like fiction writers, scientists must have a feeling for their audience in order to make convincing arguments and persuade their readers of the truth they want to convey. Scientific knowledge-making is a social process. One has to acquire an understanding of the field and make judgments about who is important and what one's audience might accept as convincing evidence. This is a skill that has to be acquired to suit the social and intellectual contexts of each generation. It requires interaction among one's peers.

The Moewus case is exemplary. First, the hierarchy of authority in German science has to be considered. Young scientists relied on the favor of powerful professors for obtaining academic positions. In this situation, Moewus might have submitted too much to the power of his patron and mentor, Hartmann, in providing the kind of answers that would be acceptable, and in producing the kind of evidence that might be convincing. In the context of the heated controversy over sex determination and relative sexuality, between biologists in Prague and those in Berlin, Moewus remained very faithful to his "father professor." Moewus's results gained their strength and legitimacy through the institutional power and scientific authority of Hartmann and later Kuhn, in Germany. However, in the context of the biochemical genetics of microorganisms which blossomed after World War II, Moewus was an outsider.

Moewus's work on sexuality evolved out of a search for fundamental and unified theories; he was trying to illustrate general concepts with large, all-embracing synthetic explanations. His program of research in Germany contrasted strikingly with that which had come into prominence in the United States. The search for universal laws and unified theories, characteristic of European science prior to World War II, stood in considerable contrast to that in the United States, where experimental biology was based on a more detailed, step-by-step approach to solving specific problems appropriated by more or less distinct specialties. Genetics in the United States tended to place greater priority on accurate descriptions and more controlled experiments than on large, all-encompassing theories. The evidence Moewus provided for his grand theory of sex determination was criticized in its details from the points of view of various disciplines and research interests.

Moewus's work had come into prominence at the very moment when genetic research in the United States was beginning to consider and embrace problems of the nature of the gene and gene action. This was a time when disciplinary and national styles were beginning to break down owing to intellectual migration across disciplines and nations. It was a time when microorganisms began to be widely domesticated for genetic use, when new techniques borrowed from the physical sciences began to be deployed, and when mycologists, algologists, bacteriologists, and protozoologists came together, with geneticists and chemists. The result was a more internationally based science, highly technified and essentially reductionistic.

Sonneborn, as we have seen, was one of the few American geneticists, prior to World War II, who appreciated the novelty and synthetic nature of Moewus's work and helped bring it to the forefront of American genetics after World War II. Sonneborn's approach to genetic work had been exceptional in the United States; he worked on difficult problems of gene expression and cellular differentiation, examining the complex relations among nucleus, cytoplasm, and the environment in the determination of hereditary characteristics. He also fully realized how difficult it was to learn how to domesticate unicellular organisms for genetic use, and saw how Moewus's techniques had preceded those developed by Beadle and Tatum and others in the United States. Sonneborn's experimental work was careful and meticulous. He coached Moewus in terms of what he judged to be his mistakes in both reporting and analyzing his results.

In passing, it might be noted that this feature of the controversy surrounding Moewus has characteristics in common with the story of another German, Alfred Wegener, who in 1912 proposed a unified theory of continental drift (see, e.g., Le Grand, 1988). Between the two world wars many geologists attacked Wegener's theory as being the result of grand speculation. They could find many flaws in the details of his arguments and claimed that he was selecting from the geological literature only those data that best supported his theory and ignoring the rest. His critics charged that he was pushing a cause as opposed to seeking the truth; he was ignoring the facts and not proceeding scientifically. Indeed, the criticisms launched against Wegener by his opponents in the Anglo-American geological communities amounted to everything entailed in fraud charges. As in the case of Moewus, the attacks against Wegener and his unified theory, between the two world wars, occurred against a background of anti-German hostility. It wasn't until the 1960s when specialties in geology began to break down, and anti-German sentiment subsided, that a modified version of Wegener's theory was erected. Wegener was vindicated, whereas Moewus was not, even though some continued to see great value in his ideas and technical insights.

The validity of Moewus's results was not tested against a background of objective truth. Nor was the recognition he might receive tested against a background of individual honesty. Those who believed that Moewus deserved recognition did so on the basis of what they perceived to be the usefulness of the information he provided, not on the basis of "moral norms." As Murneek (1941: 618) put it, "Moewus may belong to the category of scientists who like to make their studies 'beautifully complete.' But even if a small part of the results be verified, it would be a notable advance. . . ." What degree of error or "deliberate falsification" one was willing to tolerate depended on one's own interests in the work. To Sonneborn, Watson, and others, Moewus's original ideas and methodology, and the qualitative aspects of his results, outweighed the quantitatively dubious side of his results. Moewus's work may have been "too good to be true," but it was also too true to be dismissed. Indeed, Watson still had great admiration for Moewus and his work, even though he believed Moewus to be guilty on all three counts of "fraud" in Babbage's terminology – "cooking," "trimming," and "forging." Clearly what was important to Sonneborn and Watson was not whether Moewus had violated "moral norms," but whether he produced original work that was significant and useful.

To understand Moewus's demise, we have had to put aside our prejudices about whether or not he was honest and follow the scientist from the origin of the controversy to its conclusion. Before examining the steps leading to Moewus's demise, it should be made clear that denying the existence of institutionalized ethical rules of science does not entail condoning the deliberate faking of results. I am suggesting, however, that fraud, like truth, "lies" less in the statement than it does in the context in which the statement is made. For example, fudging the statistics a little may or may not be considered a serious misdemeanor; it depends on the context.

If one is caught fudging the statistics in a routine genetic experiment on *Drosophila*, it may not be considered a serious offense. It may be a low-risk strategy. As most young science students and their instructors well know, falsifying results in "laboratory practices" to obtain the correct answer is not the exception, but the rule. More often than not, "disciplined" young scientists are rewarded for obtaining the "correct" answer. Their version of "truth" is closely associated with getting an "A." For the young science student, fiction often holds more truth than fact. If this is fraud, then fraud is fully institutionalized in science. On the other hand, protests and demonstrations may ensue if one is charged with fudging the data concerning the safety of certain contraceptive pills.

We can see the same process at work deep within academic science as scientists negotiated what was to count as evidence in the contest over the validity of Moewus's results. One cannot understand Moewus's demise in terms of whether or not he intended to deceive. First, we have no special resources to know what his intentions were. All we could do is follow the scientists as they negotiated the validity of his work. Second, one could not reduce Moewus's fraud to his intentions, even if we could know them. Moewus's fraud did not emerge from his mind, no more than the discoveries of Mendel or Garrod. As discussed in Chapter 2, the discoveries of Mendel and Garrod were attributed to them in particular contexts. In this sense, we can understand Moewus's fraud as an attribution like "discovery" itself. In this case, it is a failed discovery. First, let us examine the context of the "discovery."

Prior to World War II, Moewus was simply one of a few researchers investigating problems of sexuality in microorganisms. He was an actor in a dispute over sex determination among biologists at Berlin and those at Prague. He was simply a researcher whose work was judged by competitors to be faulty in one way or another, as is often the case in science. However, after World War II, the meaning of Moewus's work was transformed. He was thrown into the center of a large ring of biologists who worked on the biochemical genetics of microorganisms. The genetics of microorganisms came to mark something of a revolution in biological research. Moewus's work, which had fallen outside the main lines of genetic research prior to World War II, now seemed to have been ahead of the game all along. Sonneborn put up Moewus as a contender in a priority dispute over the foundations of microbial – biochemical genetics. Sonneborn precisely redefined the stakes concerning the meaning of Moewus's work: Who was the first to domesticate microorganisms for genetic use? Who was the first to show that microorganisms were very useful for studying the process by which genes control biochemical reactions? And, who was the first to give a detailed biochemical description of genetic processes in terms of genes and enzymes?

This was the most crucial step leading to Moewus's demise. Sonneborn, in popularizing Moewus's work, raised the stakes in the controversy. If Moewus won, Beadle and Tatum's priority would be jeopardized, and Sonneborn's own early contributions on showing how to use microorganisms for genetic analysis would be raised to the level of an important discovery. One would have to place Moewus in the forefront of *Chlamydomonas* genetics, along with Lewin, Sager, Smith, and several others as technological offspring. And finally, one would have to celebrate and interpolate the history of modern biology with work done in Nazi Germany. Indeed, as we have seen, Moewus had other traits that helped to discredit him. Most of the work in question was carried out during the Nazi regime. As a scientist working in Germany during World War II, Moewus already had come up on the losing side. Allied with nazism, many scientists outside Germany already judged him to be morally incorrect; they were predisposed to exclude him from recognition. On the other hand, if Moewus lost, Beadle and Tatum's priority would not be jeopardized in any way. The first demonstrations of genetic principles in microorganisms would be overshadowed by other contributions; we would punctuate history differently. And Nazi Germany could continue to be seen comfortably for what it was: opposed to the "cold light of reason."

Moewus was trapped in this context; he could not escape. The penalty for losing the controversy would match the potential rewards to be gained from winning it. This would be a difficult contest to win. But Moewus had some resources. He had influential scientists in his corner: Hartmann, Kuhn, and Sonneborn. Sonneborn trivialized the existing criticisms made against Moewus's work. As we have seen, the first well-known public attack on Moewus's research came in the form of statistical criticisms pertaining to his numerical results published in *Nature* by Philip and Haldane. These criticisms contributed greatly to the initial skepticism toward Moewus and his work.

Yet, as they pertained to the reliability of the experimentalist, the statistical critiques were quite weak. Their lack of strength lay in their naive empiricist premise and their tendency to reify scientific results. The statistical criticisms ignored the human and social aspect of experimentation, the judgments required when reporting data, the process by which scientific results are transcribed, and the conscious and unconscious selection of data which underlie scientific reports. But Sonneborn reasoned from historical precedent and analogy, comparing the criticisms of Moewus's results to those of Mendel. Mendel's data were also statistically "too good to be true," but no one was willing to deny his contributions. For Sonneborn and many others, the qualitative aspects of Moewus's results were more important than the quantitative. Statistical criticisms were out of their depth when they focused on the precise biochemical pathways of sex determination and their control by genes.

Indeed, assessing the validity of Moewus's claims about the process by which genes controlled the biochemical process of sex determination was no small task. As Lewin put it, Moewus "trod the strips of no-man's land." He straddled the boundaries between what had become semiautonomous specialties: physiology, biochemistry, botany, and genetics – at a time when few scientists felt they had the breadth of background to question critically, in print, the validity of his claims. Specialization is double-edged. It allows scientists to turn inward and develop their own special techniques and language; it permits more detailed descriptions. Individually, scientists come to know more and more about less and less. As the literature becomes more technical, fewer scientists are capable of criticizing it, and thus researchers are compelled to accept claims in specialties other than their own.

Moewus's papers are exemplary, for they were technical and synthetic, bringing into association knowledge from algology, genetics, physiology, and biochemistry. It was because of their very technical and synthetic nature, Lewin argued, that few scientists were willing to question them in print. Moreover, the specialization that accompanies the growth of scientific knowledge necessitates researchers' reliance on an individual's prestige – acquired through past victories – when they evaluate a colleague's work. We have seen, for example, that Kuhn's authority as a Nobel Prize winner in biochemistry was an important influence for those who were reluctant to dismiss Moewus early. In order to debunk Moewus's claims and close the controversy, scientists had to bring together more and more evidence from the perspectives of various specialties. They had to enlist the words of many well-equipped scientists and win over Moewus's allies to combat his claims.

This procedure involved replication – what Collins (1985) has referred to as the "Supreme Court" of the scientific system. It might be noted that critics of the social organization of science contend that in all fields insufficient incentives are provided for replication. They claim that replication is seldom done in practice because there is little reward for doing it. The Moewus case contradicts this claim. The controversy surrounding Moewus represented one of those situations in science where major discoveries were being negotiated, and great prestige was at stake. The heated controversy that ensued produced a context in which there was reward for repeating Moewus's experiments.

However, as we have seen, replication, when it is attempted, is not unequivocal. It is often difficult to specify just how an experiment should be repeated and how "exact" a replication is sufficient to refute a claim. Some behavioral scientists have long been aware of the ambiguities inherent in the concept of replication and the difficulties of attributing meaning to a failed replication. Rosenthal (1966), who presented one of the most penetrating discussions of "observer effects" in the behavioral sciences, began his discussion of replication by pointing out that the same experiment can never be repeated by a different worker. Indeed, the same experiment can never be repeated even by the same experimenter. To avoid the not very helpful conclusion that there can be no replication in the behavioral sciences, Rosenthal (1966: 321) argued that one could "speak of relative replication." What Rosenthal (1966: 34) has written about the meaning of replication in the behavioral sciences bears equally heavily on the natural sciences:

> The basic control of intentional errors in science, as for other types of error, is the tradition of replication of research findings. In the sciences generally, this has sometimes led to the discovery of intentional errors. Perhaps, though in the behavioral sciences this must be less true. The reason is that whereas all are agreed on the desirability or even necessity of replication, behavioral scientists have learned that unsuccessful replication is so common that we hardly know what it means when one's data don't confirm another's. Always there are sampling differences, different subjects and different experimenters. Often there are procedural differences so trivial on the surface that no one would expect them to make a difference, yet, when the results are in, it is to these we turn in part to account for the different results. We require replication but can conclude too little

from the failure to achieve confirming data. . . . Science, it is said, is self-correcting, but in the behavioral sciences especially, it corrects only very slowly.

One might have thought that Rosenthal's remarks apply only to the so-called "soft" sciences, not to the "hard" sciences. However, perhaps the differences between "hard" sciences and "soft" sciences lie only in the extent to which methodological reflection is emphasized. Those "soft" scientists may be only more methodologically sensitive and conscientious. Collins and Pinch (1979), for example, have argued that some researchers in paranormal psychology are operating with a higher degree of rigor than many mainstream scientists. The controversy surrounding Moewus's claims lay at the feet of molecular biology, the "queen" of the life sciences, and all of the issues raised in the passage above were of crucial importance in keeping the controversy alive. Let us treat each issue in turn.

The uncertainty posed by the use of different "subjects" in attempts to replicate Moewus's experiments has been discussed repeatedly. Some scientists argued that it was necessary to use the same species of organism, because what applied to one species may not apply in the same way to another; their biological natures may be different. However, as discussed in Chapter 7, establishing that one was using the same species was not straightforward. Moreover, even when one was confident that one was using the same species, the significance of disconfirmations could be ambiguous, if one did not use the same mutant strains employed by Moewus himself. To be sure, it was best – more definitive – to obtain strains directly or indirectly from Moewus. Yet, even when one used the same strains employed by Moewus, the meaning of the experimental results remained ambiguous. Failures to confirm Moewus might result from lack of skill in carrying out the experiments or be due to slight variations in procedural differences, light conditions, temperature, etc. Different results might also have been due to changes that had occurred in the organism itself since the last successful demonstrations had been reported.

One also had to take into consideration observer expectancy when evaluating the meaning of failed replications. As Sonneborn put it, the scientist who was to repeat crucial experiments had to be both competent and above suspicion. If Moewus carried out his experiments with expectations in mind, so, too, might others who attempted to repeat his work. Their observations might have been affected by the cloud of suspicion and gossip surrounding Moewus and his work. After all, by

1950, public criticisms of Moewus and his work had come from many corners. There had been charges that his numerical results were too good to be true statistically, that he had failed to send stocks to others, and there had been theoretical objections to some of his biochemical claims as well. Could this atmosphere also have affected the outcome of the replications? The scientists in our story collectively had considered all these possibilities. The circumstantial nature of the evidence provided by failed replications continued until Moewus himself was invited to repeat some of his own experiments under the watchful eyes of special witnesses in Ryan's laboratory in New York. It came to a showdown of strength based on demonstration. Moewus had to provide demonstrations to match those of his critics. In this test of strength, Moewus lost.

But by the mid-1950s Moewus had placed himself in an especially vulnerable position. In effect, he had broken all the strategic rules of science, and we can draw lessons from them. First, he did not have a secure institutional position. He had no laboratory of his own, and he had no loyal students who knew his experimental protocols and could help him correct the flaws in his work. Instead, he continued to produce controversial results on top of controversial results. In the process, he had come to lose his influential allies – Hartmann, Sonneborn, and Kuhn. Finally, after traveling around the world with his circus animals, he was asked to make them perform to order, on the spot. In effect, he walked into a trap. Suspicions increased at Columbia University, where his organisms were neither growing well nor responding to stimulation regularly. His experimental organisms were unfaithful with respect to the phenomena under investigation. They were susceptible to modification under just about every environmental condition: too much light, not enough light; too hot, too cold; too wet, too dry. In addition, one had to observe them at the right time in their life cycle. Moewus simply lacked the kind of tight control over his organisms to which those who worked on microorganisms in the United States had become accustomed.

By the mid-1950s, it seemed clear to many participants in the controversy that the objections to using replication as a source of validating Moewus's knowledge claims were leading down a road of sophistry. However, the controversy did not reduce to an issue of either accepting sophistry or accepting the validity of experimental demonstrations. There was a plethora of other criticisms that could be woven into a uniform fabric indicting Moewus. The circumstantial evidence provided by replications became substantial in the context of other charges that had accrued against the honesty and integrity of Moewus and the validity of his reports: the statistical criticisms against the objectivity of his reports; the social accusations that he was reluctant to send stocks; the biochemical charges against the plausibility of some of his claims about reaction frequencies and chemical ratios; the theoretical objections against the possibility of certain chromatographic results; the "demonstration" at Woods Hole which looked like deliberate trickery; Moewus's own failure to repeat his results before accredited witnesses. None of these criticisms individually would have been strong enough to end the controversy, but taken together they could be woven into a consistent fabric – a burial cloth for Moewus.

The controversy over the validity of Moewus's work on sex determination was not only closed; it was also buried, never to be raised, reopened, and negotiated again – at least not for the reason of trying to rehabilitate his claims. As we followed scientists through this affair, it was necessary to keep an open mind about the truth and the outcome of the controversy as long as it lasted. To suspend disbelief when the controversy had come to a close would be to ignore the very basis of the scientific knowledge-making process. Scientific controversy is not a matter of "naked power" versus the "naked truth." Scientific truth itself is clothed in technical capacity and institutional power. One can no more see the naked truth than one can make it, defend it, or attack it naked. One has to be equipped with laboratories, experimental techniques, persuasive arguments, and significant allies. In the end, Moewus had none of these.

Bibliography

Interviews

Birch, Arthur, University of Melbourne, Parkville, Australia, October 23, 1984.
Grell, Karl, Australian Institute for Marine Sciences, Townsville, Australia, October 27–28, 1985.
Kobb, Liselotte, Heidelberg, West Germany, May 25–26, 1987.
Lederberg, Joshua, Rockefeller University, New York, May 26, 1986.
Lewin, Ralph, Scripps Institution of Oceanography, La Jolla, California, May 2, 1986.
Moore, J. A., University of California, Riverside, California, May 1, 1986.
Ryan, Elizabeth, New York, New York, May 6, 1986.
Sager, Ruth, Dana-Farber Cancer Institute, Harvard Medical School, May 15, 1986.
Watson, J. D., Cold Spring Harbor Laboratory, New York, May 13, 1986.

Letters and unpublished manuscripts

Armstrong, P. B., to T. M. Sonneborn, November 6, 1953; *Sonneborn papers*. Manuscripts Department, Lilly Library, Indiana University, Bloomington Indiana.
Beadle, G. W., and Pauling, L. 1946. *A Proposed Program of Research on the Fundamental Problems of Biology and Medicine.* Unpublished, Record Group 1.2, Series 205, Rockefeller Archive Center, Tarrytown, New York.
Beadle, G. W., to Warren Weaver, May 15, 1951; Record Group 1.1, Series 401D, Rockefeller Archive Center, Tarrytown, N.Y.
Brink, R. A., to Warren Weaver, December 4, 1947; Record Group 2.1, Series 200D. Rockefeller Archive Center, Tarrytown, N.Y.
Brodie, Harold, to T. M. Sonneborn, November 11, 1953; *Sonneborn papers.*
Brodie, Harold, to O. Winther, January 14, 1954; *Sonneborn papers.*
Bryson, Vernon, to T. M Sonneborn, October 19, 1955; *Sonneborn papers.*
Carlson, Elof, to the author, September 2, 1986.
Davis, Bernard, to the author, May 22, 1985.
Delbrück, Max, to T. M. Sonneborn, June 1, 1944; *Sonneborn papers.*
Ephrussi, Boris, to T. M. Sonneborn, October 6, 1958; *Sonneborn papers.*
Grell, Karl, to the author, March 5, 1985.
Hansen, F. B., diary, July 1, 1939; Record Group 1.1, Series 200D, Rockefeller Archive Center, Tarrytown, N.Y.
Hansen, F. B., diary, 1941; Record Group 1.1, Series 205D, Rockefeller Archive Center, Tarrytown, N.Y.

313

Hartmann, Max, to the Dean of Natural Sciences, University of Heidelberg, December 3, 1954; *Moewus file*, University of Heidelberg.

Hinreiner, Elly, to Ralph Lewin, December 20, 1954; *Lewin papers.*

Hinreiner, Elly, to Ralph Lewin, January 18, 1955; *Lewin papers.*

Hinreiner, Elly, to Ralph Lewin, November 11, 1955; *Lewin papers.*

Horowitz, Norman, to the author, August 6, 1986.

Jensen, Dean, to Franz Moewus, January 19, 1955; *Moewus file*, University of Heidelberg.

Jollos, Victor, to T. M. Sonneborn, February 6, 1940; *Sonneborn papers.*

Kaplan, Reinhard, to the author, July 5, 1986.

Kobb, Liselotte, notes to the author, May 29, 1988.

Lederberg, Joshua, to Franz Moewus, June 9, 1949; *Lederberg papers;* Rockefeller University, New York.

Lederberg, Joshua, to the author, June 26, 1985.

Lewin, R. 1950. *The Life Cycle and Genetics of Chlamydomonas moewusii Gerloff.* Unpublished doctoral dissertation; Yale University Lewin papers, Scripps Institution of Oceanography, La Jolla, California.

Lewin, Ralph, to Max Hartmann, February 2, 1956; *Lewin papers.*

Lewin, Ralph, to Elly Hinreiner, January 23, 1955; *Lewin papers.*

Lewin, Ralph, to Franz Moewus, April 17, 1951; *Lewin papers.*

Lewin, Ralph, to G. W. Prescott, May 15, 1955; *Lewin papers.*

Lewin, Ralph, to G. W. Prescott, June 13, 1955; *Lewin papers.*

Lewin, Ralph, to Otto Renner, December 12, 1958; *Lewin papers.*

Lewin, Ralph. 1976. "Epilogue." Unpublished; *Lewin papers.*

Lewin, Ralph, to the author, November 27, 1984; *Lewin papers.*

Ludwig, W., to Franz Moewus, February 3, 1951; *Moewus file*, University of Heidelberg.

Moewus, Franz, to Karl Grell, November 3, 1950; *Grell papers;* Institut für Biologie, Universität Tubingen, West Germany.

Moewus, Franz, to T. M. Sonneborn, August 7, 1947; *Sonneborn papers.*

Moewus, Franz, to T. M. Sonneborn, October 13, 1947; *Sonneborn papers.*

Moewus, Franz, to T. M. Sonneborn, February 10, 1948; *Sonneborn papers.*

Moewus, Franz, to Joshua Lederberg, August, 16, 1949; *Lederberg papers.*

Moewus, Franz, to T. M. Sonneborn, September 16, 1949; *Sonneborn papers.*

Moewus, Franz, to W. Ludwig, February 9, 1951; *Moewus file*, University of Heidelberg.

Moewus, Franz, to T. M. Sonneborn, May 20, 1951; *Sonneborn papers.*

Moewus, Franz, to T. M. Sonneborn, September 14, 1951; *Sonneborn papers.*

Moewus, Franz, to T. M. Sonneborn, January 20, 1954; *Sonneborn papers.*

Moewus, Franz, to T. M. Sonneborn, March 25, 1954; *Sonneborn papers.*

Moewus, Franz, to T. M. Sonneborn, April 20, 1954; *Sonneborn papers.*

Moewus, Franz, to T. M. Sonneborn, March 19, 1955; *Sonneborn papers.*

Moewus, Franz, to T. M. Sonneborn, April 27, 1955; *Sonneborn papers.*

Moore, J. A., to the author, February 24, 1986.

Nyborn, Nils, to Ralph Lewin, April 15, 1951; *Lewin papers.*

Pomerat, Gerard, *diary*, May 20, 1949; Record Group 2, Folder 3/27, Rockefeller Archive Center, Tarrytown, N.Y.

Pomerat, Gerard, *diary*, September 22, 1953; Record Group 2, Series 717, Rockefeller Archive Center, Tarrytown, N.Y.

Pomerat, Gerard, *diary*, December 1–2, 1953, Record Group 2, Series 717, Rockefeller Archive Center, Tarrytown, N.Y.

Pomerat, Gerard, to T. M. Sonneborn, July 30, 1951; *Sonneborn papers.*

Ryan, F. 1955. *Attempt To Reproduce with Moewus' Collaboration Some of His Experiments on Chlamydomonas and Polytoma.* Unpublished report, 7 pp. Author's copy obtained from Mrs. Elizabeth Ryan, New York, N.Y.

Seybold, August, to Franz Moewus, May 20, 1949; *Moewus file,* University of Heidelberg, Heidelberg, West Germany.

Seybold, August, to Franz Moewus, June 21, 1949; *Moewus file,* University of Heidelberg, Heidelberg, West Germany.

Sonneborn, T. M. 1948. *Syllabus on Genetics of Microorganisms.* Unpublished; *Watson papers,* Cold Spring Harbor Laboratory, N.Y. 113 pp.

Sonneborn, T. M. 1978. *My Intellectual History in Relation to My Contributions to Science.* Unpublished autobiography, *Sonneborn papers.*

Sonneborn, T. M., to P. B. Armstrong, November 13, 1953; *Sonneborn papers.*

Sonneborn, T. M., to Harold Brodie, November 13, 1953; *Sonneborn papers.*

Sonneborn, T. M., to Vernon Bryson, October 24, 1955; *Sonneborn papers.*

Sonneborn, T. M., to Boris Ephrussi, November 10, 1958; *Sonneborn papers.*

Sonneborn, T. M., to Julian Huxley, June 5, 1951; *Sonneborn papers.*

Sonneborn, T. M., to Victor Jollos, February 12, 1940; *Sonneborn papers.*

Sonneborn, T. M., to C. J. Lapp, February 25, 1949; *Sonneborn papers.*

Sonneborn, T. M., to C. J. Lapp, January 1950; *Sonneborn papers.*

Sonneborn, T. M., to Franz Moewus, September 9, 1947; *Sonneborn papers.*

Sonneborn, T. M., to Franz Moewus, January 7, 1948; *Sonneborn papers.*

Sonneborn, T. M., to Franz Moewus, November 10, 1949; *Sonneborn papers.*

Sonneborn, T. M., to Franz Moewus, April 25, 1951; *Sonneborn papers.*

Sonneborn, T. M., to Franz Moewus, May 30, 1951; *Sonneborn papers.*

Sonneborn, T. M., to Franz Moewus, June 18, 1951; *Sonneborn papers.*

Sonneborn, T. M., to Franz Moewus, November 13, 1953; *Sonneborn papers.*

Sonneborn, T. M., to Franz Moewus, August 18, 1954; *Sonneborn papers.*

Sonneborn, T. M., to Franz Moewus, March 22, 1955; *Sonneborn papers.*

Sonneborn, T. M., to Franz Moewus, May 2, 1955; *Sonneborn papers.*

Sonneborn, T. M., to Warren Weaver, June 19, 1951; *Sonneborn papers.*

Sonneborn, T. M., to Sewall Wright, February 19, 1944; *Sonneborn papers.*

Spiegelman, Sol, to Boris Ephrussi, November 16, 1946; *Ephrussi papers,* Centre de Génétique Moléculaire du C.N.R.S., Gif-sur-Yvette, France.

Stern, Herbert, to the author, October 31, 1986.

Watson, J. D. 1948 *The Genetics of Chlamydomonas with Special Regard to Sexuality.* Unpublished manuscript, copy obtained from Robert Olby, University of Leeds.

Weaver, Warren, to G. W. Beadle, May 18, 1951; Record Group 1.1, Series 401D, Rockefeller Archive Center, Tarrytown, N.Y.

Weaver, Warren, To R. B. F., December 24, 1941; Record Group 1.1, Series 205D, Rockefeller Archive Center, Tarrytown, N.Y.

Weaver, Warren, to T. M. Sonneborn, June 15, 1951; *Sonneborn papers.*

Weaver, Warren, to T. M. Sonneborn, June 25, 1951; *Sonneborn papers.*

Wright, Sewall, to T. M. Sonneborn, January 29, 1944; *Sonneborn papers.*

Wright, Sewall, to T. M. Sonneborn, May 4, 1944; *Sonneborn papers.*

Published sources

Allen, G. E. 1978a. *Life Science in the Twentieth Century.* Cambridge: Cambridge University Press.

    1978b. *Thomas Hunt Morgan: The Man and His Science.* Princeton, N.J.: Princeton University Press.

Ash, M. G. 1980. Academic Politics in the History of Science: Experimental Psychology in Germany, 1879–41. *Central European History* 13: 255–286.

Assmuth, J., and Hull, E. R. 1915. *Haeckel's Frauds and Forgeries*. Bombay: Bombay Examiner Press.

Auerbach, C. 1967. Changes in the Concept of Mutation and the Aims of Mutation Research. In Brink, R. A., and Styles, E. D., eds., *Heritage from Mendel*. Madison: University of Wisconsin Press, pp. 67–80.

Babbage, C. 1830. *Reflections on the Decline of Science in England, and on Some of Its Causes*. London: Fellows & Booth.

Barber, B. 1962. Resistance by Scientists to Scientific Discovery. In Barber, B., and Hirsch, W., Barber, B., and Hirsch, W., eds., *The Sociology of Science*. New York: Free Press, pp. 539–556.

Barnes, S. B., and Dolby, R. G. A. 1970. The Scientific Ethos: A Deviant Viewpoint. *Archives européennes de sociologie* 11: 3–25.

Barthélemy-M. 1982. *Lamarck the Mythical Precursor: A Study of the Relations Between Science and Ideology*. Cambridge, Mass.: M.I.T. Press.

Bateson, 1909. *Mendel's Principles of Heredity*. Cambridge: Cambridge University Press.

——— 1913. *Problems of Genetics* (Reprint. New Haven: Yale University Press, 1979.)

Beadle G. W. 1945. Biochemical Genetics. *Chemical Reviews* 34: 15–88.

——— 1951. Chemical Genetics. In L. G. Dunn, ed., *Genetics in the Twentieth Century*. New York: Macmillan, pp. 221–240.

——— 1958. Genes and Chemical Reactions in *Neurospora*. *Nobel Lectures*. Amsterdam: Elsevier, pp. 147–159.

——— 1966. Biochemical Genetics: Some Recollections. In Cairns, J., Stent, G. S., and Watson. J. D., eds., *Phage and the Origins of Molecular Biology*. Cold Spring Harbor, N. Y.: Cold Spring Harbor Laboratory, pp. 23–32.

——— 1967. Mendelism, 1965. In Brink, R. A., and Styles, E. D., eds., *Heritage from Mendel*. Madison: University of Wisconsin Press, pp. 335–350.

Beadle, G. W., and Ephrussi, B. 1936. The Differentiation of Eye Pigments in *Drosophila* as Studied by Transplantation. *Genetics* 21: 225–247.

Beadle, G. W., and Tatum, E. L. 1941. Genetic Control of Biochemical Reactions in *Neurospora*. *Proceedings of the National Academy of Sciences* 27: 499–506.

Beecher, E. M. 1970. *The Sex Researchers*. London: Lowe & Brydone.

Ben-David, J. 1971. *The Scientist's Role in Society: A Comparative Study*. Englewood-Cliffs, N.J.: Prentice-Hall.

Benzer, S. 1966. Adventures in the rII Region. In Cairns, J., Stent, G. S., and Watson, J. D., eds., *Phase and the Origins of Molecular Biology*, Cold Spring Harbor, N.Y.: Cold Spring Harbor Laboratory, pp. 148–158.

Birch, A. J. 1957. Biosynthetic Relations of Some Natural Phenolic and Enolic Compounds. *Progress in the Chemistry of Organic Natural Products* 14: 186–216.

Birch, A. J., Donovan, F. W., and Moewus, F. 1953. Biogenesis of Flavonoids in *Chlamydomonas eugametos*. *Nature* 172: 902–904.

Brannigan, A. 1979. The Reification of Gregor Mendel. *Social Studies of Science* 9: 432–454.

——— 1981. *The Social Basis of Scientific Discovery*. Cambridge: Cambridge University Press.

Bridges, C. B. 1923. Aberrations in Chromosomal Materials. In *Eugenics, Genetics and the Family: 2nd International Congress of Eugenics* 1: 76–81.

Bridgstock, M. 1982. A Sociological Approach to Fraud in Science. *Australian and New Zealand Journal of Sociology* 18: 364–381.

Broad, W., and Wade, N. 1982. *Betrayers of the Truth: Fraud and Deceit in the Halls of Science*. New York: Touchstone.

Brush, S. G. 1974. Should the History of Science Be Rated X? *Science* 183: 1164–1172.

Burk, D. 1973. Kuhn, Richard. *Dictionary of Scientific Biography VII*. New York: Scribner, pp. 517–518.

Callender, L. A. 1988. Gregor Mendel: An Opponent of Descent with Modification. *History of Science* 26: 41–75.

Campbell, M. 1976. Explanations of Mendel's Results. *Centaurus* 20: 159–174.

Canguilhem, G. 1979. *Études d'histoire et de philosophie des sciences*. Paris: J. Vrin, pp. 20–22.

Chapman, V. J. 1962. *The Algae*. London: Macmillan Press.

1969. *The Algae*. London: Macmillan Press.

Chapman, V. J., and Chapman, D. J. 1973. *The Algae*, 2nd ed. London: Macmillan Press.

Child, C. M. 1915. *Senescence and Rejuvenescence*. Chicago: University of Chicago Press.

Chubin, D. E. 1985. Misconduct in Research: An Issue of Science Policy and Practice. *Minerva* 23: 175–202.

Churchhill, F. B. 1979. Sex and the Single Organism: Biological Theories of Sexuality in Mid-Nineteenth Century. *Studies in History of Biology* 3: 139–178.

Cioffi, F. 1976. Was Freud a Liar? *Journal of Orthomolecular Psychiatry* 5: 275–280.

Clark, J. T. 1959. The Philosophy of Science and History of Science. In Clagett, M., ed., *Critical Problems in the History of Science*. Madison: University of Wisconsin Press, pp. 103–140.

Clark, R. 1968. *J. B. S.: The Life and Work of J. B. S. Haldane*. Oxford: Oxford University Press.

Coleman, W. 1970. Bateson and Chromosomes: Conservative Thought in Science. *Centaurus* 15: 228–314.

Collins, H. M. 1974. The TEA Set: Tacit Knowledge and Scientific Networks. *Social Studies of Science* 4: 165–186.

1985. *Changing Order: Replication and Induction in Scientific Practice*. Beverly Hills, Cal.: Sage.

Collins, H. M., and Pinch, T. J. 1979. The Construction of the Paranormal. In Wallis, R., eds., *On the Margins of Science. Sociological Review Monographs* 27, pp. 237–270.

Cook, A. H. 1945. Algal Pigments and Their Significance. In Munro, F., ed., *Biological Reviews*. Cambridge: Cambridge University Press, pp. 115–132.

Crampton, H. E. 1950. A History of the Department of Zoology of Columbia University. *BIOS* 21: 218–246.

Cuénot, L. 1902. La loi de Mendel et l'hérédité de la pigmentation chez les souris. *Arch. zool. expér. gén.* [3rd séries] 10: 27–30.

Czurda, V. 1933. Über einege Grundbegriffe der Sexualitätstheorie. *Botanisches Zentralblatt* 50 (1): 196–210.

Czurda, V. 1935. Über die "Variabilität" von *Chlamydomonas eugametos Moewus*. *Botanisches Zentralblatt [Suppl.]* 53A: 133–157.

Darlington, C. D. 1939. *The Evolution of Genetic Systems*. Cambridge: Cambridge University Press.

Delbrück, M. 1946. Discussion Following David Bonner, "Biochemical Mutations in *Neurospora*." *Cold Spring Harbor Symposia on Quantitative Biology* 11: 14–24.

1949. A Physicist's Look at Biology. Reprinted in Cairns, J., Stent, G. S.,

Watson, J. D., eds., *Phage and the Origin of Molecular Biology*. Cold Spring Harbor, N.Y.: Cold Spring Harbor Laboratory, 1966, pp. 9–22.

Demerec, M. 1951. Foreword to *Genes and Mutations*. *Cold Spring Harbor Symposia on Quantitative Biology* 16: v.

Dodge, B. O. 1927. Nuclear Phenomena Associated with Heterothallism and Homothallism in the Ascomycete *Neurospora*. *Journal of Agricultural Research* 35: 289–305.

Driesch, H. 1894. *Analytische Theorie der organischen Entwicklung*. Leipzig: Wilhelm Engelman.

Dunn, L. C. 1965. *A Short History of Genetics*. New York: McGraw-Hill.

Ephrussi, B. 1938. Aspects of the Physiology of Gene Action. *American Naturalist* 72: 5–23.

Esser K., and Straub, A. 1954. Das Pollenschlauchwachstum bei *Forsythia*, eine Stellungnahme zu der Moewusschen Hemmstoff-Ferment Hypothese. *Biologisches Zentralblatt* 73: 449–455.

Feigl, H. 1970. Beyond Peaceful Coexistence. In Stuewer, R. H., ed., *Historical and Philosophical Perspectives of Science*. Minneapolis: University of Minnesota Press, pp. 3–11.

Fisher, R. A. 1936. Has Mendel's Work Been Re-discovered? *Annals of Science* 1: 115–137.

Fleming, D. 1969. Emigré Physicists and the Biological Revolution. In Fleming, D., and Bailyn, B., eds., *The Intelectual Migration: Europe and America, 1930–1960*. Cambridge, Mass.: Harvard University Press, pp. 152–189.

Forman, P. 1969. The Discovery of the Diffraction of X-Rays by Crystals: A Critique of the Myths. *Archive for History of Exact Sciences* 5: 38–71.

Förster, H., and Wiese, L. 1954a. Untersuchungen zur Kopulationsfähigkeit von *Chlamydomonas eugametos*. *Zeitschrift für Naturforschung* 9b: 470–471.

　　1954b. Gamonwirkungen bei *Chlamydomonas eugametos*. *Zeitschrift für Naturforschung* 9b: 548–550.

Garrod, A. E. 1908. Inborn Errors of Metabolism. *Lancet* 2 (July 4): 1–7.

　　1923. *Inborn Errors of Metabolism*, 2nd ed. Oxford: Oxford University Press.

Geddes, P., and Thomson, J. A. 1889. *The Evolution of Sex*. London: Walter Scott.

Gerloff, J. 1940. Bieträge zur Kenntnis der Variabilität und Systematik der Gaffung *Chlamydomonas*. *Archiv für Protistenkunde* 94: 318–502.

Glass, B. 1965. A Century of Biochemical Genetics. *Proceedings of the American Philosophical Society* 109: 227–236.

Goldschmidt, R. 1916. Genetic Factors and Enzyme Reaction. *Science* 43: 98–100.

Gould, S. J. 1978. Morton's Ranking of Races by Cranial Capacity. *Science* 200: 503–509.

　　1981. *The Mismeasure of Man*. New York: Norton.

Gowans, C. S. 1976. Publications by Franz Moewus on the Genetics of Algae. In Lewin, R., ed., *The Genetics of Algae, Botanical Monographs*, Vol. 12. Oxford: Blackwell Scientific, pp. 310–332.

Greenberg, D. S. 1967. *The Politics of Pure Science*. New York: World Publishing.

Hagen-Seyfferth, M. 1959. Zur Kenntnis der Geisseln und der Chemotaxis von *Chlamydomonas eugametos* Moewus (*Chl. Moewussi* Gerloff). *Planta* 53: 376–401.

Haldane, J. B. S. 1920. Some Recent Work on Heredity. *Transactions of the Oxford University Junior Scientific Club* [Series 3, No. 1], 3–11.

　　1937. The Biochemistry of the Individual. In Needham, J., and Green, D. E., eds., *Perspectives in Biochemistry*. Cambridge: Cambridge University Press, pp. 1–10.

1942. *New Paths in Genetics*. New York: Harper & Brothers.

1954. *The Biochemistry of Genetics*. London: Allen & Unwin.

Hamburger, V. 1980. Evolutionary Theory in Germany: A Comment. In Mayr, E., and Provine, W., eds., *The Evolutionary Synthesis*. Cambridge, Mass.: Harvard University Press, pp. 303–308.

Harris, H. 1963. *Garrod's Inborn Errors of Metabolism*. Oxford: Oxford University Press.

Harte, C. 1948. Das Crossing-over bei *Chlamydomonas*. *Biologisches Zentralblatt* 67: 504–510.

Hartmann, M. 1927. *Allgemeine Biologie: Eine Einfuhrung in die Lehre vom Leben*. Stuttgart: Fischer.

1929. Verteilung, Bestimmung und Vererbung des Geschlechtes bei den Protisten und Thallophyten. *Handbuch der Vererbungswissenschaft II*. Berlin: Bornträger.

1932. Neue Ergebnisse zum Befrunchtungs- und Sexualitätsproblem. (Nach Untersuchungen von M. Hartmann, J. Hämmerling und F. Moewus.) *Naturwissenschaften* 20: 567–73.

1934. Beiträge zur Sexualitätstheorie: Mit besonderer Berücksichtigung neuer Ergebnisse von Fr. Moewus. *Sitzungsberichte der Preussen Akademie der Wissenschaften zu Berlin Physik-Mathematisch Klasse* 20: 1–23.

1943. *Die Sexualität*. Stuttgart: Fischer.

1955a. Sexualitätsprobleme bei Algen, Pilzen und Protozoen. (Eine kritische Darstellung im Auschluss on einen Bericht von R. A. Lewin). *Biologisches Zentralblatt* 74: 311–334.

1955b. Sex Problems in Algae, Fungi and Protozoa. *American Naturalist* 89: 321–346.

Hartmann, M., and Schartau, O. 1939. Untersuchungen über Befruchtungsstoffe der Seeigel. *Biologisches Zentralblatt* 59: 571–587.

Hayes, W. 1984. Max Delbrück and the Birth of Molecular Biology. *Social Research* 51: 641–673.

Hearnshaw, L. S. 1979. *Cyril Burt, Psychologist*. London: Hodder & Stoughton.

Hixson, J. 1976. *The Patchwork Mouse*. New York: Doubleday.

Holton, G. 1978. Subelectrons, Presuppositions, and the Millikan–Ehrenhoft Dispute. *Historical Studies in the Physical Sciences* 9: 166–224.

Hopkins, F. G. 1913. The Dynamic Side of Biochemistry. In Needham, J., and Baldwin, E., eds., *Hopkins and Biochemistry*. (Reprint. Cambridge: Heffer, 1949, pp. 136–159.)

1938. Archibald Edward Garrod, 1857–1936. *Obituary Notices of Fellows of the Royal Society* 2: 225–228.

Huxley, J. 1942. *Evolution: The Modern Synthesis*. London: Allen & Unwin.

Iltis, H. 1966. *Life of Mendel*. New York: Norton.

Jennings, H. S. 1941. Inheritance in Protozoa. In Calkins G. N., and Summers, F. M., eds., *Protozoa in Biological Research*. [Reprint. New York: Macmillan (Hafner Press), 1964, pp. 710–771.]

Johannsen, W. 1911. The Genotype Conception of Heredity. *American Naturalist* 45: 129–159.

Jollos, V. 1926. Untersuchungen über die Sexualitätverhältnisse von *Dasycladus clavaeformis*. *Biologisches Zentralblatt* 46: 279–295.

Joravsky, D. 1970. *The Lysenko Affair*. Cambridge, Mass.: Harvard University Press.

Judson, H. F. 1979. *The Eighth Day of Creation: The Makers of the Revolution in Biology*. New York: Simon & Schuster.

320 Bibliography

Kamin, L. 1974. *The Science and Politics of I.Q.* Potomac, Md.: Erlbaum.

Kay, L. 1985. Conceptual Models and Analytic Tools: The Biology of Physicist Max Delbrück. *Journal of the History of Biology* 18: 207–246.

——— 1988. *Selling Pure Science in Wartime: The Biochemical Genetics of George Beadle.* In press, 33 pp.

Kevles, D. J. 1980. Genetics in the United States and Great Britain, 1890–1930: A Review with Speculations. *Isis* 71: 441–455.

Kimmelman, B. A. 1983. The American Breeder's Association: Genetics and Eugenics in an Agricultural Context, 1903–13. *Social Studies of Science* 13: 163–204.

Kniep, H. 1928. *Die Sexualität der niederen Pflanzen. Differenzierung, Verteilung, Bestimmung und Vererbung des Geschlechts bei den Thallophyten.* Jena: Fischer.

Knightly, P., Evans, J., Potter, E., and Wallace, M. 1979. *Suffer the Children: The Story of Thalidomide.* London: Archie Deutsch.

Koestler, A. 1971. *The Case of the Midwife Toad.* London: Hutchinson.

Kohler, R. E. 1976. The Management of Science: The Experience of Warren Weaver and the Rockefeller Foundation Program in Molecular Biology. *Minerva* 14: 279–306.

——— 1978. A Policy for the Advancement of Science: The Rockefeller Foundation, 1924–29. *Minerva* 16: 480–515.

——— 1982. *From Medical Chemistry to Biochemistry: The Making of a Biomedical Discipline.* Cambridge: Cambridge University Press.

Kohn, A. 1986. *False Prophets.* Oxford: Blackwell Publisher.

Koyré, A. 1968. *Metaphysics and Measurement: Essays in Scientific Revolution.* Cambridge, Mass.: Harvard University Press

——— 1973. *The Astronomical Revolution.* Paris: Hermann.

Kuhn, R., and Löw, I. 1949. Über Kristallisiertes Paonin aus der *Chlamydomonas* – Mutante Nr. 4. *Chemische Berichte* 80: 481–484.

——— 1960. Über Flavonolglylcoside von *Forsythia* und über Inhaltsstoffe von *Chlamydomonas. Chemische Berichte* 93: 1009–1010.

Kuhn, R., and Moewus, F. 1940. Über die chemische Wirkungsweise der Gene *Mot*, M_D und *Gathe* bei *Chlamydomonas. Deutsche chemische Gesellschaft, Berichte* 73: 547–559.

Kuhn, R., Moewus, F., and Jerchel, D. 1938. Über die chemische Natur der Stoffe, welche die Kopulation der männlichen und weiblichen Gameten von *Chlamydonomas eugametos* am Lichte bewirken. *Deutsche chemische Gesellschaft, Berichte* 71: 1541–1547.

Kuhn, R., Moewus, F., and Wendt, G. 1939. Über die geschlechtsbestimmenden Stoffe einer Grünalge. *Deutsche chemische Gesellschaft, Berichte* 72: 1702–1707.

Kuhn, R., Moewus, F., and Löw, I. 1942a. Über ein hockwirksames Glykosid aus dem Pollen von Crocus. *Naturwissenschaften* 30: 373.

——— 1942b. Über die Wirkungsweise eines geschlechtsbestimmenden Stoffes (Borsäure). *Naturwissenschaften* 30: 407.

——— 1944. Über die pflanzenphysiologische Spezifität von Quercetinderivaten. *Deutsche chemische Gesellschaft, Berichte* 77: 219–220.

Latour, B. 1987. *Science in Action. How To Follow Scientists and Engineers Through Society.* Cambridge, Mass.: Harvard University Press.

Latour, B., and Woolgar, S. 1979. *Laboratory Life: The Social Construction of Scientific Facts.* Beverly Hills, Cal.: Sage.

Lederberg, J. 1948. Problems in Microbial Genetics. *Heredity* 2 (Part 2): 145–198.

——— 1986. Memoir on Edward L. Tatum. *Biographical Memoirs of the National Academy of Sciences,* 13 pp.

Lederberg, J., and Tatum, E. L. 1946a. Novel Genotypes in Mixed Cultures of Biochemical Mutants of Bacteria. *Cold Spring Harbor Symposia on Quantitative Biology* 11: 113–114.

——— 1946b. Gene Recombination in *Escherichia coli*. *Nature* 158: 558.

Le Grand, H. E. 1988. *Drifting Continents and Shifting Theories*. Cambridge: Cambridge University Press.

Lewin, R. A. 1952. Ultraviolet-Induced Mutations in *Chlamydomonas moewusii* Gerloff. *Journal of General Microbiology* 6: 233–248.

——— 1954. Sex in Unicellular Algae. In Wenrich, D. H., Lewis, I. F., and Raper, J. R., eds., *Sex in Microorganisms*. Washington, D.C.: American Association for the Advancement of Science, pp. 100–133.

——— 1956. Problems of Sex in Unicellular Algae. *American Naturalist* 90: 331.

——— 1957. The Zygote of *Chlamydomonas moewusii*. *Canadian Journal of Botany* 35: 793–804.

Lewin, R. A., ed. 1976. *The Genetics of Algae*. Oxford: Blackwell Scientific Publication, pp. 310–332.

Lewis, D. 1954. Comparative Incompatibility in Angiosperms and Fungi. *Advances in Genetics* 6: 235–285.

——— 1983. Cyril Dean Darlington. *Biographical Memoirs of Fellows of the Royal Society* 29: 113–157.

Lindegren, C. C. 1932. The Genetics of *Neurospora* I and II. *Bulletin of the Torrey Botanical Club* 59: 85–102.

——— 1934. The Genetics of *Neurospora* IV. The Inheritance of *Tan* Versus *Normal*. *American Journal Botany* 21: 55–63.

Loeb, J. 1916. *The Organism as a Whole*. New York: Putnam.

Long, D. E. 1987. Physiological Identity of American Sex Researchers Between the Two World Wars. In Geison, G. L., ed., *Physiology in the American Context*. Bethesda, Md.: American Physiological Society, pp. 263–278.

Ludwig, W. 1942. Notiz zu der unternormalen Streung in den Moewusschen *Chlamydomonas*-Versuchen. *Zeitschrift für induktive Abstammungs- und Vererbungslehre* 80: 612–615.

Lwoff, A. 1947. Some Aspects of the Problem of Growth Factors for Protozoa. *Annual Review of Biochemistry* 16: 101–114.

——— 1957. The Concept of a Virus. *Journal of General Microbiology* 17: 239–253.

MacKenzie, D. 1981. Sociobiologies in Competition: The Biometrician – Mendelian Debate. In Webster, C., ed., *Biology, Medicine and Society 1840–1940*. Cambridge: Cambridge University Press, pp. 243–288.

Mainx, F. 1933. *Die Sexualität als Problem der Genetik*. Jena: Fischer.

——— 1937. Besprechungen [Discussion of] Moewus, Franz, Über die Vererbung des Geschlechts bei *Polytoma Pascheri* und bei *Polytoma uvella*. *Zeitschrift für induktive Abstammungs- und Vererbungslehre* 1935. 69: 374–417; ___, Die Vererbung des Geschlechts bei verschieden Rassen von *Protosiphon botryoides*. *Archiv für Protistenkunde*. 1936. 86:1–57.

Manwell, C., and Baker, C. M. A. 1981. Honesty in Science: A Partial Test of a Sociobiological Model of the Social Structure of Science. *Search* 126: 151–160.

Mathews, A. P. 1916. *Physiological Chemistry*. London: Ballière, Tindall & Cox.

Mayr, E. 1982. *The Growth of Biological Thought*. Cambridge, Mass.: Harvard University Press.

Medawar, P. B. 1963. Is the Scientific Paper a Fraud? *The Listener*, September 12, pp. 377–378.

Medvedev, Z. A. 1971. *The Rise and Fall of T. D. Lysenko*, trans. I. M. Lerner. New York: Doubleday (Anchor Books).

Merton, R. K. 1957. Priorities in Scientific Discovery: A Chapter in the Sociology of Science. *American Sociological Review* 22: 635–659.

1961. Singletons and Multiples in Scientific Discovery: A Chapter in the Sociology of Science. *Proceedings of the American Philosophical Society* 105: 470–494.

1973. The Normative Structure of Science (1942). In Storer, N. W., ed., *The Sociology of Science: Theoretical and Empirical Investigation*. Chicago: University of Chicago Press, pp. 267–278.

Minchin, E. A. 1916. The Evolution of the Cell. *The American Naturalist* 50: 5–39.

Mitroff, I. 1974. *The Subjective Side of Science*. New York: Elsevier.

Moewus, F. 1931. Neue Chlamydomonaden. *Archiv für Protistenkunde* 75: 284–296.

1933a. Untersuchungen über die Variabilität von Chlamydomonaden. *Archiv für Protistenkunde* 80: 128–171.

1933b. Untersuchungen über die Sexualität und Entwicklung von Chlorophyceen. *Archiv für Protistenkunde* 80: 469–526.

1934. Über Dauermodifikationen bei Chlamydomonaden. *Archiv für Protistenkunde* 83: 220–240.

1935a. Über die Vererbung des Geschlechts bei *Polytoma Pascheri* und bei *Polytoma uvella*. *Zeitschrift für Vererbungslehre* 69: 374–417.

1935b. Über den Einfluss äusserer Faktoren auf die Geschlechtsbestimmung bei *Protosiphon*. *Biologisches Zentralblatt* 55: 293–309.

1936. Faktorenaustausch, insbesondere der Realisatoren bei *Chlamydomonas*-Kreuzungen. *Berichte über die gesamte Biologie [A]: Deutsche botanische Gesellschaft* 54: 45–57.

1937. Methodik und Nachträge zu den Kreuzungen zwischen *Polytoma*-Arten und zwischen *Protosiphon*-Rassen. *Zeitschrift für Vererbungslehre* 73: 63–107.

1938a. Carotinoide als Sexualstoffe von Algen. *Jahrbucher für wissenschaftliche Botanik* 86: 753–783.

1938b. Vererbung des Geschlechts bei *Chlamydomonas eugametos* und verwandten Arten. *Biologisches Zentralblatt* 58: 516–536.

1939a. Carotinoide als Sexualstoffe von Algen. *Forschungen und Fortschritte: Nachrichtenblatt der Deutschen Wissenschaft und Technik* 15: 39–40.

1939b. Carotenoids as Sexual Substances of Algae. *Research and Progress: Bimonthly Review of German Science* (Engl. ed. of *Forschungen und Fortschritte*) 5: 370–376.

1939c. Carotinoide als Sexualstoffe von Algen. *Naturwissenschaften* 27: 97–104.

1939d. Über die Chemotaxis von Algengameten. *Archiv für Protistenkunde* 92: 485–526.

1940a. Über Mutationen der Sexual-Gene bei *Chlamydomonas*. *Biologisches Zentralblatt* 60: 597–626.

1940b. Die Analyse von 42 erblichen Eigenschaften der *Chlamydomonas eugametos*-Gruppe. 1. Teil: Zellform, Membran, Geisseln, Chloroplast, Pyrenoid, Augenfleck, Zellteilung. *Zeitschrift für Vererbungslehre* 78: 418–462.

1940c. Die Analyse . . . 2. Teil: Zellresistenz, Sexualität, Zygoten Besprechung der Ergebnisse. *Zeitschrift für Vererbungslehre* 78: 463–500.

1940d. Die Analyse . . . 3. Teil: Die 10 Koppelungsgruppen. *Zeitschrift für Vererbungslehre* 78: 501–522.

1940e. Carotinoid-Derivate als geschlechtsbestimmende Stoffe von Algen. *Biologisches Zentralblatt* 60: 143–166.

1941. Zur Sexualität der neideren Organismen. I. Flagellaten und Algen. *Ergebnisse der Biologie* 18: 287–356.

1943. Statistische Auswertung einiger physiologischer und genetischer Versuche an *Protosiphon* und *Chlamydomonas*. *Biologisches Zentralblatt* 63: 169–203.

1949. Der Kressewurzeltest, ein neuer qualitativer Wuchsstofftest. *Biologisches Zentralblatt* 68: 118–140.

1950. Die Bedeutung von Farbstoffen bei den Sexualprozessen der Algen und Blütenpflanzen. *Angewandte Chemie* 62: 496–502.

1954. On Inherited and Adapted Rutin-Resistence in *Chlamydomonas*. *Woods Hole, Massachusetts, Marine Biological Laboratory. Biological Bulletin* 107: 293.

1955. Biogenesis of the Flavonoids. *Annals of the New York Academy of Sciences* 61: 660–664.

1959. Stimulation of Mitotic Activity by Benzidine, and Kinetin in *Polytoma uvella*. *Transactions of the American Microscopical Society* 78: 295–304.

Moewus, F., and Deulofeu, V. 1954. An Antagonist of the Sterility Hormone Rutin in the Green Algae *Chlamydomonas eugametos:* Ombuoside = 7.4′-Dimethyl-rutin. *Nature* 173: 218.

Moewus, F., and Moewus, L. 1955. Growth Pattern of a Sexual Strain of *Polytoma uvella*. *Microbial Genetics Bulletin* 12: 17–18.

Moore, A. R. 1912. On Mendelian Dominance. *Archiv für Entuicklungsmechanik* 34: 168–175.

Moore, J. 1986. Socializing Darwinism: Historiography and the Fortunes of a Phrase. In Levidon, L., ed., *Science as Politics*. London: Free Association Books, pp. 38–80.

Morgan, T. H. 1919. *The Physical Basis of Heredity*. Philadelphia: Lippincott.

1926. *The Theory of the Gene*. New Haven: Yale University Press.

Mulkay, M. J. 1976. Norms and Ideology in Science. *Social Science Information* 15: 637–656.

Mulkay, M. J., and Gilbert, G. N. 1982. Accounting for Error: How Scientists Construct Their Social World When They Account for Correct and Incorrect Belief. *Sociology* 16: 165–183.

Muller, H. J. 1922. Variations Due to Change in the Individual Gene. *American Naturalist* 56: 32–50.

1929. The Gene as the Basis of Life. *Proceedings of the International Congress of Plant Science, Ithaca* 1: 897–921.

Murneek, A. E. 1941. Sexual Reproduction and the Carotenoid Pigments in Plants. *American Naturalist* 75: 614–619.

Nanney, D. L. 1954. Mating Type Determination in *Paramecium aurelia*. A study in Cellular Heredity. In Wenrich, D. W., ed., *Sex in Microorganisms*. Washington, D.C.: American Association for the Advancement of Science, pp. 266–283.

1958. Epigenetic Factors Affecting Mating Type Expression in Certain Ciliates. *Cold Spring Harbor Symposia on Quantitative Biology* 23: 327–335.

Nash, L. K. 1956. The Origin of Dalton's Chemical Atomic Theory. *Isis* 47: 101–116.

Nemec, B. 1965. Before Mendel. In Krizenecky, J., ed., *Fundamenta genetica*. Oosterhout, the Netherlands: Anthropological Publications, pp. 7–13.

Nordenskiöld, E. 1929. *The History of Biology*. London: Kegan Paul.

Olby, R. C. 1966. *Origins of Mendelism*. New York: Schocken Books.

1974. *The Path to the Double Helix*. London: Macmillan.

1979. Mendel No Mendelian? *History of Science* 17: 53–72.

Pascher, A. 1916. Über die Kreuzung einzelliger haploider Organismen: *Chlamy-domonas. Berichte der Deutsche botanische Gesellschaft* 34: 228–242.

1918. Über die Beziehung der Ruduktionsteilung zur Mendelschen Spaltung. *Berichte der Deutsche botanische Gesellschaft* 36: 163–168.

1931. Über einen neuen einzelligen und einkernigen Organismus mit Eibe-fruchtung. *Botanisches Zentralblatt* 48: 466–480.

Pätau, K. 1941. Eine Statistische Bemerkung zu Moewus' Arbeit die 'Analyse von 42 Erblichen Eigenschaften der *Chlamydomonas-eugametos*-Gruppe. III. Teil.' *Zeitschrift für Induktive Abstammungs- und Verebungslehre* 80: 317–319.

Philip, U., and Haldane, J. B. S. 1939. Relative Sexuality in Unicellular Algae. *Nature* 143: 334.

Pilgrim, I. 1984. The Too-Good-To-Be-True Paradox and Gregor Mendel. *The Journal of Heredity* 75: 501–502.

Pirie, N. W. 1966. John Burdon Sanderson Haldane. *Biographical Memoirs of Fellows of the Royal Society* 12: 219–249.

Polanyi, M. 1958. *Personal Knowledge.* London: Routledge & Kegan Paul.

1967. *The Tacit Dimension.* New York: Doubleday (Anchor Books).

Pringsheim, E. G., and Ondratschek, K. 1939. Geschlechtsvorgange bei *Poly-toma. Botanisches Zentralblatt* 59a: 117–172.

Provine, W. B. 1971. *The Origins of Theoretical Population Genetics.* Chicago: University of Chicago Press.

1986. *Sewall Wright and Evolutionary Biology.* Chicago: University of Chicago Press.

Radl, E. 1930. *The History of Biological Theories,* trans. E. J. Hatfield. Oxford: Oxford University Press.

Raper, J. R. 1952. Chemical Regulation of Sexual Processes in the Thallophytes. *Botanical Review* 18: 447–545.

1957. Hormones and Sexuality in Lower Plants. In Porter, H. K., ed., *Symposia of the Society for Experimental Biology No. 11: The Biological Action of Growth Substance.* Cambridge: Cambridge University Press, pp. 143–165.

Ravetz, J. R. 1971. *Scientific Knowledge and Its Social Problems.* Oxford: Oxford University Press.

Renner, von, O. 1958. Auch etwas über F. Moewus, *Forsythia* und *Chlamydomonas. Zeitschrift für Naturforschung,* 13b: 399–403.

Rensberger, B. 1977. Fraud in Research Is a Rising Problem in Science. *New York Times,* January 23, pp. 1,44.

Riddle, O. 1908. Our knowledge of Melanin Color Formation and Its Bearing on the Mendelian Description of Heredity. *Biological Bulletin* 16: 316–328.

Rosenberg, C. E. 1976. *No Other Gods.* Baltimore: Johns Hopkins University Press.

Rosenthal, R. 1966. *Experimental Effects in Behavioral Research.* New York: Appleton-Century-Crofts.

Roth, J. A. 1966. Hired-Hand Research. *American Sociologist* 1: 190–196.

Ryan, F. J. 1955. Attempts to Reproduce Some of Moewus' Experiments on *Chlamydomonas* and *Polytoma. Science* 122: 470.

Sager, R. 1972. *Cytoplasmic Genes and Organelles.* New York: Academic Press.

Sager, R., and Ryan, F. 1961. *Cell Heredity: An Analysis of the Mechanisms of Heredity at the Cellular Level.* New York: Wiley.

Saha, M. S. 1984. *Carl Correns and an Alternative Approach to Genetics: The Study of Heredity in Germany Between 1880 and 1930.* Ph.D. dissertation, Michigan State University.

Sapp, J. 1983. The Struggle for Authority in the Field of Heredity, 1900–1932:

New Perspectives on the Rise of Genetics. *Journal of the History of Biology* 16: 311–342.

1986. Inside the Cell: Genetic Methodology and the Case of the Cytoplasm. In Schuster, J. A., and Yeo, R., eds., *The Politics and Rhetoric of Scientific Method.* Dordrecht: Reidel, pp. 311–342.

1987a. *Beyond the Gene: Cytoplasmic Inheritance and the Struggle for Authority in Genetics.* Oxford: Oxford University Press.

1987b. What Counts as Evidence, or Who Was Franz Moewus and Why Was Everybody Saying Such Terrible Things about Him? *History and Philosophy of the Life Sciences* 9: 277–308.

1989. The Nine Lives of Gregor Mendel. In Le Grand, H. E., ed., *Experimental Inquires.* Dordrecht: Reidel, in press.

Sayre, A. 1975. *Rosalind Franklin and DNA.* New York: Norton.

Schrödinger, E. 1944. *What Is Life?* Cambridge: Cambridge University Press.

Schuster, J. A., and Yeo, R. R., eds. 1986. *The Politics and Rhetoric of Scientific Method.* Dordrecht, Holland: Reidel.

Shapin, S. 1984. Pump and Circumstance: Robert Boyle's Literary Technology. *Social Studies of Science* 14: 481–520.

Shapin, S., and Barnes, B. 1979. Darwin and Social Darwinism: Purity and History. In Barnes, B., and Shapin, S., eds., *Natural Order.* Beverly Hills, Cal.: Sage, pp. 125–139.

Smart, J. J. C. 1972. Science, History and Methodology. *British Journal for the Philosophy of Science* 23: 266–274.

Smith, G. M. 1946. The Nature of Sexuality in *Chlamydomonas. American Journal of Botany* 33: 625–630.

1947. On the Reproduction of Some Pacific Coast Species of *Ulva. American Journal of Botany* 33: 625–630.

1951. Sexuality of Algae. In Smith, G. M., ed., *Manual of Phycology.* Waltham, Mass.: Chronica Botanica, pp. 229–241.

Smith, G. M., and Regnery, D. C. 1950. Inheritance of Sexuality in *Chlamydomonas reinhardi. Proceedings of the National Academy of Science* 36: 246–248.

Sonneborn, T. M. 1937. Sex, Sex Inheritance and Sex Determination in *Paramecium aurelia. Proceedings of the National Academy of Sciences* 23: 378–395.

1941. Sexuality in Unicellular Organisms. In Calkins, G. N., and Summers, F. M., eds., *Protozoa in Biological Research.* New York: Macmillan (Hafner Press), pp. 666–709.

1942. Sex Hormones in Unicellular Organisms. *Cold Spring Harbor Symposia on Quantitative Biology* 10: 111–125.

1950. Heredity, Environment, and Politics. *Science* 111: 529–539.

1951. Some Current Problems of Genetics in the Light of Investigations on *Chlamydomonas* and *Paramecium. Cold Spring Harbor Symposia on Quantitative Biology* 16: 483–503.

Starr, R. C. 1964. The Culture Collection of Algae at Indiana University. *American Journal of Botany* 51: 1013–1044.

Sturtevant, A. H. 1965. *A History of Genetics.* New York: Harper & Row.

Swinburne, R. G. 1962. The Presence-and-Absence Theory. *Annals of Science* 18: 131–146.

Telonicher, F., and Smith, G. M. Periodicity in the Reproduction of *Cladophora trichotoma.* (Unpublished)

Thimann, K. U. 1940. Sexual Substances in the Algae. *Chronica Botanica* 6: 31–32.

Troland, L. 1917. Biological Enigmas and the Theory of Enzyme Action. *American Naturalist* 51: 321–350.

van Niel, C. B. 1940. The Biochemistry of Micro-organisms: An Approach to General and Comparative Biochemistry. In Moulten, R., ed., *The Cell and Protoplasm*. Washington, D.C.: Science Press, pp. 106–119.

Watson, J. D. 1966. Growing Up in the Phage Group. In Cairns, J., Stent, G. S., and Watson, J. P., eds., *Phage and the Origins of Molecular Biology*. Cold Spring Harbor, N.Y.: Cold Spring Harbor Laboratory, pp. 239–245.

1969. *The Double Helix*. New York: Signet Books.

Watson, J. D., and Crick, F. H. C. 1953. Genetical Implications of the Structure of Deoxyribonucleic Acid. *Nature* 171: 964–967.

Weiner, J. S. 1955. *The Piltdown Forgery*. Oxford: Oxford University Press.

Weinstein, D. 1979. Fraud in Science. *Social Science Quarterly* 59: 639–652.

Wenrich, D. H. 1954. Comments on the Origin and Evolution of Sex. In Wenrich, D. H., Lewis, I. F., and Raper, J. R., eds., *Sex in Microorganisms*. Washington, D.C.: American Association for the Advancement of Science, pp. 335–346.

Westfall, R. 1973. Newton and the Fudge Factor. *Science* 179: 751–758.

Williams, R. J. 1931. *An Introduction to Biochemistry*. London: Chapman and Hall.

Williams, R. J., and Beerstecher, E. 1948. *An Introduction to Biochemistry*, 2nd ed. New York: Van Nostrand.

Wilson, E. B. 1923. The Physical Basis of Life. *Science* 57: 277–286.

Winge, O. 1935. On Haplophase and Diplophase in Some Saccharomycetes. *Comptes-rendus des travaux du Laboratorie Carlsberg [Série physiologie]* 21: 77–111.

Winge, O., and Lausten, O. 1938. Artificial Species Hybridization in Yeast. *Comptes-rendus des travaux du Laboratoire Carlsberg [Série physiologie]* 22: 235–244.

Wright, S. 1917. Color Inheritance in Mammals. *Journal of Heredity* 8: 224–235.

1941. The Physiology of the Gene. *Physiological Reviews* 21: 487–527.

1945. Genes as Physiological Agents. *American Naturalist* 79: 289–303.

1966. Mendel's Ratios. In Stern, C., and Sherwood, E., eds., *The Origin of Genetics*. San Francisco: Freeman, pp. 173–175.

Young, R. 1971. Darwin's Metaphor: Does Nature Select? *The Monist* 55: 442–503.

Zuckerman, H. 1977. Deviant Behaviour and Social Control in Science. In Sagarin, E., ed., *Deviance and Social Change*. London: Sage, pp. 87–138.

Publications of Franz Moewus

1931

Neue Chlamydomonaden. *Archiv für Protistenkunde* 75: 284–296.

1932

Volvocales-Literaturverzeichnis. *Botanisches Zentralblatt* [Suppl.] 49: 369–412.

1933

Untersuchungen über die Variabilität von Chlamydomonaden. *Archiv für Protistenkunde* 80: 128–171.

Untersuchungen über die Sexualität und Entwicklung von Chlorophyceen. *Archiv für Protistenkunde* 80: 469–526.

1933/34

Über einige Volvocalen aus dem Georgenfelder Moor (Erzgebirge). *Naturwissenschaftliche Gesellschaft Isis, Dresden. Sitzungsberichte und Abhandlungen,* pp. 45–51.

1934

Über Subheterözie bei *Chlamydomonas eugametos*. *Archiv für Protistenkunde* 83: 98–109.

1934. Über Dauermodifikationen bei Chlamydomonaden. *Archiv für Protistenkunde* 83: 220–240.

Neue Volvocalen aus der Umgebung von Coimbra (Portugal). *Sociedade Broteriana, Coimbra. Boletim. [Ano X, 2. Ser.],* 1–14.

1935

Über die Vererbung des Geschlechts bei *Polytoma Pascheri* und bei *Polytoma uvella. Zeitschrift für Vererbungslehre* 69: 374–417.

Über den Einfluss äusserer Faktoren auf die Geschlechtsbestimmung bei *Protosiphon. Biologisches Zentralblatt* 55: 293–309.

Die Vererbung des Geschlechts bei verschiedenen Rassen von *Protosiphon botryoides. Archiv für Protistenkunde* 86: 1–57.

1936

Faktorenaustausch, insbesondere der Realisatoren bei *Chlamydomonas*-Kreuzungen. *Berichte über die gesamte Biologie [A]: Deutsche botanische Gesellschaft* 54: 45–57.

1937

Faktorenaustausch, insbesondere der Realisatoren bei *Chlamydomonas*-Kreuzungen. *Forschung und Fortschritt* 13:25–26.

Die allgemeinen Grundlagen der Sexualität. *Biologe* 6: 145–151.

Methodik und Nachträge zu den Kreuzungen zwischen *Polytoma*-Arten und zwischen *Protosiphon*-Rassen. *Zeitschrift für Vererbungslehre* 73: 63–107.

1938

Carotinoide als Sexualstoffe von Algen. *Jahrbucher für wissenschaftliche Botanik* 86: 753–783.

Vererbung des Geschlechts bei *Chlamydomonas eugametos* und verwandten Arten. *Biologisches Zentralblatt* 58: 516–536.

Sexualstoffe der Pflanzen. *Umschau in Wissenschaft und Technik* 42: 1147–1148.

Die Sexualität und der Generationswechsel der Ulvaceen und Untersuchungen über die Parthenogenese der Gameten. *Archiv für Protistenkunde* 91: 357–441.

Kuhn, R., Moewus, F., and Jergel, D. Über die chemische Natur der Stoffe, welche die Kopulation der männlichen und weiblichen Gameten von *Chlamydomonas eugametos* am Lichte bewirken. *Deutsche chemische Gesellschaft Berichte* 71: 1541–1547.

1939

Carotinoide als Sexualstoffe von Algen. *Forschungen und Fortschritte: Nachrichtenblatt der Deutschen Wissenschaft und Technik* 15: 39–40.

Carotenoids as Sexual Substances of Algae. *Research and Progress: Bimonthly Review of German Science* (Engl. ed. of *Forschungen und Fortschritte*) 5: 370–376.

Carotinoide als Sexualstoffe von Algen. *Naturwissenschaften* 27: 97–104.

Untersuchungen über die relative Sexualität von Algen. *Biologisches Zentralblatt* 59: 40–58.

Volvocales-Literaturverzeichnis. 1. Nachtrag (1932–1937). *Botanisches Zentralblatt* [A] 59: 225–234.

Über die Chemotaxis von Algengameten. *Archiv für Protistenkunde* 92: 485–526.

Kuhn, R., Moewus, F., and Wendt, G. Über die geschlechtsbestimmenden Stoffe einer Grünalge. *Deutsche chemische Gesellschaft, Berichte* 72: 1702–1707.

1940

Die chemischen Grundlagen der Sexualvorgänge bei Algen. *Natur und Volk* 70: 131–136.

Über Mutationen der Sexual-Gene bei *Chlamydomonas*. *Biologisches Zentralblatt* 60: 597–626.

Die Analyse von 42 erblichen Eigenschaften der *Chlamydomonas eugametos*-Gruppe. 1. Teil: Zellform, Membran, Geisseln, Chloroplast, Pyrenoid, Augenfleck, Zellteilung. *Zeitschrift für Vererbungslehre* 78: 418–462.

Die Analyse . . . 2. Teil: Zellresistenz, Sexualität, Zygoten Besprechung der Ergebnisse. *Zeitschrift für Vererbungslehre* 78: 463–500.

Die Analyse . . . 3. Teil: Die 10 Koppelungsgruppen. *Zeitschrift für Vererbungslehre* 78: 501–522.

Carotinoid-Derivate als geschlechtsbestimmende Stoffe von Algen. *Biologisches Zentralblatt* 60: 143–166.

Über Zoosporenkopulationen bei *Monostroma*. *Biologisches Zentralblatt* 60: 225–238.

Über die Sexualität von *Botrydium granulatum*. *Biologisches Zentralblatt* 60: 484–498.

Los Carotinoides como substancias sexuales de las algas. *Investigation y Progreso* 11: 177–182.

Kuhn, R., and Moewus, F. Über die chemische Wirkungsweise der Gene *Mot*, *M_D* und *Gathe* bei *Chlamydomonas*. *Deutsche chemische Gesellschaft, Berichte* 73: 547–559.

Kuhn, R., and Moewus, F. Wie kommen die Verhältniszahlen *cis:trans*-Crocetin-dimethylester bei den getrenntgeschlechtlichen Rassen von *Chlamydomonas* zustande? *Deutsche chemische Gesellschaft, Berichte* 73: 559–562.

1941

Über die Wirkungsweise von Sexualgenen, *Erbarzt* 9: 145–156.

Polyploidie. *Frankfurter Zeitung* (*Naturwissenschaftlicher Bericht* 33).

Zur Sexualität der niederen Organismen. I. Flagellaten und Algen. *Ergebnisse der Biologie* 18: 287–356.

1942

Kuhn, R., Löw, I., and Moewus, F. Über ein hochwirksames Glykosid aus dem Pollen von Crocus. *Naturwissenschaften* 30: 373.

Kuhn, R., Löw, L., and Moewus, F. Über die Wirkungsweise eines geschlechts-bestimmenden Stoffes (Borsäure). *Naturwissenschaften* 30: 407.

1943

Zur Sexualität der niederen Organismen. II. Myxomyceten und Phycomyceten. *Ergebnisse der Biologie* 19: 82–142.

Die Erforschung des Erbgefüges. *Rheinisch Westfälische Zeitung*.

Die pflanzlichen Symbiosen. *Rheinisch Westfälische Zeitung*.

Das entdeckte Geheimnis der Natur. *Rheinisch Westfälische Zeitung*.

Geisselbildung und Beweglichkeit bei *Chlamydomonas*. *Naturwissenschaften* 31: 420.

Statistische Auswertung einiger physiologischer und genetischer Versuche an *Protosiphon* und *Chlamydomonas*. *Biologisches Zentralblatt* 63: 169–203.

Pflanzen ohne Frucht. Die Probleme der Selbststerilität. *Rheinisch Westfälische Zeitung*.

Kuhn, R., Jerchel, D., Moewus, F., Möller, E. F., and Lettré, H. Über die chem-ische Natur der Blastokoline und ihre Einwirkung auf keimende Samen, Pollenkörner, Hefen, Bakterien, Epithelgewebe und Fibroblasten. *Natur-wissenschaften* 31: 468.

1944

Kuhn, R., Moewus, F., and Löw, I. Über die pflanzenphysiologische Spezifität von Quercetinderivaten. *Deutsche chemische Gesellschaft, Berichte* 77: 219–220.

1946

Über die Temperatur-Resistenz von *Chlamydomonas*-Zygoten. *Field Information Agency Technical Report* No. 964: 1–11.

Über die durch eine Molekel verschiedener Wirkstoffe ausgelösten Vorgänge in der *Chlamydomonas*-Zelle. *Biologisches Zentralblatt* 65: 18–29.

1947
Hemmstoffe der Pflazen. *Natur und Volk* 77: 98–101.
Über morphologische Geschlechtsunterschiede bei Valeriana dioica. *Zeitschrift für Naturforschung [B]: Anorganische, organische und biologische Chemie, Botanik und verwandte Gebiete* 2: 313–316.
Über nicht kopulierende *Chlamydomonas*-Zellen. *Naturwissenschaften* 34: 282–283.

1948
Zur Genetik und Physiologie der Kern- und Zellteilung. I. Die Apomiktosis von Enteromorpha-Gameten. *Biologisches Zentralblatt* 67: 277–293.
Gebundener und freier Wuchsstoff in der Kartoffelknolle. *Zeitschrift für Naturforschung [B]* 3: 135–136.
Ein neuer quantitativer Test für pflanzliche Wuchsstoffe. *Naturwissenschaften* 35: 124–125.
Über die Erblichkeit des Kopulationsverhaltens bei *Chlamydomonas*. *Zeitschrift für Naturforschung [B]* 3: 279–290.
Koppelung und Austausch der Realisatoren bei *Brachiomonas*. *Beiträge zur Biologie der Pflanzen* 27: 297–338.
Über den Stand der Erforschung pflanzlicher Wuchsstoffe. *Angewandte Chemie [A]* 60: 336.
Bestimmung des Wuchsstoff- und Hemmstoffgehaltes von Pflanzenextrakten. *Züchter* 19: 108–115.
Fortpflanzung und Sexualität der Pflanzen. *FIAT Review of German Science 1939–46: Biology [II]* 117–122.
Blastokoline. *FIAT Review of German Science 1939–46: Biochemistry [II]* 185–205.
Wieland, Th., Moewus, F., and Fischer, E. Natrium-skatylsulfonat, ein Antagonist von Heteroauxin beim Wurzelwachstum. *Justus Liebig's Annalen der Chemie* 561: 47–52.

1949
Zur Genetik und Physiologie der Kern- und Zellteilung. *Forschungen und Fortschritt* 25: 67–68.
Die Wirkung von Wuchs- und Hemmstoffen auf die Kressewurzel. *Biologisches Zentralblatt* 68: 58–72.
Der Kressewurzeltest, ein neuer pflanzlicher Wuchsstofftest. *Biologisches Zentralblatt* 68: 118–140.
Zur Genetik und Physiologie der Kern- und Zellteilung. III. Über die Erblichkeit der Parthenogenese bei *Enteromorpha compressa*. *Biologisches Zentralblatt* 68: 232–243.
Das Rätsel der Blattform. *Umschau in Wissenschaft und Technik* 49: 529–531.
Gebundener und freier Wuchsstoff in fleischigen Früchten. *Planta* 37: 413–430.
Zur biochemischen Genetik des Rutins. *Portugaliae Acta Biologica Richard Goldschmidt Volumen* 161–199.

1950
Die Bedeutung von Farbstoffen bei den Sexualprozessen der Algen und Blütenpflanzen. *Angewandte Chemie* 62: 496–502.

Zur Physiologie und Biochemie der Selbststerilität bei *Forsythia*. *Biologisches Zentralblatt* 69: 181–197.

Zur Genetik und Physiologie der Kern- und Zellteilung. II. Über den Synchronismus der Kernteilungen bei *Protosiphon botryoides*. *Beiträge zur Biologie der Pflanzen* 28: 36–63.

Sexualität und Sexualstoffe bei einem einzelligen Organismus. *Zeitschrift für Sexualforschung* 1: 17–41.

Der heutige Stand der Termonforschung bei Algen. *Zeitschrift für Vitamin-, Hormon- und Fermentforschung* 3: 139–147.

Beiträge zur Systematik der Volvocales-Gattung *Polytoma*. I. Die Arten mit Kugeligen Formen. *Archiv für Mikrobiologie* 14: 542–553.

Über die physiologischen und biochemischen Grundlagen der Selbststerilität bei *Forsythia*. *Forschung und Fortschritt* 26: 101–102.

Über ein Blastokolin der *Chlamydomonas*-Zygoten. *Zeitschrift für Naturforschung [B]* 5: 196–202.

Über die Zahlenverhältnisse der Geschlechter bei *Valeriana diocica* L. *Zeitschrift für Naturforschung [B]* 5: 380–383.

Zur Physiologie und Biochemie der Selbststerilität bei *Forsythia*. In Osvald, H., and Alberg, E., eds., *Proceedings of the 7th International Botanical Congress, Stockholm*: Almqvist and Wicksell, p. 777.

1951

Die Sexualstoffe von *Chlamydomonas eugametos*. *Ergebnisse der Enzymforschung* 12: 173–206.

Die Reaktionskette Gen – Ferment – Wirkstoff – Merkmal für die Rutinbildung bei *Chlamydomonas*. *Berichte über die gesamte Biologie [A]: Deutsche botanische Gesellschaft* 63: 11–12.

Über die Anwendbarkeit des Kressewurzel-Testes. *Berichte über die gesamte Biologie [A]: Deutsche botanische Gesellschaft* 64: 213–215.

Moewus, F., and Banerjee, B. Effect of *cis*-Cinnamic Acid and Some Isomeric Compounds on the Germination of Zygotes of *Chlamydomonas*. *Nature* 168: 561–562.

Moewus, F., and Banerjee, B. Über die Wirkung von *cis*-Zimtsäire und einigen isomeren Verbindungen auf *Chlamydomonas*-Zygoten. *Zeitschrift für Naturforschung [B]* 6: 270–273.

Moewus, F., and Schader, E. Die Wirkung von Cumarin und Parasorbinsäure auf das Austreiben von Kartoffelknollen. *Zeitschrift für Naturforschung [B]* 6: 112–115.

Moewus, F., and Schader, E. Uber die Keimungs- und wachstumshemmende Wirkung einiger Phthalide. *Berichte über die gesamte Biologie. [A], Deutsche botanische Gesellschaft* 64: 124–129.

Moewus, F., Moewus, L., and Schader, E. Vorkommen und Bedeutung von Blastokolinen in fleischigen Früchten. *Zeitschrift für Naturforschung [B]* 6: 261–270.

1952

Moewus, F., and Moewus, L. Sensitivity of Cress Roots to Indoleacetic Acid. *Nature* 170: 372.

Moewus, F., and Schader, E. Über den Einfluss von Wuchs- und Hemmstoffen auf das Rhizoidwachstum von *Marchantia*-Brutkörpern. *Beitrage zur Biologie der Pflanzen* 29: 171–184.

Moewus, F., and Wolfschlag, U. Über Grössenunterschiede von Pollenkörnern und Narbenpapillen bei einigen heterostylen Pflanzen. *Zeitschrift für Naturforschung [B]* 7: 196–197.

Moewus, F., Moewus, L., and Skwarra, H. Nachweis von zwei Wuchsstoffen in Samen und Wurzeln der Kresse *(Lepidium sativum)*. *Planta*. 40: 254–264.

1953

Über die Biosynthese des Quercetins. *Angewanmdte Chemie* 65.

Biosynthesis of Pigments in a Unicellular Plant. *International Congress of Microbiology, 6th, Rome* 1: 292–295.

Birch, A. J., Donovan, F. W., and Moewus, F. Biogenesis of Flavonoids in *Chlamydomonas eugametos. Nature* 172: 902–903.

1954

On Inherited and Adapted Rutin-Resistance in *Chlamydomonas. Woods Hole, Massachusetts, Marine Biological Laboratory. Biological Bulletin* 107: 293.

About the Hormone Control in the Life Cycle of the Green Alga *Chlamydomonas eugametos. International Botanical Congress, 8th, Paris, Proceedings [Sect 17],* 46–47.

The Action of 2.4-Dichlorophenoxyacetic Acid on Deaminating Enzymes. *International Botanical Congress, 8th, Paris, Proceedings,* [Sect. 11/12], 149–150.

Moewus, F., and Deulofeu, V. An Antagonist of the Sterility Hormone Rutin in the Green Alga *Chlamydomonas eugametos:* Ombuoside = 7.4'-Dimethylrutin. *Nature* 173: 218.

1955

Biogenesis of the Flavonoids. *Annals of the New York Academy of Sciences* 61: 660–664.

Interrelations Between Growth and Sexuality in a Homothallic Strain of *Polytoma uvella. Journal of Protozoology* 2 (Suppl.): 7.

Moewus, F., and Moewus, L. Growth Pattern of a Sexual Strain of *Polytoma uvella. Microbial Genetics Bulletin* 12: 17–18.

1956

Moewus, F., and Moewus, L. First Approach to UV-Induced Mutations in *Polytoma uvella. Microbial Genetics Bulletin* 13: 19–21.

Moewus, F., and Moewus, L. Utilization of Fatty Acids for Induction of Sexuality in Populations of *Polytoma uvella. Microbial Genetics Bulletin* 14: 20–21.

1957

The Manifestation of the Homothallic Sex Behavior in Algae. *Transactions of the American Microscopical Society* 76: 337–344.

1959

Competitive Antagonism Between Kinetin and 8-Azaguanine in *Polytoma uvella. Science* 130: 921–922.

Stimulation of Mitotic Activity by Benzidine and Kinetin in *Polytoma uvella. Transactions of the American Microscopical Society* 78: 295–304.

Index

Entries under personal names are arranged in chronological order; all other entries are in alphabetical sequence.